The CORAF's 2023 Symposium on Processing and Transformation of Agricultural Products in West and Central Africa

The CORAF's 2023 Symposium on Processing and Transformation of Agricultural Products in West and Central Africa

Volume Editors

G. Esaïe Kpadonou
Kyky Komla Ganyo
Ayoni Ogunbayo
Emmanuel Njukwe
Niéyidouba Lamien
Aissatou Drame Yaye

MDPI

Basel • Beijing • Wuhan • Barcelona • Belgrade • Novi Sad • Cluj • Manchester

Volume Editors
G. Esaïe Kpadonou
CORAF
Dakar
Senegal

Kyky Komla Ganyo
CORAF
Dakar
Senegal

Ayoni Ogunbayo
CORAF
Dakar
Senegal

Emmanuel Njukwe
CORAF
Dakar
Senegal

Niéyidouba Lamien
CORAF
Dakar
Senegal

Aissatou Drame Yaye
Université Abdou Moumouni
Niamey
Niger

Editorial Office
MDPI AG
Grosspeteranlage 5
4052 Basel, Switzerland

This is a reprint of the Proceedings, published open access by the journal *Proceedings* (ISSN 2504-3900), freely accessible at: https://www.mdpi.com/2504-3900/118/1.

For citation purposes, cite each article independently as indicated on the article page online and as indicated below:

Lastname, A.A.; Lastname, B.B. Article Title. *Journal Name* **Year**, *Volume Number*, Page Range.

ISBN 978-3-7258-5041-9 (Hbk)
ISBN 978-3-7258-5042-6 (PDF)
https://doi.org/10.3390/books978-3-7258-5042-6

Contents

About the Editors

G. Esaïe Kpadonou

G. Esaïe Kpadonou is an agricultural systems analysist with a PhD in Soil Science. He currently works as a scientist at the West and Central African Council for Agricultural Research and Development (CORAF), conducting and facilitating research activities on climate smart agriculture, climate change, nutrition sensitive agriculture and nutrition smart agriculture. He has expertise in biophysical modelling and is dedicated to valorizing research findings for the improvement of smallholders' livelihoods, particularly in African countries. His works focus on the evaluation of the smartness and effectiveness of agricultural technologies and innovations (including agri-food and nutrition technologies) as well as appropriate scaling mechanisms to facilitate the dissemination and up-take of the best-bet of these technologies to improve agricultural productivity, nutrition and food security.

Komla Kyky Ganyo

Komla Kyky Ganyo is an Agricultural Engineer (With PhD in Soil Science) and expert in natural resource management and climate change working with West and Central African Council for Agricultural Research and Development (CORAF). He is dedicated to conducting and facilitating activities to (i) strengthen national and regional research centers, (ii) deepen and expand regional and national R&D networks, (iii) modernize national extension services, (iv) promote technology access and exchange and (vi) strengthen regional food security through integrated landscape management. His main research interests include soil science, soil fertility and fertilization, climate smart agriculture, climate change, nutrition sensitive agriculture and nutrition smart agriculture. Prior to joining CORAF, he was a staff of Togolese Institute of Agricultural Research where he coordinated a national project on the elaboration of the fertility map of agricultural soils in Togo and contributed to the implementation of project improvement of agricultural resilience to the salinity and promotion of technologies adapted to vulnerable producers (RESADE), capacitating producers in food conservation and processing.

Ayoni Ogunbayo

Ayoni Ogunbayo holds a PhD degree in Plant Breeding from the Federal University of Agriculture, Abeokuta (FUNAAB), Nigeria. He has a prolific career spanning more than two decades in agricultural research and development at five reputable organizations (IITA, AfricaRice, ICRISAT, CIAT and CORAF), supported by immense scientific knowledge production, technology development and dissemination. Prior to joining ICRISAT, he served as the TARSPro Regional Project Coordinator for West and Central Africa at CORAF, Senegal with a special focus on out-scaling technologies and innovations with focus on those related to nutrition and foods security. He has authored and co-authored over 50 publications that are published in reputable scientific journals, proceedings, books and chapters.

Emmanuel Njukwe

Emmanuel Njukwe holds a PhD in agronomy with over 30 years of experience in Agricultural Research and Development and works for CORAF as Director of Research and Innovation. Prior to joining CORAF, Emmanuel was IITA country representative in Rwanda and in Burundi. Emmanuel promotes the development of sustainable agriculture and seed systems as an agricultural enterprise, and for this he was awarded certificate of excellence and recognition by several institutions and organizations including the World Bank, and the UK body of Farming & Agriculture. Emmanuel's research activities are also oriented towards improving agri-food and nutrition systems in West and Central Africa. He is a member of the External Scientific Council of UMR-AGAP Institute, CIRAD, University of Montpellier and a board member of the African Growth and Development Policy (AGRODEP) Modeling Consortium.

Niéyidouba Lamien

Niéyidouba Lamien is a PhD holder and researcher in agroforestry and ecology. With over 30 years' experience in Agricultural research and development, he has strong experience in regional projects and programs management, including those related to food and nutrition. He currently serves as program manager at CORAF with activities related to promoting the inclusion of food and nutrition in agricultural programs. Examples include the development of (i) the regional nutrition sensitive agricultural research strategy in West and Central Africa and (ii) a tool (app) for evaluating the nutrition smartness and sensitivity of agricultural technologies and innovations.

Aissetou Drame Yaye

Drame Yaye Aissetou is a married female lecturer and scientist working at the Department of Forestry at the Faculty of Agriculture of Abdou Moumouni University of Niamey in the Republic of Niger. She is specialized in Agroforestry and Forestry, particularly on the issues of parklands and their impacts on the entomofauna and on above ground carbon sequestration, as well as their effective use in human nutrition. Having worked for nine years (2007–2016) as the Executive Secretary of the African Network for Agriculture, Agroforestry and Natural Resources Education (ANAFE) at the World Agroforestry Centre (ICRAF) in Nairobi, Kenya, she has expertise in agroforestry landscape management, projects development and led the elaboration and implementation of several projects supporting agriculture, agroforestry, climate change and agribusiness curricula development for universities. Since 2019, she is the Chair of the Scientific and Technical Committee (STC) of CORAF where she leads a multidisciplinary team of twelve scientific leaders working as the technical arm of the CORAF Board of Directors.

Preface

This second edition of the West and Central African Council for Agricultural Research and Development (CORAF) Symposium focuses on a subject of growing importance for the region: the processing and transformation of agricultural products. With agriculture serving as a backbone of many economies in West and Central Africa, the shift from raw commodity production to value-added processing represents a strategic pathway to enhance food security, generate employment, increase farmers' incomes, and stimulate regional trade.

The aim of this edition is to document achievements, identify emerging challenges, and highlight promising opportunities for enhancing the transformation of local agricultural products for value addition. It responds to growing interest in unlocking the potential of agricultural value chains and ensuring that research outcomes are translated into concrete economic benefits. This Reprint is addressed to a broad audience: researchers, policymakers, agribusinesses, development partners, investors, farmers' organizations, and all those committed to transforming agriculture into a driver of inclusive and sustainable growth. It is our hope that this work will inspire renewed collaboration and sustained engagement across sectors and borders.

Dr Moumini Savadogo
Executive Director, CORAF

G. Esaïe Kpadonou, Kyky Komla Ganyo, Ayoni Ogunbayo, Emmanuel Njukwe, Niéyidouba Lamien, and Aissatou Drame Yaye
Volume Editors

proceedings

MDPI

Editorial

Preface: 2nd Edition of the West and Central African Council for Agricultural Research and Development's Symposium on "Processing and Transformation of Agricultural Products in West and Central Africa: Achievements and Opportunities for Private Sector Engagement"

Abdou Tenkouano [1], Emmanuel Njukwe [1], G. Esaïe Kpadonou [1,*], Kyky Komla Ganyo [1], Niéyidouba Lamien [1], Christel Kenou [1], Ayoni Ogunbayo [1] and Aissatou Drame Yaye [2]

1 West and Central African Council for Agricultural Research and Development (CORAF), Dakar BP 48, Senegal; atenkouano@gmail.com (A.T.); e.njukwe@coraf.org (E.N.); k.ganyo@coraf.org (K.K.G.); n.lamien@coraf.org (N.L.); c.kenoul@coraf.org (C.K.); bogunbayo@hotmail.com (A.O.)

2 Faculté d'Agronomie, Université Abdou Moumouni, Niamey BP 10896, Niger; adrameyaye@gmail.com

* Correspondence: e.kpadonou@coraf.org

1. Introduction

1.1. Context

The agricultural sector faces new challenges, including pest and disease outbreaks, land degradation, and the effects of climate change. These challenges can be addressed through new technologies and practices, combined with appropriate tools and approaches for communication and knowledge management. These innovative technologies, practices, and approaches have the potential to increase production and productivity, reduce post-harvest losses, improve the quality of products and services, and make food systems more resilient and sustainable. Moreover, they provide a critical foundation for ensuring that agriculture remains a viable and thriving sector in the face of ongoing and future challenges.

In recognition of the pivotal role of research and development in addressing these challenges, CORAF organized an international symposium from 21st to 23rd November 2023, in Lomé, Togo. The theme of the symposium was, "Processing and transformation of Agricultural Products: Achievements and Opportunities Available to the Private Sector for Job and Wealth Creation in West and Central Africa". The symposium, which was attended by 147 participants, started with a call for abstracts, and 265 abstracts were received from authors spread across the West, Central, and Eastern Africa region, out of which 31 papers were accepted as full research papers with 18 manuscripts finally accepted for publication after the peer-review process.

This symposium was more than just a gathering of minds; it was a confluence of ideas, a testament to the power of collaboration in advancing agricultural research and development. It provided a unique opportunity for plant breeders, researchers, students, private companies, national agricultural research systems, and policymakers to discuss their findings, share their achievements, and forge new partnerships. These interactions are the seeds from which future innovations will grow, paving the way for a more prosperous and resilient agricultural sector in West and Central Africa region and beyond.

The symposium's outcomes and discussions, presented in this *Proceedings* volume, reflect the commitment of CORAF to advancing agriculture in the region, addressing contemporary challenges, and fostering innovative solutions for the benefit of West and Central Africa. Let it spark new ideas, foster deeper understandings, and ignite the drive to continue exploring and innovating in the field of agriculture.

1.2. Organizers

The Symposium is an annual regional scientific meeting in West and Central Africa focusing on current and important topics in the agricultural research and development areas. It is organized by the West and Central African Council for Agricultural Research and Development (CORAF) and its partners. CORAF is a sub-regional organization made up of the national agricultural research systems (NARS) of 23 countries in West and Central Africa. It was established in 1987, with a mandate to coordinate and facilitate innovative and cutting-edge research products needed to unlock the agricultural potential of West and Central Africa.

1.3. Scopes and Topics

The 2023 symposium focused on "Processing and transformation of Agricultural Products: Achievements and Opportunities for Private Sector for Job and Wealth Creation in West and Central Africa". For this proceeding, it covered three topics:

- Foods processing and nutrition;
- Postharvest management and conservation;
- Foods security.

2. Committee Members

2.1. Organizing Committee

- Dr. Emmanuel NJUKWE, CORAF, Senegal
- Dr. Ayoni OGUNBAYO, CORAF, Senegal
- Dr. Niéyidouba LAMIEN, CORAF, Senegal
- Dr. G. Esaïe KPADONOU, CORAF, Senegal
- Dr. Kyky Komlan GANYO, CORAF, Senegal
- Dr. Caroline MAKAMTO SOBGUI, CORAF, Senegal
- Dr. Réana FALL DIAGNE, CORAF, Senegal
- Dr Fatou DIENG GUEYE, CORAF, Senegal
- Dr. Yao LOMBO, Institut Togolais de Recherche Agronomique (ITRA), Togo
- Dr. Kossi Koudjega, ITRA, Togo
- Mr. Christel KENOU, CORAF, Senegal
- Mr. Etsri ANANI, ITRA, Togo
- Ms. Haby TAMBEDOU, CORAF, Senegal
- Ms. Poko Alida NADINGA, CORAF, Senegal
- Mrs. Pauline Ngandoul DIOUF, CORAF, Senegal
- Ms. Henriette KABA, CORAF, Senegal
- Mr. Mouhamadou LAMINE NIASS, CORAF, Senegal
- Mrs. Sophie Yandé Ngning KAMONY, CORAF, Senegal

2.2. Scientific Committee

- Prof. Aissatou DRAME YAYE, Université Abdou Moumouni, Niger
- Prof. Mauricette OUALI-N'GORAN, Université Félix Houphouët-Boigny, Côte d'Ivoire
- Prof. Fonteh Florence ANYANGWE, University of Dschang, Cameroon
- Prof. Isaac Oludayo DANIEL, University of Manitoba, Canada
- Prof. Emmanuel K. AJANI, University of Ibadan, Nigeria
- Prof. Shehu Garki ADO, Ahmadu Bello University Zaria, Nigeria

- Prof. Emmanuel MBETID-BESSANE, Université de Bangui, Centrafrique
- Prof. Isaac Roger TCHOUAMO (retd), University of Dschang, Cameroon
- Dr. Rachidatou SIKIROU, Institut National des Recherches Agricoles du Benin, Benin
- Dr. Robert ZOUGMORE, Alliance of Bioversity International and CIAT, Senegal
- Dr. Pauline LIMI-MOUNJOUENPOU, Institut de Recherche Agricole pour le Développement, Cameroun
- Dr. Maïmouna CISSE, Institut Sénégalais de Recherche Agricole, Senegal
- Dr. Emmanuel NJUKWE, West and Central African Council for Agricultural Research and Development, Senegal
- Dr. Ekpetsi O. Chantal BOUKA GOTO, Institut Togolais de Recherche Agronomique, Togo
- Prof. Kwasi Dzola AYISAH, Université de Lomé, Togo
- Prof. Fousseni FOLEGA, Université de Lomé, Togo
- Dr. Ayoni OGUNBAYO, West and Central African Council for Agricultural Research and Development, Senegal
- Dr. Caroline MAKAMTO SOBGUI, West and Central African Council for Agricultural Research and Development, Senegal
- Dr. Kyky Komlan GANYO, West and Central African Council for Agricultural Research and Development, Senegal
- Dr. G. Esaïe KPADONOU, West and Central African Council for Agricultural Research and Development, Senegal

2.3. Advisory Committee

- Dr. Moumini SAVADOGO, CORAF, Senegal
- Dr. Abdou TENKOUANO, CORAF, Senegal
- Dr. Alioune FALL, CORAF, Senegal
- Dr. Ousmane NDOYE, CORAF, Senegal
- Dr. Yacouba DIALLO, CORAF, Senegal

2.4. Independent Reviewers

- Prof. Rasak OLAJIDE, University of Ibadan, Nigeria
- Prof. Sylvain DABADE, Université d'Abomey Calavi, Benin
- Prof. Akeem Babatunde DAUDA, Federal University Dutsin-Ma, Nigeria
- Dr. Felix Kwashie MADILO, Ho Technical University, Ghana
- Dr. Paul HOUSSOU, Institut National des Recherches Agricoles du Benin, Benin
- Dr. Oluwafunmilayo Olarewaju AMINU, Olusegun Agagu University of Science and Technology, Nigeria
- Dr. Adegbite S. ADEYEMI, Federal Institute of Industrial Research, Nigeria
- Dr. Sedo Eudes Landry ANIHOUVI, Université d'Abomey Calavi, Benin
- Dr. Victoria NWADILI, National Root Crops Research Institute, Nigeria
- Dr. Vincent Awe EZIN, Université d'Abomey Calavi, Benin
- Dr. Christian ADJIBA, Université de Parakou, Benin
- Dr. Ekpetsi O. Chantal BOUKA GOTO, Institut Togolais de Recherche Agronomique, Togo
- Dr. Mahyao ADOLPHE, Centre National de Recherche Agronomique, Côte d'ivoire
- Dr. Oluwadiran Olufunmilola OLUDOYIN, Emmanuel Alayande University of Education, Nigeria
- Dr. André ADJOGBOTO, Université de Parakou, Benin
- Dr. Elysée M. HOUEDJOFONON, Institut National des Recherches Agricoles du Benin, Benin
- Dr. Adrien DOGO, Université de Parakou, Benin

- Dr. François ASSOGBA, Faculté des sciences agronomiques, Université d'Abomey Calavi, Benin
- Dr. Herbert Afe IKOH, Faculté des sciences agronomiques, Université d'Abomey Calavi, Benin
- Prof. Ahmed M. ABDEL-AZEEM, Botany and Microbiology Department, Faculty of Science, Suez Canal University, Egypt
- Dr. Amarachi OSI, Université polytechnique de Bucharest, Roumanie

3. Sponsors

CORAF appreciates support of the **Swiss Cooperation** in funding the implementation of Technologies and Innovations for Increasing the Resilience of Production Systems and Family Farms in West and Central Africa (**TARSPro**) Program in West and Central Africa region.

Acknowledgments: Our deepest gratitude goes to the Swiss Cooperation, all the authors, reviewers, and contributors who made this symposium and the resulting publication possible. We are grateful to Dr. Moumini Savadogo for his guidance and encouragement. We also acknowledge the invaluable support of our partners, stakeholders, and collaborators. We appreciate our partners from the National Agricultural Research System (NARS) in the 23 countries, your unwavering belief in the transformative power of research fuels our collective mission and propels us forward.

Conflicts of Interest: The authors declare no conflicts of interest.

proceedings

Editorial

Statement of Peer Review

G. Esaïe Kpadonou [1,*], Kyky Komla Ganyo [1], Ayoni Ogunbayo [1], Emmanuel Njukwe [1], Niéyidouba Lamien [1] and Aissatou Drame Yaye [2]

1 West and Central African Council for Agricultural Research and Development (CORAF), Dakar BP 48, Senegal
2 Faculté d'Agronomie, Université Abdou Moumouni, Niamey BP 10896, Niger
* Correspondence: e.kpadonou@coraf.org

In submitting conference proceedings to *Proceedings*, the volume editors of the proceedings certify to the publisher that all papers published in this volume have been subjected to peer review administered by the volume editors. Reviews were conducted by expert referees to the professional and scientific standards expected of a proceedings journal.

- Type of peer review: The single-blind style was used at the first level of the review process. Full papers were evaluated by the scientific committee of CORAF. Successful papers were then sent to independent reviewers using the double-anonymized peer review style.
- Conference submission management system: Institutional email.
- Number of submissions sent for review: 265 abstracts were received, out of which 31 papers were accepted as full research papers for the first stage.
- Number of submissions accepted: 18 papers were accepted for publication after the peer-review process.
- Number of submission published: 17 papers were finally published online.
- Acceptance rate (number of submissions accepted/number of submissions received): 54%.
- Average number of reviews per paper: 2 reviews.
- Total number of reviewers involved: 39 reviewers.
- Review process and criteria:
    - ○ Clarity of the research problem and objectives as presented in the manuscript.
    - ○ Clarity and rigor of the methodology adopted with respect to the type of research.
    - ○ Shows evidence in advancing knowledge related to the topics.
    - ○ Quality of writing.

Acknowledgments: The authors would like to express their gratitude to all the independent reviewers for their time and invaluable support in reviewing the symposium papers.

Conflicts of Interest: The authors declare no conflicts of interest.

proceedings

MDPI

Proceeding Paper

Assessment of Local Rice Processing, Packaging and Storage Among Rice Processors in Southwestern Nigeria, West Africa †

Sikiru Banjo [1], Timothy Olawumi [1], Abiala Abiala [2,*], John Jolayemi [2], Oye Ogunyanwo [2] and Yemisi Asamu [2]

1 Department of Tourism and Hospitality Management, College of Agriculture and Hospitality Management, Tai Solarin University of Education, Ijagun 120103, Nigeria; banjoos@tasued.edu.ng (S.B.); aremutim@gmail.com (T.O.)

2 Department of Agricultural Economics and Extension, College of Agriculture and Hospitality Management, Tai Solarin University of Education, Ijagun 120103, Nigeria; jjohn9@gmail.com (J.J.); ogunyanwoo@gmail.com (O.O.); yemas200@gmail.com (Y.A.)

* Correspondence: abialaalatise@gmail.com

† Presented at the CORAF's 2023 Symposium on Processing and Transformation of Agricultural Products in West and Central Africa: Achievements and Opportunities for Private Sector Engagement, Lome, Togo, 21–23 November 2023.

Abstract: Among the factors threatening food security in Nigeria are poor access to credit facilities, the high cost of inputs, and poor processing and storage. Cereals and grains are among the staple food crops highly consumed by Nigerians. In this study, multi-stage sampling procedure was used to select 1200 registered rice processors from Agricultural Development Programme zones in the Nigerian states of Lagos, Oyo, Ogun, Ondo, Osun, and Ekiti, and a structured questionnaire was used to obtain data on primary, secondary, and tertiary postharvest operations. The data were analyzed using descriptive statistics and Pearson Product Moment Correlation. The results showed that the majority (65.1%) of the respondents were male, 54.5% were 30–60 years old, 86.9% were married, 96.3% had been formally educated, and 99.9% processed, 71.5% packaged, and 79.4% stored more than 5001 kg of rice monthly. The majority (85.9%) of the respondents had no knowledge of rice moisture content and still used local means of rice processing, while 14.1% of the respondents used modern means of rice processing. We concluded that stored local rice was still subject to more wastage, spoilage, and losses due to the poor processing, packaging, and storage methods used in the study area. We recommend the adoption of modern and suitable rice technologies for processing, packaging, and storage. Furthermore, credit facilities should be made available, and inputs should be subsidized for rice farmers and processors.

Keywords: rice; postharvest; processing; packaging; storage

1. Introduction

The second Sustainable Development Goal (SDG) highlights the significance of ensuring access to food and achieving food security, aiming to eliminate hunger by 2030 [1]. Enhancing the production of locally grown cereals and grains is a key strategy for addressing hunger. Rice (*Oryza sativa*) serves as a primary food source for over half of the global population and stands as one of the most vital commercial crops, with an annual global yield of approximately 700 million tons. It is cultivated in at least 114 countries, with China and India contributing 50% of the total production [2], Nigeria's rice sub-sector currently faces challenges, including weak and inefficient producer–market linkages and suboptimal rice quality. In parts of Southwestern Nigeria, paddy rice is predominantly harvested manually using traditional tools like sickles, knives, scythes, and cutters, whereas in developed nations, almost all rice harvesting is mechanized through combine harvesters [3].

Nigeria has a dynamic rice-farming sector, employing various techniques to process harvested paddy into finished rice. The process begins with threshing, which separates the grain from the straw, followed by milling, which removes the husk and bran layers, leaving the white endosperm. Polishing is then carried out to eliminate any residual bran layers and enhance the rice's appearance. After being polished, the rice is sorted into grades based on size, shape, and color. Broken rice, a by-product of milling, is separated and processed differently. The rice supply chain in Nigeria is crucial for linking farmers with millers and processors, a process focusing on improving product quality, minimizing waste, and reducing processing time [4,5]. Timing and harvesting methods are critical factors affecting rice production. Significant losses can occur if harvesting is not conducted when the paddies are appropriately mature and moist. Premature harvesting when rice has a high moisture content increases drying costs, promotes mold growth and insect infestation, and leads to high breakage rates and low milling yields [6]. A study on Southeastern Nigeria revealed that delays in harvesting increased paddy losses by 67% due to scattering, rising from 1.5% to 2.5%. Similarly, a study from India reported a 10.3% rise in paddy-harvesting losses, increasing from 1.74% to 1.92%, attributed to delayed harvesting and insufficient harvesting equipment [7]. In Southwestern Nigeria, threshing is primarily performed manually, although mechanical threshing is employed on large, irrigated farms. Farmers often beat the harvested straw against metal drums, cemented vessels, or tree trunks using wooden sticks. After threshing, cleaning is carried out to separate whole grains from broken grains and foreign materials like straw, stones, sand, chaff, and weed seeds. Winnowing is the most common cleaning method in developing countries, though screening or sifting, conducted manually or mechanically, is also prevalent. [8], increases in rice production can enhance farmers' income, reduce poverty, and improve living standards. However, in Nigeria, rice production and processing are hindered by numerous challenges, with limited access to agricultural credit being a significant factor [9].

2. Statement of the Research Problem

Cereal and grain production in Nigeria is hindered by factors such as limited access to credit facilities, high input costs, and inadequate processing, packaging, and storage. These crops are staple foods for most Nigerians, yet postharvest grain losses (PHGLs) and quality losses (PGQLs) remain substantial. Efforts to mitigate these losses and improve production have been constrained by the absence of simple, cost-effective, and adaptable postharvest management practices and technologies [3]. In Southwestern Nigeria, rice processing is hindered by issues such as a reliance on traditional threshing methods, which result in significant husk breakage. The uneven distribution of heat during parboiling, due to rudimentary handling techniques, further exacerbates quality issues. The inadequate drying methods practiced by local farmers also lead to milling inefficiencies and increased rice breakage. These challenges underscore the need for a study to evaluate local rice-processing, -packaging, and storage practices among rice processors in Southwestern Nigeria, West Africa [10].

3. Research Objectives

1. Describe the socio-economic characteristics of local rice processors in the study area.
2. Assess the processing, packaging, and storage methods adopted by rice processors in the study area.

3. Identify the constraints faced by rice processors during rice processing, packaging, and storage in the study area.

4. Methodology

4.1. Study Area

This study was conducted in the Southwestern geopolitical zone of Nigeria, encompassing the states of Oyo, Ogun, Osun, Ondo, Ekiti, and Lagos. This zone, located between latitudes 5° N and 9° N, spans an area of 114,271 square kilometers, accounting for 12% of the nation's total land area. According to the 2006 Census, the region's population was 21,974,678 [11]. The zone comprises four distinct sub-ecologies: swamp, mangrove forests, moist and dry lowland forests, and derived savannah. The states of Lagos, Ondo, and Ogun fall within the tropical rainforest zone, which includes swamp forests along the coastal areas, while parts of Osun, Oyo, and Ekiti lie in the derived savannah [12]. The climate is equatorial, featuring distinct wet and dry seasons. The rainy season spans up to nine months, with peaks in July and September. Annual rainfall averages between 2600 mm and 2900 mm in the southern coastal areas, such as the states of Lagos and Ogun, with a zonal mean of 480 mm. During the dry season, average monthly temperatures range from 18 °C to 24 °C. Agriculture remains the primary livelihood in this zone, characterized by arable cropping systems involving rice, maize, millet, yam, and cassava, alongside common tree crops such as cocoa, oil palm, cashew, and kolanut. Both small- and large-scale rice farmers operate in the zone, engaging in rice processing, packaging, storage, and sales.

4.2. Data Collection Method

The procedure for data collection adhered to the ethical guidelines for human research with approval from Tai Solarin University of Education Human Research Ethics Committee (Approval No. TASUED-HREC/2023/045). Informed content was obtained from all respondents, ensuring their voluntary participation and confidentiality. Data collection involved direct engagement with rice farmers, processors, and preservers through field reconnaissance surveys, personal observations, and structured questionnaires. A total of 1200 questionnaires were randomly distributed to rice processors across the six states covered in the study.

4.3. Sampling Procedure and Data Analysis Method

A multi-stage sampling procedure was adopted to select the study sample, chosen for its effectiveness in simplifying data collection across large and geographically dispersed populations while enabling probability sampling without requiring a complete sampling frame. The process involved five stages: Firstly, the six southwestern states, namely, Ogun, Oyo, Ondo, Osun, Ekiti, and Lagos, were purposively selected due to their prominence in rice processing [12]. Secondly, registered rice processors in the extension blocks of these states were identified, with assistance from the Agricultural Development Programme (ADP) and relevant official reports. Thirdly, 10% of the identified extension blocks were randomly selected using a random sampling technique. Fourthly, another 10% of the extension cells within the selected blocks were also randomly sampled. Fifthly, a minimum of 200 registered rice processors were randomly selected from each of the six states, culminating in a total sample size of 1200 respondents. We categorized local rice processing, packaging, and storage methods based on the pooled mean scores of respondents, distinguishing between manual and mechanical approaches. Descriptive statistics such as frequency counts, percentages, means, and standard deviations were used to describe the respondents' socio-economic characteristics.

5. Results

Table 1 reveals that the rice processors in the region are predominantly male (65.1%), with females accounting for 34.9%. Regarding marital status, 86.9% were married, 9.2% were single, 2.5% were divorced, and 1.4% were widowed. The average age of the respondents was 30 years, the average education level was 7 years, the average household size was 13 persons, and the average monthly income was NGN 55,000, and the respondents' processed rice output averaged 5001 kg. Notably, 79.3% had over ten years of experience in rice processing, and 80.4% were members of cooperative societies.

Table 1. Socio-economic characteristics of the rice processors (n = 1200).

Variables	Frequency	Percentage	Mean	SD
Age (years)				
<30	221	18.4	30	10
31–40	343	28.6		
41–50	480	40.0		
51–60	156	13.0		
Sex				
Male	781	65.1		
Female	419	34.9		
Marital status				
Single	110	9.2		
Married	1043	86.9		
Divorced	30	2.5		
Widowed	17	1.4		
Education (years)				
No Formal Education	44	3.7	7	3
1–6	900	75.0		
7–12	216	18.0		
>13	40	3.3		
Household size (persons)				
0–4	58	4. 8	13	5
5–9	192	16.0		
>10	950	79.2		
Processing experience (years)				
1–5 yrs	48	4.0	14	10
6–10 yrs	200	16.7		
>10 yrs	952	79.3		
Areas of Specialization				
Rice Processing Only	320	26.7		
Rice Packaging and Storage	300	25.0		
Rice Storage Only	305	25.4		
All of the above	275	22.9		
Processed Rice Output (kg)			5001	2000
<3000	228	19.0		
3001–6000	672	56.0		
6001–9000	180	15.0		
>9001	120	10.0		
Monthly Income (NGN)			55,000	20,000
<20,000	50	4.2		
20,000–40,000	60	5.0		
40,001–60,000	460	38.3		
60,001–80,000	360	30.0		
>80,001	270	22.5		

Table 1. *Cont.*

Variables	Frequency	Percentage	Mean	SD
Member of a Cooperative				
Yes	965	80.4		
No	235	19.6		

Source: Field Survey, 2023.

Table 2 shows that the majority of rice processors employed manual techniques for threshing (82.5%), parboiling (83.3%), milling (83.3%), sun-drying (83.3%), packaging (70.8%), and storage (83.3%), leading to corresponding spoilage (13.6%), wastage (20.4%), losses (13.0%), reduced income (30.5%), and profit (22.5%), while few rice processors employed mechanical methods.

Table 2. Processing, packaging, and storage methods adopted by rice processors.

Processing Methods	Manual (%)	Mechanical (%)
Threshing	82.5	17.5
Parboiling	83.3	16.7
Milling	83.3	16.7
Drying	83.3	16.7
Packaging	70.8	29.2
Storage	83.3	16.7
Spoilage	13.6	3.0
Wastage	20.4	3.5
Losses	13.0	3.0
Income reduction	30.5	3.5
Profit	22.5	87.0

Source: Field survey, 2023.

Table 3 shows the major constraints faced by rice processors, among which a lack of modern equipment ranked first, high labor costs ranked second, high input costs ranked third, inadequate harvesting tools occupied the fourth spot, poor processing ranked fifth, poor packaging ranked sixth, and poor storage equipment ranked seventh.

Table 3. Distribution and ranking of major constraints faced by rice processors.

Items	Mean	Rank
Lack of modern equipment	1.901	1st
High cost of labor	1.860	2nd
High cost of inputs	1.800	3rd
Inadequate harvesting equipment	1.750	4th
Poor processing equipment	1.745	5th
Poor packaging equipment	1.741	6th
Poor storage equipment	1.735	7th
Poor marketing system	1.715	8th
Government policy	1.700	9th
Limited farmland	1.659	10th
Poor quality of extension services	1.640	11th
Late disbursement of loans	1.605	12th
Underfunding	1.600	13th
Lack of improved seeds	1.590	14th

Source: Field survey, 2023 Note: Means $\geq$ 1.4 severe, mild, and not a constraint.

6. Discussion

In the light of the results obtained, it can be seen that rice farmers and processors with higher education and smaller household size were able to enjoy increased income. These results corroborate the findings reported [13], who identified factors such as income, education, farm size, household size, experience, and cooperative membership as significant drivers of food security among rice farmers and processors. Similarly, [14] highlighted a positive relationship between socio-economic characteristics and the performance of rice farmers and processors in Nigeria.

Regarding methods of processing rice by farmers and processors, these findings align with those reported [5], who noted that a reliance on manual methods for rice processing often results in food spoilage, waste, and losses, stressing the need for modern alternatives. This trend mirrors the observations made [6], who reported similar packaging practices among rice farmers and processors in Northern Nigeria. These findings are consistent with those reported [15], which emphasized the importance of transitioning to modern storage methods to minimize food spoilage and waste. This demonstrates the decreased profitability of mechanical methods, in line with the findings of [10] who emphasized the role of modern technologies in boosting food production and security.

The results obtained on the constraints align with the finding of [8] who highlighted similar barriers, such as inadequate technologies, late loan disbursement, and limited financial support. [16] reported that lack of fund to purchase modern rice equipment is a major constraint to local rice farmers and processors. [17] also revealed that poor processing, packaging and storage equipment contribute greatly to the rice losses and wastages during production process.

7. Conclusions

We concluded that local rice storage in the study area continued to be hindered by significant waste, spoilage, losses, and income reduction due to inadequate processing, packaging, and storage methods. To enhance local rice production and minimize these issues, we recommend that modern technologies appropriate for rice processing, packaging, and storage, alongside credit facilities and inputs, should be made available for rice farmers and processors. The techniques used in rice farming and processing in Nigeria remain vital to its rice industry's overall success. With governmental support and the implementation of innovative practices, rice processing, packaging, and storage in Southwestern Nigeria can become more efficient and profitable.

Author Contributions: Conceptualization of the study work and data analysis, A.A.; methodology and resources, S.B.; data collection and writing—original draft preparation, T.O.; preparation and writing—review, J.J.; editing and supervision, O.O.; administration of questionnaire which involved gathering of local rice processors together and to collect needed information directly from them through the use of structured questionnaire with the help of all the authors and Association of Nigerian Rice Processors in Southwest Nigeria, Y.A. All authors have read and agreed to the published version of the manuscript.

Funding: The study received no external funding.

Institutional Review Board Statement: Not applicable.

Informed Consent Statement: The procedure for data collection adhered to the ethical guidelines for human research with approval from Tai Solarin University of Education Human Research Ethics Committee (Approval No. TASUED-HREC/2023/045). Informed content was obtained from all respondents, ensuring their voluntary participation and confidentiality.

Data Availability Statement: The datasets generated during and/or analyzed during the current study are included in the manuscript.

Acknowledgments: The researchers would like to specially thank CORAF for financing all the expenses incurred to attend and present at the 2023 symposium held in Lome, Togo. The researchers are also grateful to the state coordinators of the Association of Nigerian Rice Processors in Southwest Nigeria for supplying the data required for this study and connecting us directly with rice processors in their various local areas.

Conflicts of Interest: The authors declare that there are no conflicts of interest.

References

1. World Food Programme. *Fighting Hunger Worldwide;* World Food Programme Annual Report 2022; Series 10; World Food Programme: Rome, Italy, 2022; pp. 35–42.
2. United States Agency for International Development (USAID). *Food Insecurity and Nutritional Inadequacy in West African Countries;* USAID Publication: Washington, DC, USA, 2020; pp. 163–172.
3. Ugwu, K.C. Investigation on local rice processing, packaging and storage among farmers in some southeastern Nigeria. *SSRG Int. J. Recent Eng. Sci.* **2022**, *9*, 8–15.
4. Amours, D.T. Analysis of Food Production in Nigeria. *The Guardian.* 31 December 2022. Available online: https://guardian.ng/saturday-magazine/cover/analysis/food/production (accessed on 10 August 2023).
5. Odusina, O.A.; Banjo, J.O.S.; Akano, R.A. Assessment of the response of rural farming household's food security status to health and other idiosyncratic shocks. *Afr. J. Gend. Dev.* **2017**, *4*, 10–21.
6. Babatunde, R.O.; Omotesho, O.A.; Sholotan, O.S. Socio-economic characteristics and food security status of farming households in Kwara State, North-Central Nigeria. *Pak. J. Nutr.* **2021**, *6*, 49.
7. Danbaba, N. Rice Postharvest Technology in Nigeria: An Overview. 2019. Available online: https://scirp.org/journal/paperinformation?paperid=94313 (accessed on 5 July 2023).
8. Balana, B.B.; Oyeyemi, M.A. Agricultural credit constraints in Smallholder farming in developing countries. Evidence from Nigeria. *J. World Dev. Sustain.* **2022**, *22*, 100–112. [CrossRef]
9. Saheed, Z.; Alexander, A.; Isa, A.; Adeney, O. Anchor Borrowers' Programme on agricultural commodity price and employment generation in Kebbi State, Nigeria. *Eur. Sci. J.* **2018**, *14*, 213–256.
10. Belachew, D.G. Measuring food insecurity: The frequency and severity of "coping strategies" (PDF). *Food Policy* **2022**, *21*, 291–303.
11. National Population Commission. *Federal Government of Nigeria: 2006 Census Record;* National Population Commission: Abuja, Nigeria, 2006; pp. 61–65.
12. Adeoye, W.A.; Ademolu, A.D. *Research Methodologies and Models;* Tonad Publishers Limited: Lagos, Nigeria, 2022; pp. 141–145.
13. Nwahia, O.C. Analysis of the cost and economic returns in rice production in Ebonyi State, Nigeria. *Indones. J. Agric. Res.* **2020**, *3*, 205–214. [CrossRef]
14. Balogun, O.; Ayo-Bello, T.A.; Abasilim, C.; Abimbola, O.; Afodu, O.; Akinwole, O. Assessment of the performance of Anchor Borrowers' Programme Beneficiary and Non-Beneficiary rice farmers in Badagry Local Government Area, Lagos State, Nigeria. *Ife J. Agric.* **2021**, *33*, 62–76.
15. United Nations (UN). *Quantitative Indicators from Food Expenditure Survey Can Be Used to Target the Food Insecure in Africa;* UN Publications: New York, NY, USA, 2022; pp. 223–246.
16. Kiros, S.; Meshasha, G.B. Factors affecting farmers' access to formal financial credit in Basona Worana District, North Showa Zone, Amhara Regional State, Ethiopia. *J. Cogent Econ. Financ.* **2022**, *10*, 1–22. [CrossRef]
17. Sennuga, S.O.; Baines, R.N.; Conway, J.S.; Angba, C.W. Awareness and Adoption of Good Agricultural Practices among smallholder famers in relation to the Adopted Villages programme: The Case Study of Northern Nigeria. *Int. J. Biol. Agric. Healthc.* **2020**, *10*, 34–49.

proceedings

MDPI

Proceeding Paper

Postharvest Handling Practices of Sweet Orange Marketers in Benue State, Nigeria †

Oluwafunmilayo Olarewaju Aminu

Department of Agricultural Economics and Extension, Olusegun Agagu University of Science and Technology, Okitipupa 350105, Ondo State, Nigeria; funmiaminu83@gmail.com or oo.aminu@oaustech.edu.ng; Tel.: +234-8060990906

† Presented at the CORAF's 2023 Symposium on Processing and Transformation of Agricultural Products in West and Central Africa: Achievements and Opportunities for Private Sector Engagement, Lome, Togo, 21–23 November 2023.

Abstract: Successful marketing of fresh fruits requires careful attention to the details involved in postharvest handling (PH), which protects fruit's quality and nutritive and economic value. This study assessed the PH practices of sweet orange marketers in Benue State. Six (6) major orange markets were purposively selected. Thereafter, a total of 104 marketers were randomly sampled. The data were collected using an interview schedule and analysed using the frequencies, percentages, means, Chi-squares, and Pearson's Product Moment Correlations at $\alpha_{0.05}$. The knowledge of PH practices and the level of PH practice utilisation were high. Marketers' age ($r = -0.220$, $p < 0.05$), the use of sacks for packaging ($\chi^2 = 23.314$, $p < 0.01$), knowledge of PH ($r = 0.272$, $p < 0.01$), and attitudinal disposition towards PH ($r = 0.492$, $p < 0.01$) significantly influence the utilisation of PH practices. This study recommends the generation of awareness and sensitisation among marketers by horticultural crop institutions on improved preservation and packaging methods for sweet oranges.

Keywords: postharvest handling practices; knowledge; attitude; utilisation; marketers; packaging; sorting

1. Introduction

Sweet orange (*Citrus sinensis* (L.) Osbeck) is a citrus fruit notable for its wide cultivation and consumption globally. Sweet oranges account for about 50% of the citruses produced worldwide [1]. The global sweet orange production was 47.40 million metric tons in the 2023/2024 marketing season, 46.88 million metric tons in the 2022/2023 marketing season, and 50.32 million metric tons in the 2021/2022 marketing season [2], indicating a downward trend in production from the 2022/2023 to 2021/2022 marketing seasons and then a minor increase in production in the 2023/2024 marketing season.

According to the Food and Agriculture Organization of the United Nations, Brazil is the global leader in sweet orange production [3]. In the 2023/2024 marketing year, Brazil led globally in its orange production, with an annual output of about 16.5 million metric tons within the global production of fresh oranges for this marketing season [4]. Thus, Brazil's sweet orange production accounts for about 30% of the sweet oranges produced globally. There was a decline in sweet orange production in the 2023/2024 marketing year (48.82 million metric tons) compared to that in the 2022/2023 marketing year (50.41 million metric tons). Declines in orange production in countries such as Brazil and the United States of America have been attributed to bacterial disease [1]. In Africa, the FAO statistics in 2021 revealed that the leading producer of oranges was Egypt, followed by South Africa, Algeria,

Morocco, and Ghana [5]. Over the years, Egypt has been among the top 10 producers of oranges in the world [3,4,6], accounting for about 8% of global orange production.

Sweet oranges are also cultivated in Nigeria, where the climate is favourable for growing. Tropical rainforest and guinea savannah ecological zones are the most suitable for sweet orange cultivation in Nigeria [7]. Its annual production of sweet oranges is nearly 930,000 tons, cultivated on approximately 3 million hectares of land in Nigeria [8]. In Nigeria, Benue State is the leading producer of sweet oranges [9,10]. In 2019, an average production of 22,164.6 bags of sweet oranges was reported, out of which 16,672.3 bags were sold and 5492.3 bags were lost in selected LGAs in Benue State, Nigeria; the post-harvest losses of sweet oranges can be as high as 21–30 percent [11]. In a study conducted among roadside orange sellers in Oyo State, Nigeria, the losses incurred were reportedly up to 14.2% [12]. Previous research [8] noted that sweet oranges contribute to the development of agriculture and the nation's economy. However, the maximisation of their potential is limited by challenges such as pests, diseases, inadequate capital, price volatility, and climate change, among others [7,8]. Sweet oranges are common fruits consumed widely by the populace in Nigeria, especially because of their sweet taste and health benefits, contributing to nutrition security. The common varieties grown in Benue State, Nigeria, are mostly Ibadan Sweet and Valencia [11].

Sweet oranges are important sources of vitamins, minerals, and dietary fibre, and they play vital roles in improving human nutritional status. The vitamins and flavonoids present in citrus fruits such as sweet oranges enhance their nutritional and healing properties [13]. Sweet oranges are essential to an individual's normal body growth and development. They are very nutritious and low in calories and aid in lowering the risk of many health conditions [14]. Sweet orange fruits are quite sweet because they contain high concentrations of fruit sugar. Sweet oranges are usually harvested when they are fully ripe or half-ripe. Orange juice is rich in vitamin C and antioxidants; hence, sweet orange fruits have health benefits. Sweet oranges are characterised by having a distinctive flavour and being a good source of carbohydrates and many B vitamins, minerals, and biologically active phytochemicals, such as carotenoids and flavonoids, which provide provitamin A and purported antioxidant benefits [15].

Fresh fruits like sweet oranges are inherently perishable and are prone to deterioration, especially when injured, which limits the optimisation of their health benefits by the human body. Their deterioration after harvest contributes to most of this loss and quality degradation, thus contributing to both cost and nutritive losses [13]. Due to the highly perishable nature of sweet oranges, losses are incurred during postharvest handling and their distribution from production areas to consumption centres. The losses in citrus fruits can be as high as 30% of the total output and can reach up to 50% in low-income countries [13]. Hence, successful marketing of sweet oranges requires careful attention to the details involved in postharvest handling, which can protect their quality and nutritive and economic value and assure food safety. Good postharvest handling practices will reduce postharvest losses, improve customers' satisfaction with the produce, attract good prices, and increase marketers' incomes.

Efforts to reduce postharvest losses in perishable fruits such as sweet oranges are crucial to avoiding wastage and also reducing the cost of preventing food losses. The causes of post-harvest losses of sweet oranges in Nigeria include mechanical damage, weather conditions, physiological deterioration, inadequate handling skills, insect and pest infestations, poor storage, and poor market patronage [16,17]. In Ethiopia, fruit losses are caused by factors such as the harvesting techniques, improper storage, improper packaging methods, and improper transportation/marketing [18]. Likewise, in Zimbabwe, the contributing factors to postharvest losses include poor production practices, pest infestations, diseases, poor storage facilities, and poor physical infrastructure [19].

Thus, proper postharvest handling of sweet oranges is of great importance to reducing losses and ensuring their continuous availability and increases in marketers' incomes. The level of profit that will be realised by marketers from oranges sold is, to an extent, dependent on their employment of good postharvest handling practices. A viable orange marketing enterprise is crucial to the sustainability of crop production, hence the need to examine the postharvest handling practices of sweet orange marketers. During postharvest handling, oranges must be sorted to remove bruised or diseased fruits and avoid contamination of wholesome oranges. Also, protecting oranges from the sun and rain is crucial to enhancing their market value. How oranges are loaded and offloaded from vehicles is equally important.

Previous studies among sweet orange marketers in Benue State, Nigeria, have focused on postharvest losses [10,20]. However, empirical data on the postharvest handling practices of sweet orange marketers in Benue State, Nigeria, is sparse. Hence, this study examined the postharvest handling practices (preservation, sorting, packaging, transportation, and marketing) of sweet orange marketers in this study area. Also, the sweet orange marketers' knowledge of postharvest handling practices and their attitudinal dispositions were examined. This study hypothesised that the sweet orange marketers' knowledge and attitudes would affect their utilisation of postharvest handling practices.

2. Materials and Methods

Benue is a state in Central Nigeria, with its capital situated in Makurdi. Its geographic coordinates are a longitude $7°47'$ and $10°0'$ east and a latitude $6°25'$ and $8°8'$ north. The state shares a common boundary with the Republic of Cameroon in the southeast. Benue State is acclaimed to be the nation's "food basket" because of its rich and diverse agricultural produce, which includes oranges, mangoes, pears, garden eggs, yams, rice, beans, cassava, potatoes, and maize, among others. The population of this study was sweet orange marketers in Benue State, Nigeria. Major sweet orange markets were purposively selected in Benue State. The selected markets were the Ihugh, Gboko, Lessel, Ikyobo, Aliade, and Korinya markets, with populations of 104, 118, 30, 26, 36, and 33 sweet orange marketers, respectively, giving a total of 347 marketers from the sampled markets. Using simple random sampling, 30% (104) of the marketers were selected. The marketers in the selected sweet orange markets mainly buy oranges in bulk from sweet orange farmers and transport them to different states/cities in Nigeria such as Abuja, Kano, Kaduna, Portharcourt, Jos, Onitsha, Lagos, Maiduguri, Lafia, Aba, Benin, Ibadan, Bauchi, Sokoto, Yobe, and Zaria, among others (this was discovered during the data collection). The sampled sweet orange marketers in this study mainly sell them wholesale to other marketers in these states/cities because they deal with a large volume of oranges. A well-structured interview schedule was used for the data collection.

Knowledge of postharvest handling practices was assessed by presenting a list of 17 knowledge items consisting of positive and negative statements to the sweet orange marketers. This was measured using a 2-point scale of true and false, with a score of 1 assigned to the correct response and 0 to the wrong response. The marketers' knowledge scores were added up, and their mean scores were computed. The minimum and maximum knowledge scores were 13 and 16, respectively, and the mean score was 14.72 ± 0.79. The mean score was used to categorise respondents as either having a very high or a high knowledge level of postharvest handling. This categorisation was based on the fact that out of the 17 knowledge items presented to the sweet orange marketers, the minimum score was 13, indicating that all the marketers had a high knowledge of postharvest handling practices.

Attitudes towards postharvest handling were assessed by presenting a list of 10 attitudinal items consisting of positive and negative statements to the sweet orange marketers. This was measured using a 5-point Likert-type scale of Strongly Agree, Agree, Undecided, Disagree, and Strongly Disagree, with scores of 5, 4, 3, 2, and 1 assigned, respectively, for positively worded statements and the reverse for negatively worded statements. The marketers' scores on their attitudes were added up, and the mean score was computed. The minimum and maximum attitudinal scores were 33 and 46, respectively, and the mean score was 39.53 ± 3.57. The mean score was used to categorise the marketers as either having a favourable or unfavourable attitude towards postharvest handling.

Postharvest sweet orange handling practices were assessed by presenting the marketers with a list of 20 items on their postharvest handling practices. They were measured using a 3-point scale of always, sometimes, and never, with scores of 2, 1, and 0 assigned, respectively, indicating the frequency of use of postharvest handling practices. The marketers' scores on their postharvest handling practices were added up, and the mean score was computed. The minimum and maximum postharvest handling practice utilisation scores were 14 and 26, respectively, and the mean score was 19.91 ± 2.54. The mean score was used to categorise the respondents as either having a high or low level of utilisation of postharvest handling practices.

The descriptive statistics used for the data analysis were the frequency, percentages, means, and standard deviation, and the inferential statistics used to analyse the data collected were Chi-squares and Pearson's Product Moment Correlations.

3. Results

3.1. The Personal Characteristics of Sweet Orange Marketers

Table 1 presents the distribution of the marketers based on their personal characteristics. The results show that the average age of the marketers was 39.89 ± 7.56 years. Most of the marketers were within the age range of 31–40 years (40.4%) and 41–50 years (38.5%). It was found that the marketers were mostly male (95.2%), the majority were married (85.6%), and very few (10.6%) were single. It was found that most of the marketers (96.1%) had formal education, while very few (3.8%) had no formal education. The majority of the marketers were primarily engaged in trading (90.4%), while a few marketers were primarily engaged in farming (9.6%). More than half of the marketers (55.8%) had secondary occupations, while 44.2% had no secondary occupation. The main secondary occupation of the marketers was farming (38.5%).

Table 1. Distribution of marketers based on personal characteristics (n = 104).

Variable	Freq.	%	Mean	Std. Dev.
Age (in years)				
<31	12	11.5	39.89	7.56
31–40	42	40.4		
41–50	40	38.5		
51–60	10	9.6		
Gender				
Male	99	95.2		
Female	5	4.8		

Table 1. *Cont.*

Variable	Freq.	%	Mean	Std. Dev.
Marital status				
Single	11	10.6		
Married	89	85.6		
Widowed	3	2.9		
Separated	1	1.0		
Education				
No formal education	4	3.8		
Secondary education	15	14.4		
Tertiary education	85	81.7		
Primary occupation				
Farming	10	9.6		
Trading	94	90.4		
Secondary occupation *				
None	46	44.2		
Farming	40	38.5		
Trading	7	6.7		
Crop processing	4	3.8		
Renting out taxis and motorcycles	4	3.8		
Driving	3	2.9		
Cycling	3	2.9		
Transport	2	1.9		

* multiple response set.

3.2. The Enterprise Characteristics of the Marketers

Table 2 shows the distribution of the marketers based on their enterprise characteristics. The findings from this study show that the average marketing experience was 12.29 ± 6.54 years. Personal savings (100.0%) were a source of capital for all of the marketers, while nearly a quarter (24.0%) of the marketers' source of capital was friends. Very few of the marketers' sources of capital were microfinance banks (6.7%), cooperatives (3.8%), or customers (3.8%). The results showed that all of the marketers used hired labour (100.0%). Furthermore, the maturity stages at which oranges were purchased were the mature green (90.4%), fully ripe (87.5%), and half-ripe (80.8%) stages. The marketers transport oranges at any time of the day, whether morning (100.0%), afternoon (100.0%), evening (100.0%), or night (100.0%). The major vehicles used for transportation were *911* lorries (87.5%). Other means of transportation were *faka* (43.3%), motorcycles (41.3%), trailers (27.9%), *12-tyre* trucks (30.8%), containers (14.4%), and J5 vans (18.3%). The marketers purchase the oranges while they are still on the trees (100.0%). Oranges are preserved under tree shade (100.0%), and the marketers do not process the oranges into any products (100.0%). All of the marketers patronised urban markets (100.0%) and sold them to wholesalers (100.0%). Oranges are transported by the marketers with and without (67.3%) packaging materials. Sack bags (86.5%) were used as the packaging materials.

Table 2. Distribution of marketers based on other enterprise characteristics (n = 104).

Variables	Freq	%	Mean	SD
Marketing experience				
<11	52	50.0	12.29	6.54
11–20	40	38.5		
21–30	11	10.6		
31–40	1	1.0		

Table 2. *Cont.*

Variables	Freq	%	Mean	SD
Sources of capital *				
Personal savings	104	100.0		
Cooperatives	4	3.8		
Customers	4	3.8		
Friends	25	24.0		
Microfinance banks	7	6.7		
Labour employed				
Hired	104	100.0		
Maturity stage at which you purchased oranges *				
Mature green	94	90.4		
Fully ripe	91	87.5		
Half-ripe	84	80.8		
Time of transport of oranges *				
Morning	104	100.0		
Afternoon	104	100.0		
Evening	104	100.0		
Overnight	104	100.0		
Anytime	104	100.0		
Means of transportation of oranges *				
911 lorries	91	87.5		
Trailers	29	27.9		
Motorcycles	43	41.3		
Faka	45	43.3		
12-tyre trucks	32	30.8		
Containers	15	14.4		
J5 vans	19	18.3		
Point of purchase of oranges				
While still on the trees	104	100.0		
Preservation of oranges				
Under shade of trees	104	100.0		
Processing of oranges				
No	104	100.0		
Marketing outlets				
Urban markets	104	100.0		
Category of buyers				
Wholesalers	104	100.0		
Method of loading				
Without packaging material	14	13.5		
With packaging material	20	19.2		
Both (with and without packaging material)	70	67.3		
Packaging material used				
Sack bags	90	86.5		

* Multiple-response.

3.3. The Marketers' Knowledge on Postharvest Handling

The distribution of the marketers based on their knowledge of postharvest handling is shown in Table 3. The results of this study show that all of the marketers (100.0%) knew that orange deterioration can be caused by vehicle breakdowns; sorting bruised oranges helps prevent contamination; protecting oranges from the sun after harvesting is good; orange bruising is not good; separating damaged oranges from good ones when packaging is important; poor ventilation results in orange deterioration during storage; mature and immature oranges should not be packed together; poor loading can lead to postharvest losses; bacterial diseases can be spread through poor field sanitation; and damaged oranges have no market value. Also, the majority of the marketers knew that throwing packaged oranges (in sacks) around can lead to bursting of the packaging materials (86.5%) and that oranges become overripe or softened when they are exposed to heat (97.1%).

Table 3. The distribution of the marketers based on their knowledge of postharvest handling.

S/N	Knowledge Statements	Correct Responses	
		F	%
1	A breakdown of vehicles while moving oranges to the market increases orange deterioration	104	100.0
2	After harvest, oranges should be protected from the sun	104	100.0
3	Sorting of diseased oranges helps prevent contamination of healthy oranges	104	100.0
4	Oranges should be prevented from getting bruised	104	100.0
5	The high cost of transportation contributes to postharvest loss	76	73.1
6	Damaged oranges should not be packed with good ones	104	100.0
7	Poor ventilation during storage encourages deterioration	104	100.0
8	Packing mature and immature oranges together is not good	104	100.0
9	Poor loading contributes to postharvest loss	104	100.0
10	Packaging oranges properly prevents mechanical damage	13	12.5
11	Poor field sanitation promotes the spread of bacterial diseases	104	100.0
12	Oranges lose value due to mechanical damage during the marketing period	104	100.0
13	Produce loses value due to exposure to high temperatures if it cannot be sold right away	104	100.0
14	The use of shade is important for oranges	101	97.1
15	Injuries to oranges increase water loss in oranges	103	99.0
16	Throwing packaged oranges in sacks can result in bursting of the packaging materials	90	86.5
17	Oranges become overripe or soften when exposed to heat	101	97.1

3.4. Level of Knowledge of Postharvest Handling and Loss

The results in Table 4 reveal that the majority of the marketers (60.6%) had a very high level of knowledge on postharvest handling of sweet oranges, while 39.4% of the marketers had a high level of knowledge on postharvest handling.

Table 4. The distribution of the marketers based on their level of knowledge.

Knowledge	Freq.	%	Minimum	Maximum	Mean	S Dev.
High (<mean)	41	39.4	13.00	16.00	14.72	0.79
Very high (≥mean)	63	60.6				

3.5. The Marketers' Attitudes Towards Postharvest Handling

According to the results in Table 5, the marketers' responses to attitudinal statements are shown. The majority of the marketers agreed that preservation of oranges under shade is good (90.4%) and well-ventilated materials should be used to pack oranges (71.1%), and all of the marketers (100.0%) agreed that postharvest handling of oranges is very important. Sorting damaged oranges was considered an important exercise (99.0%), and most of the marketers (92.3%) attested that damaged oranges are prone to attacks by decay mechanisms. All of the marketers (100.0%) agreed that it is important to ensure good sanitation of the collection centres to maintain fruit's quality and market and economic value. Furthermore, about 40.0% of the marketers believed that oranges cannot be packaged with any type of packaging material, while 60.5% were indifferent. Most of the marketers (97.1%) believed that careful handling of oranges would prevent spoilage, and 76.9% agreed that oranges should be packaged to protect them from being roughly handled. Almost all of the marketers (99.0%) opined that washing oranges may not be necessary.

Table 5. Percentage distribution of marketers based on attitudes towards postharvest handling.

S/N	Attitudinal Statements	SA	A	U	D	SD	Mean
1	Preserving oranges under shade is simply not needed	1.0	5.8	2.9	67.3	23.1	4.06
2	It is necessary to pack oranges in well-ventilated materials	26.9	44.2	27.9	0.0	1.0	3.96
3	Postharvest handling of oranges is a waste of time	0.0	0.0	0.0	59.6	40.4	4.40
4	Sorting of damaged oranges is not just an important exercise	0.0	1.0	0.0	44.2	54.8	4.53
5	Washing of oranges may not be necessary	90.4	8.7	0.0	0.0	1.0	1.13
6	Damage caused during postharvest handling renders oranges liable to attacks by decay	77.9	14.4	0.0	1.0	6.7	4.56
7	Any type of packaging material is suitable for oranges	0.0	0.0	60.5	33.7	5.8	3.45
8	Good sanitation of the collection centre does not matter	0.0	0.0	0.0	41.3	58.7	4.59
9	Careful handling of oranges may not necessarily prevent spoilage/contamination	1.0	1.9	0.0	20.2	76.9	4.70
10	Packaging of oranges is necessary to protect the produce against rough handling during loading and unloading	41.3	35.6	20.2	2.9	0.0	4.15

SA = Strongly Agree, A = Agree, U = Undecided, D = Disagree, and SD = Strongly Disagree.

3.6. Levels of Attitudes Towards Postharvest Handling

The results, as presented in Table 6, reveal the marketers' attitudes. Based on the attitudinal mean values computed, the marketers were categorised as either having a favourable or an unfavourable level of attitudinal disposition towards postharvest handling. The findings from this study show that 53.8% of the marketers had an unfavourable attitudinal disposition towards postharvest handling, while 46.2% had a favourable attitudinal disposition.

Table 6. The distribution of the marketers based on their attitudinal dispositions.

Attitudinal Disposition	Freq.	%	Minimum	Maximum	Mean	S Dev.
Favourable (≥mean)	48	46.2	33.00	46.00	39.53	3.57
Unfavourable (<mean)	56	53.8				

3.7. Postharvest Handling Practices of the Orange Marketers

Table 7 shows the distribution of the marketers based on their frequency of use of postharvest handling practices. It was found that 76.9% of the marketers always preserved their oranges under the shade of trees, and the majority of the marketers sometimes sorted oranges by bruises (68.5%), by stage of ripeness (71.3%), and by size (70.4%). Sorting of oranges before packaging them or loading them into vehicles (95.7%), the use of packaging materials for oranges before loading (75.3%), and packaging oranges immediately after harvesting (97.8%) were applied on an occasional basis. Some of the marketers (43.0%) always ensure that only oranges in good condition are packaged. Most marketers occasionally avoid overloading of vehicles (87.0%) and throwing produce from vehicles when loading and offloading it (67.6%). The majority of the marketers always sell their oranges even if the price is not favourable (63.9%), and 73.1% do not guarantee that oranges will be sold on the same day that they arrive to the market.

Table 7. The distribution of the marketers based on their postharvest handling practices.

S/N	Postharvest Handling Practices	Frequency of Use			Mean
		Always	Sometimes	Never	
	Preservation				
1	Store in a cold room	0.0	0.0	100.0	0.00
2	Store under shade of trees	76.9	23.1	0.0	1.77
3	Store in sheds/tents	0.0	0.0	100.0	0.00
	Sorting				
4	By bruises	15.7	68.5	15.7	1.00
5	By stage of ripeness	12.0	71.3	16.7	0.95
6	By size	13.9	70.4	15.7	0.98
	Packaging				
7	Sort oranges before packaging/loading	4.3	95.7	0.0	1.04
8	Use packaging materials for oranges before loading	24.7	75.3	0.0	1.26
9	Package orange immediately after harvesting	2.2	97.8	0.0	1.02
10	Ensure packaging materials are clean to prevent contamination	100.0	0.0	0.0	2.00
11	Package only good-quality oranges	43.0	57.0	0.0	1.45
	Transportation				
12	Avoid overloading the vehicle	13.0	87.0	0.0	1.13
13	Ensure the load is stable	88.9	11.1	0.0	1.88
14	Keep vehicles under shade when loading to minimise heat	1.9	1.9	96.3	0.04
15	Avoid throwing produce from the vehicle when loading and offloading	20.4	67.6	12.0	1.09
	Marketing				
16	Sell oranges even if the price is not favourable	63.9	36.1	0.0	1.65
17	Sort diseased oranges before marketing	25.0	61.1	13.9	1.12
18	Sell oranges the same day on which they arrive at the market	0.0	26.9	73.1	0.24
19	Do not leave oranges for days before selling	24.1	75.9	0.0	1.25
20	Always keep collection centres clean	0.0	0.9	99.1	1.01

3.8. The Level of Utilisation of Postharvest Handling Practices

Table 8 shows the distribution of the marketers based on their level of utilisation of postharvest handling practices. The results show that 62.5% of the marketers had a high level of utilisation of postharvest handling practices, while 37.5% had a low level of utilisation of postharvest handling practices.

Table 8. The distribution of the marketers based on their level of postharvest handling practices utilisation.

Postharvest Handling Practice Utilisation Levels	Freq.	%	Minimum	Maximum	Mean	S Dev.
Low (<mean)	39	37.5	14.00	26.00	19.91	2.54
High (≥mean)	65	62.5				

3.9. Hypothesis Testing

3.9.1. The Chi-Square Relationship Between the Independent Variables and the Utilisation of Postharvest Handling Practices

The results in Table 9 show that there was no significant relationship between the marketers' sex ($\chi^2 = 3.152$, $p > 0.05$), marital status ($\chi^2 = 4.781$, $p > 0.05$), education ($\chi^2 = 2.550$, $p > 0.05$), or primary occupation ($\chi^2 = 1.446$, $p > 0.05$) and their utilisation of postharvest handling practices. However, there was a significant relationship between the use of sacks for packaging and postharvest handling practices ($\chi^2 = 23.314$, $p < 0.01$).

Table 9. Chi-square relationships between independent variables and postharvest handling practice utilisation.

Variables	χ^2	Df	*p*-Value	Decision
Sex	3.152	1	0.076	Not significant
Marital status	4.781	3	0.189	Not significant
Education	2.550	2	0.279	Not significant
Primary occupation	1.446	1	0.229	Not significant
Packaging with sacks	23.314 **	1	0.000	Significant

** Significant at 0.01.

3.9.2. Pearson's Product Moment Correlation (PPMC) Between Independent Variables and Utilisation of Postharvest Handling Practices

From the results in Table 10, a significant association exists between the marketers' age (r = −0.220, $p < 0.05$), knowledge of postharvest handling (r = 0.272, $p < 0.01$), and attitudinal disposition towards postharvest handling (r = 0.492, $p < 0.01$) and the marketers' utilisation of postharvest handling practices. But there was no significant association between years of experience marketing sweet oranges (r = −0.115, $p > 0.05$) and the marketers' utilisation of postharvest handling practices.

Table 10. PPMC relationship between independent variables and postharvest handling practices utilisation.

Variables	r-Value	*p*-Value	Decision
Age	−0.220 *	0.025	Significant
Years of marketing experience	−0.115	0.246	Not significant
Knowledge	0.272 **	0.005	Significant
Attitude	0.492 **	0.000	Significant

* = significant at 0.05, ** = significant at 0.01.

4. Discussion

4.1. The Personal Characteristics of the Marketers

The average age of the marketers obtained in this study shows that they are agile and active, as they are still productive. The average age of 39.89 ± 7.56 years indicates the youthfulness of the marketers in Benue State, Nigeria. Younger marketers are expected to understand the importance of postharvest handling practices better. The result of this study aligns with previous research [10] which reported a mean age of 39.0 years among sweet orange marketers in Benue State. Also, in a study [21] conducted among fruit marketers in Bahir Dar, Ethiopia, it was found that fruit marketing is mainly handled by young people. Marketing of oranges was found to be predominantly dominated by men in the study area. This finding might be attributed to the long distance associated with orange marketing in Benue. During the field survey, it was found that the marketers spent days on the road when transporting oranges. Hence, women may not be able to combine housework with the rigour of travelling long distances for days, unlike men. This conflicts with a previous finding [22] that showed that the majority of the fruit marketers in Lagos, Ogun, and Osun States, Nigeria, were female. Likewise, another study reported that sweet orange marketing is dominated by women in Oyo State, Nigeria [23]. The higher number of marketers who are married shows that they have a measure of household responsibilities to attend to. So, ensuring good postharvest handling practices might translate into a higher income to meet their household needs. The marketers were found to be formally educated in this study. Most of the marketers had a tertiary education, indicating that the majority were graduates. This high level of education is expected to enhance the marketers' acceptance of any initiative that will enhance their postharvest handling practices. This result agrees with a previous study [21] that found that fruit marketing is mainly handled by relatively educated individuals compared to fruit production. Contrastingly, the result on education disagrees with a previous study [24] which reported that the majority of fruit marketers are formally educated to the level of a primary education. The marketing of oranges was the primary means of livelihood for almost all of the marketers. In addition, more than 50.0% of the marketers engage in other livelihood-generating activities, implying multiple streams of income for some of the marketers. More than 40.0% of the marketers engaged solely in orange marketing.

4.2. The Enterprise Characteristics of the Marketers

The average years of marketing experience of 12.5 years indicated that the marketers were not just beginners or novices in the orange marketing business. Marketing experience spanning over 10 years should have allowed the marketers to understand the dangers of poor postharvest handling practices and how to guide against them. Similarly, a past study [7] reported average marketing experience of 15 years among sweet orange marketers in Benue State, Nigeria. Marketers finance their orange businesses primarily by using their savings, while one-quarter of the marketers receive financial assistance from friends. Financial assistance from cooperatives, customers, and microfinance banks was very sparse. This finding is in tandem with a past study [23] that found that very few of the sweet orange marketers in Oyo State received finance from cooperatives. Hence, marketers might not be able to expand their orange businesses beyond their personal financial strength. Also, poor postharvest handling practices leading to the loss of produce can result in a collapse in the business for marketers in the absence of an alternate source of finance. All of the marketers employed the services of hired labourers for their marketing activities. This might be because of the high volume of trade in the study area, as found during the field survey. The findings from this study align with previous research [25] that found that all marketers relied on the use of hired labourers for their orange enterprises regarding transportation. Oranges are usually purchased by marketers while they are still on the trees, either at the mature green or fully or half-ripe stages. The most important factor that determines the postharvest life and final quality of fruits, such as their appearance, texture, flavour, and nutritive value, is maturity at harvest [26]. After harvesting, oranges are temporarily preserved under the shade of trees before loading them into vehicles. Marketers do not process the oranges into any products. All oranges bought are sold in urban markets to wholesalers. The oranges are transported using various means, such as 911 lorries, trailers, motorcycles, 12-tyre trucks, containers, and J5 vans. But in Lagos, Osun, and Ogun States, the use of open lorries and pick-up vans for the transportation of oranges is common [22]. As stated by the marketers during the field survey, vehicles are left open most of the time during transportation. The use of motorcycles is to facilitate the movement of the oranges from the farmers' farms to collection points, where they are loaded into vehicles for onward transportation to urban markets in other states in Nigeria. Oranges are transported at any time of the day, whether morning, afternoon, or evening, or even at night. In essence, the transportation of oranges is not restricted to a particular time of the day. During the field survey, it was discovered that the marketers travelled long distances to urban markets in both far and near states, such as Kano, Kaduna, Nasarawa, Oyo, Rivers, Abia, Sokoto, Borno, Yobe, Bauchi, Jos, Edo, Abuja, and Lagos, among others, to sell their produce. This might account for the reason why credence is not given to transporting oranges during the coolest part of the day in Benue State, Nigeria. Before loading the oranges into vehicles, marketers can choose to package them or not. Thus, packaging oranges is not compulsory, but whenever marketers want to package their oranges, sacks are used. In Oyo State, Nigeria, it was reported that none of the orange marketers sampled (both wholesalers and retailers) used any packaging materials for their oranges [23].

4.3. Knowledge of Postharvest Handling Practices

The marketers displayed high knowledge of postharvest handling practices. They knew that oranges deteriorate whenever vehicles break down. This reduces the market value of oranges, which invariably lowers the marketers' incomes. Good roads are essential to ensuring timelines in the transportation of oranges. This helps to ensure that oranges arrive at the markets or distribution centres in good condition. Bad roads are a severe cause of postharvest losses of sweet oranges [27]. Marketers know the negative effects of heat on oranges. So, oranges are gathered under trees to protect them from the sun. An increase in temperature reduces the shelf life of horticultural crops, as most factors that cause produce to deteriorate or reduce its quality usually occur at a faster rate as the temperature increases [28]. Another postharvest handling practice that marketers know about is the sorting of oranges. Sorting is carried out to remove diseased oranges so that healthy ones will not be contaminated. Sorting is also carried out to remove bruised oranges. Bruised oranges spoil quickly, and if they are not removed, they contaminate and quicken the deterioration of healthy ones. No customer will want to buy oranges that have gone bad. Among roadside orange sellers in Oyo State, Nigeria, a high premium is placed on the cleaning and sorting of oranges [12]. Marketers know that oranges lose value due to exposure to high temperatures if they cannot be sold right away. After harvesting, the longer oranges remain unsold, the greater the reduction in their market value and the lower the profit made. Since oranges are perishable, leaving them for days without preservation increases their rate of deterioration.

4.4. The Marketers' Attitudes Towards Postharvest Handling

The marketers were positively inclined towards placing the oranges in the shade. This helps to reduce the effects of heat on oranges. The marketers agreed that it is necessary to pack oranges in well-ventilated materials. This is crucial to reducing the heat generation within packaged oranges. In the opinion of the marketers, postharvest handling of oranges is not a waste of time. This implies that the marketers have knowledge of the benefits accruing from good postharvest handling and are positively inclined towards ensuring good practices. Careful handling of oranges helps prevent spoilage. The marketers' attitudinal disposition towards sorting out damaged oranges was favourable, as it was considered an important exercise. Damage caused during postharvest handling makes oranges susceptible to attacks by decay. The marketers were indifferent towards the attitudinal statement "any type of packaging material is suitable for oranges". This might have been because the only packaging materials known by the marketers were sacks. Also, packaging oranges is not mandatory, as the marketers in the study area sometimes load their oranges into vehicles without using any packaging materials. The marketers show a favourable attitude towards maintaining good sanitation in collection centres. The cleanliness of the collection centres is crucial to reductions in losses, as contamination of healthy fruits results in more losses.

4.5. The Postharvest Handling Practices of the Orange Marketers

The preservation method used by marketers is keeping the oranges under shade before loading them into vehicles and when the oranges arrive to market. None of the marketers are using advanced methods of preservation. Poor storage facilities and a lack of processing facilities have been identified as severe causes of postharvest losses of sweet oranges [4]. Sorting was occasionally carried out by most of the marketers. Oranges are sometimes sorted by bruising, stage of ripeness, or size. It was discovered during the field survey that the marketers sell most of their oranges at urban markets in northern Nigeria, where the buyers are not so meticulous about sorting, unlike in Lagos markets. Nevertheless, sorting is important to maintaining the good quality of the oranges. Any scarring on sweet oranges serves as a point of entry for microbes, thereby quickening spoilage [29]. It was discovered that most marketers who use packaging materials do so on an occasional basis. This implies that these marketers do not use sacks exclusively during orange marketing. Oranges are occasionally packaged immediately after harvesting. This is because harvesting might take days at times, and there is no point in packaging the oranges if they are not ready for transportation. More so, packaging materials are not used by marketers all the time. Past research [21] reported that the use of packaging materials during fruit transportation would reduce fruit spoilage resulting from mechanical damage during loading and unloading and when driving on rough roads. However, this is dependent on the kind of packaging materials used because with the use of sacks in the study area, oranges can still suffer mechanical damage [21]. The cleanliness of the packaging material is considered important, as noted by the marketers who use packaging materials in the study area. There is no guarantee that all packaged oranges are of good quality. It was found during the field survey that packaging is mostly carried out by hired labourers who may not pay careful attention to sorting the oranges. Furthermore, overloading of vehicles seems to be inevitable, as most of the marketers only occasionally ensure that vehicles are not overloaded. Overloading and stacking of oranges contribute to the postharvest losses experienced by marketers [30]. However, stability of the loaded vehicles is always ensured by the marketers. When the vehicles are being loaded with oranges, they are not usually kept under shade by most marketers. Also, there is no guarantee that the oranges will be sold on the same day that they arrive to the market. There are times when the marketers have to sell their oranges even if the price is unfavourable. This is to guard against losses and shortages.

4.6. The Relationship Between the Independent Variables and Postharvest Handling

The personal characteristics of the marketers, such as their sex, marital status, education, and primary occupation, do not significantly relate to their postharvest handling practices. This implies that these aforementioned personal characteristics do not influence the marketers' postharvest handling practices. The significant relationship that exists between the use of sacks for packaging and postharvest handling practices indicates that the use of sacks as packaging materials had a significant influence on the marketers' postharvest handling practices. A significant and negative correlation was obtained between the marketers' age and their use of postharvest handling practices. This indicates that the use of postharvest handling practices decreases with age. Hence, younger marketers engage more in postharvest handling practices, thus having a higher level of use of postharvest handling practices. However, the older marketers engaged less in postharvest handling practices. The correlation between the marketers' knowledge and postharvest handling practices was significant and positive. This implies that knowledge had a positive influence on their use of postharvest handling practices. Thus, the higher their level of knowledge, the higher their use of postharvest handling practices. There was a significant and positive correlation between attitude and postharvest handling practices. This indicates that a favourable attitudinal disposition positively affects the use of postharvest handling practices. Thus, the more favourable marketers' attitudes are, the greater the increase in their use of postharvest handling practices.

5. Conclusions

A viable orange marketing enterprise is crucial to the sustainability of crop production. As found in this study, orange marketers are young and still in their active years. The marketers were predominantly male and educated, mainly graduates. Not all of the marketers engage in other income-generating activities; hence, maintaining good postharvest handling practices is a way for them to sustain their orange enterprises by achieving good returns from the sale of oranges while curtailing their losses. Oranges are transported at any time of the day to the various urban markets where they will be sold. There was no importance attached to moving the oranges during the coolest part of the day to guard against heat. Also, oranges are transported either with or without packaging materials. The marketers had good knowledge of postharvest handling practices and were positively inclined towards the use of these practices. However, their use of some of the postharvest handling practices assessed in this study is occasional. The empirical evidence from this study established that knowledge and attitude positively influence the marketers' use of postharvest handling practices. Hence, with increased knowledge and favourable attitudinal dispositions, the marketers' postharvest handling practices can be enhanced. Invariably, the postharvest losses can be minimised and the marketers' profits maximised. This study recommends the generation of awareness and sensitisation among marketers by horticultural crop institutions on improved preservation and packaging methods for sweet oranges.

Funding: This research received no external funding.

Institutional Review Board Statement: Not applicable.

Informed Consent Statement: Informed consent was obtained from all subjects involved in the study.

Data Availability Statement: The dataset is available on request from the author. The raw data supporting the conclusions of this article can be made available by the author on request.

Conflicts of Interest: The author declares that there are no conflicts of interest associated with this study.

References

1. Seminara, S.; Bennici, S.; Di Guardo, M.; Caruso, M.; Gentile, A.; La Malfa, S.; Distefano, G. Sweet Orange: Evolution, Characterization, Varieties, and Breeding Perspectives. *Agriculture* **2023**, *13*, 264. [CrossRef]
2. Villa, P.N. Orange production worldwide from 2012/2013 to 2023/2024. *Statista* 14 February 2025. Available online: https://www.statista.com/statistics/577398/world-orange-production/ (accessed on 14 April 2025).
3. Worldostats. Orange Production by Country 2024. Available online: https://worldostats.com/orange-production-by-country-2024/#google_vignette (accessed on 14 April 2025).
4. Villa, P.N. Leading Orange Producing Countries Worldwide in 2023/2024 (in *million* metric tons). *Statista* 31 March 2025. Available online: https://www.statista.com/statistics/1044840/major-orange-producers-worldwide/ (accessed on 14 April 2025).
5. Africa View Facts. Top orange producers in Africa. 2 May 2024. Available online: https://africaviewfacts.com/stats/top-orange-producers-in-africa/ (accessed on 14 April 2025).
6. United States Department of Agriculture. Production-Oranges. USDA. 2024. Available online: https://fas.usda.gov/data/production/commodity/0571120 (accessed on 14 April 2025).
7. Ogunbode, T.O.; Esan, V.I.; Ayegboyin, M.H.; Ogunlaran, O.M.; Sangoyomi, E.T.; Akande, J.A. Analysis of farmers' perceptions on sustainable sweet orange farming in Nigeria amid climate change. *Sci. Rep.* **2025**, *15*, 5205. [CrossRef] [PubMed]
8. Attah, A.J.; Mbah, F.N.; Okeke, M.N. Assessment of factors limiting production of Citrus among Smallholder Farmers in Benue State, Nigeria. *J. Agric. Crops* **2018**, *4*, 170–175. [CrossRef]
9. DailyTrust. Oranges Everywhere in Benue but no Buyers. 30 January 2022. Available online: https://dailytrust.com/oranges-everywhere-in-benue-but-no-buyers/ (accessed on 15 April 2025).
10. Okpe, P.C.; Atagher, M.M.; Iheanacho, A.C. Effect of Socio-Economic Characteristics on postharvest losses among sweet orange marketers in Benue State, Nigeria. *J. Agripreneurship Sustain. Dev.* **2022**, *5*, 130–141. [CrossRef]
11. Ikwuba, A.A.; Wegh, F.S.; Agwaza, T.A.; Angera, R.I. A study on sweet orange production and its post-harvest losses among farmers in northeastern zone of Benue State, Nigeria. *J. Biol. Sci. Bioconservation* **2019**, *11*, 64–105.

12. Bolarin, O.; Adesiji, G.B.; Adebayo, S.A.; Komolafe, S.E. Postharvest management practices by roadside orange sellers in Oyo State, Nigeria. *J. Hortic. Postharvest Res.* **2024**, *7*, 69–80. [CrossRef]
13. Richa, R.; Kohli, D.; Vishwakarma, D.; Mishra, A.; Kabdal, B.; Kothakota, A.; Richa, S.; Sirohi, R.; Kumar, R.; Naik, B. Citrus fruit: Classification, value addition, nutritional and medicinal values, and relation with pandemic and hidden hunger. *J. Agric. Food Res.* **2023**, *14*, 100718. [CrossRef]
14. Ware, M. What to Know About Oranges. Medical News Today. 2023. Available online: https://www.medicalnewstoday.com/articles/272782 (accessed on 15 April 2025).
15. YuQiu, L.; Emily, H.; Sherry, A.T. History, Global Distribution, and Nutritional Importance of Citrus Fruits. *Compr. Rev. Food Sci. Food Saf.* **2012**, *11*, 530–545. [CrossRef]
16. Odeyemi, O.M. Postharvest handling practices and losses of some major fruits sold in Abeokuta, Southwestern Nigeria. *Niger. J. Hortic. Sci.* **2019**, *24*, 113–122.
17. James, D.; Abba, A.; Aisha, M.; Shamsuddeen, J.; Oni, O.O. Determination of causes of post-harvest losses in orange marketing in selected markets in Kano State, Nigeria. *Int. J. Agric. Earth Sci.* **2018**, *4*, 23–31.
18. Etefa, O.F.; Forsido, S.F.; Kebede, M.T. Postharvesr loss, causes and handling practices of fruits and vegetables in Ethiopia: Scoping review. *J. Hortic. Res.* **2022**, *30*, 1–10. [CrossRef]
19. Musasa, S.T.; Mvumi, B.M.; Manditsera, F.A.; Chinhanga, J.; Musiyandaka, S.; Chigwedere, C. Chinho Postharvest orange losses and small-scale farmers' perceptions on the loss causes in the fruit value chain: A case study of Rusitu Valley, Zimbabwe. *Food Sci. Qual. Manag.* **2013**, *18*, 1–9.
20. Doki, N.O.; Eya, C.I.; Tughgba, M.; Gabriel, A.; Ameh, A. Determinants of post-harvest losses of orange in selected local government areas of Benue State. *Int. J. New Econ. Soc. Sci.* **2019**, *10*, 295–308. [CrossRef]
21. Yigzaw, D.; Habtemariam, A.; Teshome, D.; Amare, H. Assessment of fruit postharvest handling practices and losses in Bahir Dar, Ethiopia. *Afr. J. Agric. Res.* **2016**, *11*, 5209–5214. [CrossRef]
22. Adekalu, O.A.; Agboola, D.A.; Atanda, S.; Akande, S.A. Survey of Post-Harvest Techniques Employed in Handling and Storage of Sweet Oranges (*Citrus sinensis* Osberk) in Lagos, Ogun and Osun States, Nigeria. *Asian J. Agric. Food Sci.* **2019**, *7*, 20–30. [CrossRef]
23. Obayelu, O.A.; Dairo, D.O.; Olowe, O.O. What Factors Explain Postharvest Losses of Orange Fruit (*Citrus sinensis*) from Farm to Fork in the Tropics? *Agricultura* **2022**, *19*, 7–15. [CrossRef]
24. Busari, A.O.; Idris-Adeniyi, K.M.; Lawal, A.O. Food security and postharvess losses in fruit marketing in Lagos Metropolis, Nigeria. *J. Agric. Food Sci.* **2015**, *3*, 52–58.
25. Debela, A.; Daba, G.; Bane, D.; Tolessa, K. Identification of major causes of postharvest losses among selected fruits in Jimma zone for proffering veritable solutions. *Int. J. Curr. Res.* **2011**, *3*, 040–043.
26. El-Ramady, H.R.; Domokos-Szabolcsy, E.; Abdalla, N.A.; Taha, H.S.; Fári, M. *Postharvest Management of Fruits and Vegetables Storage*; Sustainable Agriculture Reviews; Lichtfouse, E., Ed.; Springer International Publishing: Cham, Switzerland, 2015; pp. 65–152. [CrossRef]
27. Aminu, F.O.; Balogun, S.L. Analysis of post-harvest losses in orange in Lagos metropolitan markets, Nigeria. *Niger. J. Hortic. Sci.* **2020**, *25*, 143–151.
28. Atanda, S.A.; Pessu, P.O.; Agoda, S.; Isong, I.U.; Ikotun, I. The concepts and problems of postharvest food losses in perishable crops. *Afr. J. Food Sci.* **2011**, *5*, 603–613.
29. Agriculture Nigeria. Postharvest Handling of Oranges. 2022. Available online: https://www.agriculturenigeria.com/post-harvest-handling-of-oranges/ (accessed on 16 April 2025).
30. James, D.; Bamishayi, O.M.; Shamsuddeen, J.; Williams, O.; Olukunde, T.; Zainab, M. Analysis of the Level of Post-Harvest Losses in Orange Marketing Case Study of Yanlemo Orange Market in Kano State. *Int. J. Agric. Earth Sci.* **2017**, *3*, 17–24.

proceedings

Proceeding Paper

Effect of Processing Methods on Quality Characteristics of Maize Flour †

Olufunmilola Oludoyin Oluwadiran [1,*], Olajide Philip Sobukola [2], Folake Olayinka Henshaw [2] and Wasiu Akinloye O. Afolabi [3]

1 Department of Home Economics, Emmanuel Alayande University of Education, Oyo P.M.B. 1010, Nigeria
2 Department of Food Science and Technology, Federal University of Agriculture, Abeokuta P.M.B. 2240, Nigeria; sobukolaop@funaab.edu.ng (O.P.S.); henshawfo@funaab.edu.ng (F.O.H.)
3 Department of Nutrition and Dietetics, Federal University of Agriculture, Abeokuta P.M.B. 2240, Nigeria; afolabiwao@funaab.edu.ng
* Correspondence: funmilola_akinpelu@yahoo.com; Tel.: +234-8038063499
† Presented at the CORAF's 2023 Symposium on Processing and Transformation of Agricultural Products in West and Central Africa: Achievements and Opportunities for Private Sector Engagement, Lome, Togo, 21–23 November 2023.

Abstract: Maize flour (MF) is a cheap and common cereal product that can be used in various foods. However, different processing methods can affect its quality and suitability for different purposes. This study compares the effects of three processing methods (grain non-soaking (GNSM), grain soaking (GSM), and nixtamalization (NIX)) on the nutritional, pasting, and functional properties of MF. MF samples (GNSM-MF, GSM-MF, and NIX-MF) were analyzed using standard laboratory techniques. Protein and moisture contents ranged from 9.34 to 10.12% and 5.35 to 7.05%, respectively, with NIX-MF having significantly ($p < 0.05$) higher protein content. Calcium, iron, and zinc contents ranged from 3.64 to 10.91 mg/g, 3.69 to 7.64 mg/g, and 1.45 to 1.92 mg/g, respectively. GNSM-MF had the lowest calcium content. Peak, breakdown, final, and setback viscosities were 980.25–2904.15 RVU, 91–1147 RVU, 778–1210 RVU, and 331.5–919 RVU, respectively. GSM significantly ($p < 0.05$) increased peak, breakdown, final, and setback viscosities. Water-holding capacity (WHC), emulsifying capacity (EC), and oil absorption capacity (OAC) values of MF ranged from 78.93 to 111.95%, 4.97 to 42.18%, and 160.27 to 186.13%, respectively. High protein and calcium contents in NIX-MF make it a nutritious option for pregnant women and children. GSM-MF might be incorporated into snacks, complementary foods, and puddings. This study provides insights into appropriate selection of maize flour processing methods based on food application.

Keywords: processing methods; maize flour; grain non-soaking method; grain soaking method; nixtamalization

1. Introduction

Maize, a staple cereal, is cultivated and processed for food primarily in developing countries across Africa. With the global population projected to exceed 9 billion by 2050 [1], there is an urgent need to address the growing pressure on agriculture and food processing sectors to develop affordable, accessible, and nutritious food products. Processing techniques such as soaking and nixtamalization have been established to enhance the nutritional quality of cereals by reducing antinutrient content and improving the bioavailability of key nutrients [2,3].

Soaking has been reported to significantly decrease the antinutrient content of cereals (polyphenols, phytate); it also increases protein digestibility in vitro and improves the bioavailability of zinc and iron [4]. Nixtamalization, an alkaline processing technique, holds particular promise for improving maize-based food products. By removing the pericarp through hemicellulose disruption, this method enhances nutrient bioavailability, especially that of calcium and niacin, while also improving protein quality [5]. These attributes make it especially beneficial for vulnerable populations like children and pregnant women, who require higher calcium intake. Low processing costs coupled with simplicity of technology and product diversity from the same unit operations have recently drawn the interest of researchers to alkaline cooked maize.

Despite its potential, the adoption of nixtamalization and other processing techniques in developing countries often prioritizes end-use over nutritional outcomes. This has limited the ability to fully harness their benefits. A more systematic comparison of maize flour quality derived from different processing methods is needed to guide the choice of techniques that balance simplicity and nutritional delivery. This study aims to compare three maize flour processing methods and evaluate their effects on the quality characteristics of maize flour, thereby providing insights into optimizing processing techniques for improved nutritional and functional outcomes.

2. Materials and Methods

2.1. *Material*

White maize grains were procured from a local market in Ibadan, Oyo State, Nigeria. Damaged grains and foreign materials were manually sorted out to ensure purity. The cleaned grains were stored in airtight containers to maintain freshness and prevent contamination before milling.

2.2. *Maize Flour Processing Methods*

2.2.1. Grain Non-Soaking Method (GNSM)

Cleaned whole maize grains (10 kg) were milled using a locally fabricated disc attrition mill [6], which is widely utilized in small-scale grain processing. The resulting maize flour was stored in airtight polythene bags to preserve its quality before analysis.

2.2.2. Grain Soaking Method (GSM)

A similar quantity of whole maize grains as that used for GNSM (grain non-soaking method) was soaked in water at room temperature for 3 h and then drained. The soaked grains were dried in an oven (Eurosonic, SW-XY, China) at 55 °C for 16 h. Milling was performed using the same method as GNSM, and the resulting flour was stored in airtight polythene bags to maintain quality [6].

2.2.3. Nixtamalization (NIX)

The nixtamalization process was performed following the method described in [7] (2007). Ten kilograms of maize grains were submerged in water at a 1:6 ratio, and 2% lime [$Ca(OH)_2$] was added. The mixture was cooked for 80 min and then allowed to stand for 14 h to complete the process. The nixtamalized grains were washed thoroughly under running water, with four rinses performed to ensure complete removal of residual lime. Drained grains were dried in a hot air oven at 55 °C for 12 h to reduce moisture content. The dried grains were then ground into flour.

2.3. Proximate Analysis

The proximate properties of maize flour, including ash, crude protein (determined by the Kjeldahl method), moisture, crude fibre, and crude fat (determined by solvent extraction), were analyzed following standard procedures of the Association of Official Analytical Chemists [8,9]. Carbohydrate content was calculated by difference using the formula:

Carbohydrates = (100 − (% Crude Proteins + % Lipids + % Ash + % Crude fibre + % Moisture))

2.4. Colour Properties Determination

Colour measurements of maize flours, including L* (lightness), a* (redness), and b* (yellowness), were obtained using a Hunter Lab colorimeter (Chroma meter CR-10, Osaka, Japan). The instrument was calibrated with the following standard values: L* = 96.9, a* = −0.04, and b* = 1.84. From these measurements, flour colour index, yellow index, browning index, degree of whiteness (%W), colour intensity (ΔE), hue angle ($h°$), and delta chroma (ΔC) were calculated using standard equations [10,11].

$$\Delta C = \left(a^2 + b^2\right)^{1/2} \quad (1)$$

$$h^o = \tan^{-1}(b/a) \quad (2)$$

2.5. Mineral Determination

The mineral assay of processed maize flour was conducted using an Atomic Absorption Spectrometer to determine calcium, iron, and zinc levels, following the method described by [9] (2005). This technique was selected for its accuracy and reliability in quantifying essential minerals important for nutrition.

2.6. Pasting Parameters Determination

Pasting parameters of maize flour samples were analyzed using a Rapid Visco Analyzer (RVA) (TECMASTER, Perten Instruments, Hägersten, Sweden) as described by [6] (2020). For each sample, 4 g of flour was weighed into an RVA container and mixed with 25 mL of distilled water to form a slurry. The slurry was heated and continuously stirred at 50 °C and 160 rpm for 10 s to ensure proper dispersibility. It was then held at 50 °C for 1 min before being heated to 95 °C over 8 min and held at 95 °C for an additional 5 min. The parameters measured included peak viscosity, pasting temperature, time to peak, breakdown, trough, final viscosity, and setback.

2.7. Functional Properties Analysis

The water-holding capacity (WHC), emulsion capacity (EC), and the oil absorption capacity (OAC) of maize flour were analyzed to evaluate its functional properties. For WHC, a modified centrifugation method, described in [12] (1996), was employed. Two hundred milligrams of maize flour sample were placed in a specialized tube, and 20 mL of water was added. The mixture was incubated at 25 °C with continuous agitation for 24 h. After incubation, the supernatant was discarded, and the sediment was weighed. The sediment was then oven-dried at 120 °C for up to 2 h, and its moisture content was measured. WHC was calculated as the amount of water retained per gram of the corresponding sample. The emulsion capacity (EC) and oil absorption capacity (OAC) of the maize flour samples were determined using a centrifugation method described in [9].

2.8. Statistical Analysis

Sample preparation and analyses were performed in triplicates to ensure precision and reliability of the results. The data obtained were subjected to analysis of variance (ANOVA), and mean values were separated using Duncan's Multiple Range Test at a significance level of $p < 0.05$. The independent variables consisted of the processed maize flour samples: GNSM-MF (Grain Non-Soaking Method Maize Flour), GSM-MF (Grain Soaking Method Maize Flour), and NIX-MF (Nixtamalized Maize Flour). The dependent variables included measured values for proximate composition, mineral composition, colour properties, pasting parameters and functional properties.

3. Results

3.1. Proximate Composition

The proximate composition of maize flour produced by different processing methods is summarized in Table 1. The ash content ranged from 1.12% to 1.37%, with NIX-MF exhibiting the highest value and GNSM-MF the lowest. However, there were no statistically significant differences ($p > 0.05$) among the samples. Protein content varied between 9.34% and 10.12%, with NIX-MF having the highest value, significantly higher ($p < 0.05$) than those of GSM-MF and GNSM-MF. Moisture content ranged from 5.35% to 7.05%, and significant differences ($p < 0.05$) were observed across all samples. GNSM-MF had the highest moisture content, while GSM-MF exhibited the lowest. The crude fibre content of NIX-MF was significantly lower ($p < 0.05$) compared to GSM-MF, which showed the highest fibre content. No significant differences ($p > 0.05$) were found in the crude lipid content among the samples, though GSM-MF had the highest value at 3.85%.

Table 1. Proximate composition (%) of processed maize flour.

Sample	Ash	Crude Protein	Moisture	Crude Fibre	Crude Fat	Carbohydrate
GNSM-MF	$1.12^{a} \pm 0.13$	$9.15^{b} \pm 0.18$	$7.05^{a} \pm 0.07$	$2.41^{a} \pm 0.06$	$3.60^{a} \pm 0.27$	$76.64^{a} \pm 0.60$
GSM-MF	$1.18^{a} \pm 0.13$	$9.34^{b} \pm 0.18$	$5.35^{c} \pm 0.07$	$2.89^{a} \pm 0.35$	$3.85^{a} \pm 0.04$	$77.41^{a} \pm 0.50$
NIX-MF	$1.37^{a} \pm 0.08$	$10.12^{a} \pm 0.16$	$6.30^{b} \pm 0.14$	$1.05^{b} \pm 0.03$	$3.25^{a} \pm 0.29$	$77.92^{a} \pm 0.35$

Values are means of three replicates ± standard deviation. Values in columns with the same alphabet in superscript are not significantly different from each other ($p < 0.05$). GNSM-MF grain non-soaking method processed maize flour; GSM-MF: grain soaking method processed maize flour; NIX-MF: nixtamalized maize flour.

3.2. Colour Properties

The colour properties of processed maize flour samples are presented in Table 2. The lightness (L*), redness (a*), and yellowness (b*) values ranged from 87.93 to 90.06, −0.05 to −0.09, and 14.59 to 25.00, respectively. Among the samples, GNSM-MF exhibited the lowest lightness, while NIX-MF showed the highest. No significant differences ($p > 0.05$) were observed in the lightness and redness values across the samples. However, NIX-MF was significantly different ($p < 0.05$) in yellowness, with nixtamalization notably increasing the yellowness of the maize flour sample. Hue values for all processed samples were below 90°, with NIX-MF having the highest hue angle (89.89°).

Table 2. Colour characteristics of processed maize flour.

Samples	L*	a*	b*	Deltachroma ΔC	Colour Intensity ΔE	Hue Angle h^o	Whiteness (%)	FCI	Yellowness Index	Browning Index
GNSM-MF	87.93 ^{a} ± 0.02	0.07 ^{a} ± 0.01	14.59 ^{a} ± 0.01	13.62 ^{a} ± 0.02	89.13 ^{a} ± 0.04	89.73 ^{b} ± 0.02	81.06 ^{a} ± 0.01	73.34 ^{a} ± 0.01	5.92 ^{c} ± 0.01	23.70 ^{d} ± 0.01
GSM-MF	88.83 ^{a} ± 0.01	0.09 ^{a} ± 0.01	14.68 ^{a} ± 0.01	13.71 ^{a} ± 0.01	90.04 ^{a} ± 0.01	89.64 ^{b} ± 0.01	81.61 ^{a} ± 0.01	74.15 ^{a} ± 0.03	5.90 ^{c} ± 0.02	23.61 ^{d} ± 0.02
NIX-MF	90.06 ^{a} ± 0.03	0.05 ^{a} ± 0.01	25.00 ^{b} ± 0.11	24.03 ^{b} ± 0.05	93.47 ^{b} ± 0.02	89.89 ^{b} ± 0.01	73.1 ^{b} ± 0.16	65.06 ^{b} ± 0.04	9.58 ^{b} ± 0.01	39.66 ^{e} ± 0.10

Values are means of three replicates ± standard deviation. Values in columns with the same alphabet in superscript are not significantly different from each other ($p < 0.05$). GNSM-MF: grain non-soaking method processed maize flour; GSM-MF: grain soaking method processed maize flour; NIX-MF: nixtamalized maize flour; FCI: flour colour index.

3.3. Mineral Composition

The effect of processing methods on the mineral content of maize flour samples is summarized in Table 3. Calcium levels ranged from 3.64 to 10.91 mg/g, with NIX-MF having the highest value and GNSM-MF the lowest. Significant differences ($p < 0.05$) were observed across all processing methods for calcium content. Iron content varied between 3.69 and 7.64 mg/g, with GNSM-MF exhibiting the highest level and NIX-MF the lowest. Similarly, zinc levels ranged from 1.45 to 1.92 mg/g, with GSM-MF showing the highest content and GNSM-MF the lowest. These differences highlight the impact of processing methods on the mineral composition of maize flour, which may influence its nutritional value and application.

Table 3. Mineral content of processed maize flour.

Sample	Calcium (mg/g)	Iron (mg/g)	Zinc (mg/g)
GNSM-MF	3.64 ^{c} ± 0.09	7.64 ^{a} ± 0.04	1.45 ^{c} ± 0.02
GSM-MF	5.85 ^{b} ± b0.11	3.92 ^{b} ± 0.01	1.92 ^{a} ± 0.01
NIX-MF	10.91 ^{a} ± 0.01	3.69 ^{c} ± 0.01	1.70 ^{b} ± 0.01

Values are means of three replicates ± standard deviation. Values in columns with the same alphabet in superscript are not significantly different from each other ($p < 0.05$). GNSM-MF: grain non-soaking method processed maize flour; GSM-MF: grain soaking method processed maize flour; NIX-MF: nixtamalized maize flour.

3.4. Pasting Properties

The influence of processing methods on the pasting properties of maize flour is detailed in Table 4. Peak viscosity values ranged from 980.25 to 2904.15 RVU, with GNSM-MF exhibiting the lowest and GSM-MF the highest. These differences were statistically significant ($p < 0.05$). Trough viscosity ranged from 410.5 to 1755 RVU, with NIX-MF showing the lowest value and GSM-MF the highest. Breakdown viscosity values also varied significantly ($p < 0.05$), ranging from 91 to 1147 RVU. Final viscosity values were between 778 and 1210 RVU, while setback viscosities ranged from 331.5 to 919 RVU. Both parameters differed significantly ($p < 0.05$) across the samples. Peak time ranged from 4.9 to 5.57 min, with GSM-MF having the shortest and NIX-MF the longest peak time. Pasting temperatures showed significant variation ($p < 0.05$), ranging from 73.38 °C (GSM-MF) to 84.08 °C (GNSM-MF).

Table 4. Pasting properties of processed maize flour.

Sample	Peak Viscosity (RVU)[2]	Trough (RVU)	Breakdown (RVU)	Final Viscosity (RVU)	Setback (RVU)	Peak Time (min)	Pasting Temp (°C)
GNSM-MF	$980.25^{c} \pm 0.35$	$887.5^{b} \pm 0.71$	$91^{c} \pm 1.41$	$1210^{b} \pm 14.14$	$331.5^{c} \pm 2.12$	$4.72^{c} \pm 0.07$	$84.08^{a} \pm 0.04$
GSM-MF	$2904.15^{a} \pm 0.21$	$1755^{a} \pm 1.41$	$1147^{a} \pm 1.41$	$2672^{a} \pm 2.83$	$919^{a} \pm 1.41$	$5.57^{a} \pm 0.05$	$73.38^{c} \pm 0.04$
NIX-MF	$1059^{b} \pm 1.41$	$410.5^{c} \pm 1.85$	$639^{b} \pm 1.41$	$778^{c} \pm 4.24$	$363.5^{b} \pm 2.12$	$4.9^{b} \pm 0.05$	$76.59^{b} \pm 0.01$

Values are means of three replicates ± standard deviation. Values in columns with the same alphabet in superscript are not significantly different from each other ($p < 0.05$). GNSM-MF: grain non-soaking method processed maize flour; GSM-MF: grain soaking method processed maize flour; NIX-MF: nixtamalized maize flour

3.5. Functional Properties

The functional properties of processed maize flour samples are shown in Table 5. Water-holding capacity (WHC) ranged from 78.93% to 111.95%, with NIX-MF showing the highest value and GSM-MF the lowest. Emulsion capacity (EC) and oil absorption capacity (OAC) ranged from 4.97% to 42.18% and 160.27% to 186.13%, respectively, with significant differences ($p < 0.05$) observed among all samples. GNSM significantly ($p < 0.05$) increased EC compared to the other processing methods, while nixtamalization (NIX) significantly ($p < 0.05$) reduced the EC of the maize flour. OAC values also varied significantly ($p < 0.05$) across samples.

Table 5. Functional properties of processed maize flour.

Sample	WHC (g/g)	EC (g/cm^3)	OAC (g/g)
GNSM-MF	$80.84^{b} \pm 0.88$	$42.18^{a} \pm 3.46$	$186.13^{a} \pm 1.08$
GSM-MF	$78.93^{b} \pm 1.24$	$27.33^{b} \pm 3.53$	$166.57^{b} \pm 1.17$
NIX-MF	$111.95^{a} \pm 0.14$	$4.97^{c} \pm 0.01$	$160.27^{c} \pm 0.24$

Values are means of three replicates ± standard deviation. Values in columns with the same alphabet in superscript are not significantly different from each other ($p < 0.05$). GNSM-MF grain non-soaking method processed maize flour; GSM-MF, grain soaking method processed maize flour; NIX-MF, nixtamalized maize flour; WHC, water-holding capacity; EC, emulsion capacity; OAC, oil absorption capacity.

3.6. Pearson Correlation Between Functional Properties and Colour Characteristics of Processed Maize Flour

Table 6 shows the correlation of water-holding capacity (WHC) and emulsion capacity (EC) with various colour and functional parameters. WHC exhibited significant correlations ($p < 0.05$) with b* (0.998), ΔC (0.998), %W (−0.987), and BI (0.998). Additionally, WHC showed highly significant correlations ($p < 0.01$) with ΔE (0.962), H* (−0.954), and YI (0.999). Emulsion capacity (EC) was significantly correlated ($p < 0.05$) with L* (−0.999) and highly significantly correlated ($p < 0.01$) with ΔE (−0.982).

Table 6. Pearson correlation between functional properties and colour characteristics of processed maize flour.

Variables	L*	a*	b*	ΔC	ΔE	H*	%W	FCI	YI	BI
WHC	0.888	0.838	0.998^{b}	0.998^{b}	0.962^{a}	-0.954^{a}	-0.987^{b}	-0.999^{b}	0.999^{a}	0.998^{b}
EC	-0.999^{b}	0.504	−0.919	−0.921	-0.982^{a}	−0.728	0.889	0.875	−0.917	−0.918
OAC	−0.948	0.176	−0.726	−0.731	−0.858	−0.449	0.676	0.654	−0.722	−0.724

[a]: Significant correlation at the 1% level. [b]: Significant correlation at the 5% level; L*, Lightness; a*, redness; b*, yellowness; ΔC, deltachroma; ΔE, colour intensity; H*, hue angle; %W, % whiteness; FCI, flour colour index; YI, yellowness index; BI, brownness index; WHC, water-holding capacity; EC, emulsion capacity; OAC, oil absorption capacity.

4. Discussion

4.1. Proximate Composition

The increase in protein content observed for NIX-MF in this study may be attributed to the removal of soluble starch during processing. This phenomenon results in the solubilization of corn pericarp hemicelluloses, which are subsequently lost in the nejayote. However, only trace amounts of protein (less than 5%) are lost during this process compared to other constituents [13]. Additionally, the aleurone structure of maize, which remains attached to the endosperm during nixtamalization, plays a critical role in minimizing protein loss, as demonstrated in multiple studies [14–17].

The methods of processing significantly influenced the moisture content of the maize flour samples. Contrary to the findings of [18] (2015), which associated the soaking of whole grains with increased moisture content, the GNSM showed a significant ($p < 0.05$) increase in moisture content. This discrepancy may arise from the high intrinsic moisture content of the whole maize grain used in this study. Furthermore, the moisture content of NIX-MF was notably higher compared to that of maize processed using the GSM, likely due to the increased water absorption rate during the initial stages of steeping and cooking. Steeping for approximately three hours followed by cooking has been shown to lead to about a 36% increase in moisture content, which stabilizes or progresses slowly [14]. The hydration process during NIX is influenced by the physical properties of the grain, nixtamalization variables, and the permeability of the pericarp. Maize pericarp, which typically acts as a barrier to water entry, enhances hydration when hydrolyzed [19].

The findings of this study align with [20] (2013), which compared dietary crude fibre between nixtamalized and steeped millet dough, revealing reduced crude fibre content in the former. Fibre loss during NIX may be linked to pericarp removal during processing [21,22]. Additionally, reduced ion interactions (OH^- and Ca^{2+}) within the maize grain constituents and the cooking medium may hinder the formation of indigestible products, further contributing to lower crude fibre content. However, the authors of [23] (2010) reported contrasting results, where millet cooked in lime exhibited the highest crude fibre content. These variations are likely due to differences in the raw materials used, particularly their physical properties.

None of the processing methods significantly ($p > 0.05$) influenced crude fat and carbohydrate contents. Nevertheless, maize flour produced via the GSM exhibited the highest crude fat values, whereas NIX increased carbohydrate content. The GNSM and NIX, in contrast, reduced both parameters. The high temperature achieved during NIX may facilitate fat degradation and oxidation, similar to findings reported by [20] (2013) for nixtamalized millet. Consequently, NIX-MF might be less prone to rancidity during storage compared to maize flour processed using GNSM and GSM methods. This is advantageous for maintaining preferred aroma and flavour during storage [23]. The carbohydrate content of the maize flour samples ranged higher than values documented by [18] (2015), [24] (2012), and [25] (2012) for quality protein maize and maize-based products.

4.2. Colour Properties

The significant ($p < 0.05$) increase in yellowness observed in NIX-MF can be attributed to the yellowing of the pericarp during processing, which is linked to the presence of o-glycoside flavonoid-type pigments in the pericarp. These pigments are typically colourless in acidic and neutral media; however, the alkaline environment characteristic of nixtamalization facilitates substitution of hydroxyl groups on the benzopyrone rings. This reaction results in the formation of flavonol structures, leading to the distinct yellow coloration [20]. This transformation during nixtamalization presents a potential advantage in industries such as bakery, snack production, and complementary flour manufacturing, where enhanced yellowness may be desirable. Lower yellowness values recorded for GNSM-MF and GSM-MF align with findings by [6] (2020) for whole maize meal. Hue angles, which are classified as follows: 0° for red, 90° for yellow, and 180° for green hues, further reinforce the results. NIX-MF, with its highest hue angle and corresponding yellowness ($b^* = 25.00$), exhibits a vibrant yellow hue. This is complemented by its highest saturation level (chroma = 24.03), confirming the distinctive colour characteristics of NIX-MF [26].

4.3. Mineral Content

Nixtamalization (NIX) significantly ($p < 0.05$) increased the calcium content of maize flour compared to other processing methods. This increase may be attributed to the inherent calcium content of the reagent used, which facilitates calcium diffusion into the endosperm through percolation. As steeping time and temperature increase during NIX, the solubilization and hydrolysis of the pericarp intensify, creating micro holes that promote calcium diffusion into the endosperm [27,28]. These findings are consistent with those of [29] (2002) for ground nixtamalized maize, [30] (2005) for corn tortillas, and [31] (2010) for eleven maize varieties. In contrast, the significant ($p < 0.05$) reduction in calcium content observed in GNSM-MF may result from the absence of soaking, which prevents the hydrolysis and solubilization of the maize pericarp required for calcium release into the endosperm. However, the lack of steeping in GNSM-MF favoured a significant ($p < 0.05$) increase in iron content compared to other processing methods. This aligns with the findings of [18] (2015), which documented reductions in iron content following soaking in the production of complementary foods using quality protein maize (QPM).

The grain soaking method (GSM) significantly ($p < 0.05$) increased the zinc content of maize flour compared to GNSM-MF, which exhibited significantly ($p < 0.05$) lower zinc levels. However, some studies have reported contrary findings, noting that soaking decreased the zinc content of complementary flours from finger millet and chickpea [32,33], likely due to leaching. These discrepancies may arise from differences in the raw materials used or the prolonged steeping times adopted. Supporting the current study, the authors of [34] (2009) and [35] (2010) observed that soaking enhanced zinc content in sorghum flour.

It is evident that the effects of processing methods on mineral content vary across agricultural raw materials. This variability can present both challenges and opportunities for flour processors and end users. While GSM alone may not be the optimal method for enhancing mineral bioavailability, optimizing soaking conditions and combining it with other techniques, such as NIX, could prove beneficial. Thus, integrating GSM and NIX into traditional maize processing practices holds promise for improving mineral delivery, particularly calcium, in Nigeria and other developing countries.

4.4. Pasting Parameters

The pasting profile of flour is strongly influenced by factors such as the amylopectin-to-amylose ratio, starch content, protein amylase, and lipids [36]. In this study, the steeping involved in the GSM appeared to enhance the swelling index of the maize flour samples, consistent with findings from [37] (2005), [38] (2009), and [39] (2019), which reported increased pasting profiles in pre-treated maize flour.

The cooking stage of NIX likely facilitated the formation of partially gelatinized starch granules, which may have lowered the peak viscosity. This observation aligns with reports by [40] (2008) and [41] (2014). Annealing, which strengthens intragranular forces and enhances starch granule crystallinity, also likely contributed to these changes [16,20,42]. However, a decrease in peak viscosity for NIX-MF is plausible when the rate of gelatinization surpasses the annealing effect [43].

The significantly ($p < 0.05$) lower breakdown viscosity observed in maize flour processed via the GNSM suggests strong paste stability, making it better able to withstand shear-thinning during further processing [44]. Setback viscosity, an indicator of retrogradation and associated syneresis, was lower in flours processed using GNSM, suggesting minimal retrogradation tendencies [45]. Conversely, maize flour processed via the GSM demonstrated a higher retrogradation tendency when cooked, potentially due to differences in starch structure.

The significant increase in pasting temperature observed in GNSM-MF indicates higher resistance to swelling and starch granule rupture. On the other hand, the soaking and cooking stages involved in GSM and NIX appear to have modified the starch granules' swelling behaviour, enhancing their hydration potential. These results align with those of [46] (2017), where similar findings were reported. The gelatinization temperature, which occurs due to molecular order disruption inside starch granules, typically ranges between 60 and 70 °C for ideal corn starch, but it is influenced by the adopted processing method [47]. The presence of partially pregelatinized starch granules in NIX-MF (resulting from the high cooking temperatures) may have resisted further gelatinization, elevating the pasting temperature.

The lower pasting temperature observed in GSM-MF suggests that it is likely to gelatinize more quickly than those from the other processing methods. This may be attributed to enzyme stimulation during soaking, which breaks down the starch matrix and facilitates swelling and gelatinization [48]. The reduction in peak time for GNSM-MF and NIX-MF samples indicates a potential reduction in energy and time required for processing in food formulations. However, this contrasts with the findings of [6] (2020), where maize fractions with lower peak times achieved peak viscosity faster. In the current study, GSM-MF exhibited the highest mean peak viscosity but required longer times to attain this, likely due to the use of whole maize rather than its fractions.

4.5. Functional Properties

Water absorption is the first change that occurs in the endosperm during the steeping and cooking stages of nixtamalization, as starch granules absorb water and swell. The increased water-holding capacity (WHC) observed in this study may be attributed to the enhanced ability of starch granules to absorb water during alkaline cooking, particularly within the initial three hours of steeping. This process is facilitated by the effect of lime on starch gelatinization. Calcium hydroxide ($Ca(OH)_2$) dissociates into hydroxide (OH^-) ions, which penetrate starch granules, breaking bonds between the hydroxyl groups of starch chains and water molecules. This mechanism increases water penetration into starch granules, thus enhancing their water absorption capacity [49–51]. These findings align with those reported by [23] (2010) for nixtamalized millet flour.

While GNSM exhibited a significantly ($p < 0.05$) increased effect on emulsion capacity (EC), NIX significantly ($p < 0.05$) reduced the EC of maize flour samples. This reduction could be attributed to the significantly ($p < 0.05$) lower crude lipid content of nixtamalized maize flour, as previously noted. A contrasting observation was reported by [23] (2010) for untreated and lime-cooked millets, though the mean EC value for GNSM-MF in the present study exceeded that reported for untreated millet (26%). As a result, maize flour produced via the GNSM is likely to be more easily incorporated into food preparations compared to NIX-MF. To overcome this limitation, NIX-MF may need to be blended with flours possessing higher EC.

The GNSM also significantly ($p < 0.05$) increased the oil absorption capacity (OAC) of maize flour, whereas NIX significantly ($p < 0.05$) reduced it. This disparity may be attributed to the higher fat content observed in GNSM-MF compared to the lower fat value recorded for NIX-MF. Similar findings were reported in [5] (2014), who noted that fat loss during nixtamalization was a factor contributing to reduced OAC. Oil binding capacity depends on the surface properties of hydrophobic amino acids [52], which may be more prevalent in untreated whole maize grains, facilitating the exposure of non-polar residues within interior protein molecules. Enhanced OAC in processed flour formulations is particularly important, as fat serves as a major flavour retainer [53], enhancing mouthfeel during food consumption. This suggests that GNSM-MF may have better functional properties in food applications requiring high OAC.

4.6. Pearson Correlation Between Functional Properties and Colour Characteristics of Processed Maize Flour

The strong positive correlation between WHC and b*, ΔC, ΔE, and YI suggests that yellowness and chromatic characteristics of the flour contribute significantly to its water-holding behaviour. Conversely, the strong negative correlation between WHC with variables like H, %W, and FCI, indicates that structural and hydrophobic characteristics of maize flour samples may inversely influence water absorption.

The strong negative correlation between EC and L*, implies that as flour samples become lighter, their ability to form emulsions diminishes. Negative correlations with other variables, including b*, ΔC, ΔE, and YI, further highlight the impact of flour colour and compositional changes on EC.

Negative correlation between OAC and most variables, such as L*, b*, ΔC, and YI indicates that colour properties have less influence on OAC, while other factors, such as the presence of hydrophobic amino acids, likely play a more critical role. Weak positive correlations with %W and FCI suggest slight dependencies on specific structural or physical characteristics of the flour.

These results underscore the importance of colour and structural properties in defining the functional characteristics of maize flour. Enhanced water-holding and oil absorption capacities are particularly critical in food formulations, as they influence the behaviour and quality of the final product [54]. Additionally, the high emulsion capacity observed in grain non-soaking method maize flour makes it a strong option for applications where emulsification is essential. By understanding these relationships, processors can optimize maize flour formulations for specific industrial and dietary uses.

5. Conclusions

The processing methods studied significantly influenced the proximate, colour, functional, and pasting characteristics of maize flour. Nixtamalized maize flour (NIX-MF) demonstrated the highest protein content, calcium content, and water-holding capacity, making it particularly suitable for inclusion in the diets of pregnant women and young children in developing countries, where nutritional enhancement is critical. Conversely, the grain non-soaking method maize flour (GNSM-MF) exhibited the highest oil absorption capacity, while NIX-MF had the lowest.

The grain soaking method maize flour (GSM-MF) with the highest values for pasting properties, including peak viscosity, trough, breakdown viscosity, final viscosity, setback viscosity, and peak time, making it ideal for diverse applications in food formulations, such as snacks, complementary foods, and puddings. Additionally, GNMS-MF displayed the lowest breakdown viscosity, indicative of improved paste stability under processing conditions.

In summary, each processing method presents unique functional advantages that cater to different nutritional and industrial needs. The incorporation of nixtamalization (NIX) and grain soaking methods (GSM) into traditional processing practices could significantly enhance the value and applicability of maize flour in both dietary and industrial contexts.

Author Contributions: Conceptualization, O.O.O.; methodology, O.O.O.; formal analysis, O.O.O.; writing—original draft preparation, O.O.O.; Supervision, O.P.S., F.O.H. and W.A.O.A.; writing—review and editing, O.P.S., F.O.H. and W.A.O.A. All authors have read and agreed to the published version of the manuscript.

Funding: This research received no external funding.

Institutional Review Board Statement: Not applicable.

Informed Consent Statement: Not applicable.

Data Availability Statement: The original contributions presented in this study are included in the article. Further enquiries can be directed to the corresponding author.

Acknowledgments: The authors duly acknowledge the Department of Food Science and Technology, Federal University of Agriculture for providing necessary facilities to carry out this research.

Conflicts of Interest: The authors declare no conflict of interest.

References

1. Godfray, C.; Beddington, J.R.; Crute, I.R.; Haddad, L. Food security: The challenge of feeding 9 billion people. *Science* **2010**, *327*, 812. [CrossRef] [PubMed]
2. Sarita, E.S.; Singh, E. Potential of millets: Nutrients composition and health benefits. *J. Sci. Innov. Res.* **2016**, *5*, 46–50. [CrossRef]
3. Houssou, P.; Ayernor, G.S. Appropriate processing and food functional properties of maize flour. *Afr. J. Sci. Technol.* **2002**, *3*, 15297. [CrossRef]
4. Pawar, V.; Thaker, V. In vitro efficacy of 75 essential oils against *Aspergillus niger*. *Mycoses* **2006**, *49*, 316–323. [CrossRef]
5. Jeffrey, A.G.; Maria, N.G. Processing maize flour and corn meal food products. *Ann. N.Y. Acad. Sci.* **2014**, *1312*, 66–75. [CrossRef]
6. Bolade, M.K. Effect of flour production methods on the yield, physicochemical properties of maize flour and rheological characteristics of a maize-based non-fermented food dumpling. *Afr. J. Food Sci. Res.* **2020**, *8*, 1–11.
7. Afoakwa, E.O.; Sefa-Dedeh, S.; Budu, A.G.; Sakyi-Dawson, E.; Asomaning, J. Influences of spontaneous fermentation on some quality characteristics of maize-based cowpea-fortified nixtamalized foods. *Afr. J. Food Agric. Nutr. Dev.* **2007**, *7*. eISSN: 1684-5374 print ISSN: 1684-5358.
8. AOAC. *Official Methods of Analysis of the Association of Official Analytical Chemistry*, 15th ed.; Association of Official Analytical Chemista: Arlington, VA, USA, 1990.
9. AOAC. *Official Methods of Analysis*, 18th ed.; Association of Official Analytical Chemists: Washington DC, USA, 2005.
10. Maskan, M.J. Kinetics of colour change of kiwifruits during hot air and microwave drying. *J. Food Eng.* **2001**, *48*, 169–175. [CrossRef]
11. Falade, K.O.; Ayetigbo, O.E. Effects of annealing, acid hydrolysis and citric acid modifications on physical and functional properties of starches from four yam (*Dioscorea* spp.) cultivars. *Food Hydrocoll.* **2015**, *43*, 529–539. [CrossRef]
12. Suzuki, T.; Ohsugi, Y.; Yoshie, Y.; Shirai, T.; Hirano, T. Dietary fiber content, water holding capacity and binding capacity of seaweeds. *Fish. Sci.* **1996**, *62*, 454–461. [CrossRef]
13. Gutiérrez-Cortez, E.; Rojas-Molina, I.; Rojas, A.; Arjona, J.L.; Cornejo-Villegas, M.A.; Zepeda-Benítez, Y.; Velázquez-Hernández, R.; Ibarra-Alvarado, C.; Rodríguez-García, M.E. Microstructural changes in the maize kernel pericarp during cooking stage in nixtamalization process. *J. Cereal Sci.* **2010**, *5*, 81–88. [CrossRef]
14. David, S.; Juan, D.; Figueroa, C.; Rosa, M.M.; Anayansi, E.A.; Néstor, P.C.; José, M.M. Physical and chemical changes undergone by pericarp and endosperm during corn nixtamalization—A review. *J. Cereal Sci.* **2018**, *81*, 108–117. [CrossRef]
15. Campechano, E.M.; Figueroa, C.J.; Arámbula, V.G.; Martínez Flores, H.E.; Jiménez Sandoval, S.J.; Luna Bárcenas, J.G. New ecological nixtamalization process for tortilla production and its impact on the chemical properties of whole corn flour and wastewater effluents. *Int. J. Food Sci. Technol.* **2012**, *47*, 564–571. [CrossRef]
16. Figueroa, J.; Véles Medina, J.J.; Hernández Landaverde, M.A.; Aragón Cuevas, F.; Gaytán Martínez, M.; Chávez Martínez, E.; Palacios, N.; Willcox, M. Effect of annealing from traditional nixtamalization process on the microstructural, thermal, and rheological properties of starch and quality of pozole. *J. Cereal Sci.* **2013**, *58*, 457–464. [CrossRef]
17. Muñoz Hernández, R.A.; Calderón, A.; Cruz-Orea, A.; Sánchez Sinencio, F.; Tomas, S.A.; González de la Cruz, G. Estudio de la influencia del $Ca(OH)_2$ en las películas de pericarpio de maíz nixtamalizado mediante técnicas fototérmicas. *Superf. Y Vacío* **1999**, *84*, 80–84.

18. Beruk, B.D.; Kebede, A.E. Effect of Blending Ratio and Processing Technique on Physicochemical Composition, Functional Properties and Sensory Acceptability of Quality Protein Maize (QPM) Based Complementary Food. *Int. J. Food Sci. Nutr.Eng.* **2015**, *5*, 121–129.
19. Narváez-González, E.D.; Figueroa-Cárdenas, J.; Taba, S.; Rincón Sánchez, F. Kernel microstructure of Latin American races of maize and their thermal and rheological properties. *Cereal Chem.* **2006**, *83*, 605–610. [CrossRef]
20. James, O.; Fortune, A. Applicability of Nixtamalization in the Processing of Millet-based Maasa, a Fermented Food in Ghana. *J. Food Res.* **2013**, *2*, 59.
21. Serna-Saldivar, S.O.; Gomez, M.H.; Rooney, L.M. Technology, chemistry, and nutritional value of alkaline-cooked corn products. *Adv. Cereal Sci. Technol.* **1990**, *10*, 243–306.
22. Parades-Lopez, O.; Saharopulos, M.E. Scanning electron microscopy studies of limed corn kernels for tortilla making. *J. Food Technol.* **1982**, *17*, 687–693. [CrossRef]
23. Ocheme, B.O.; Oloyede, O.O.; Mikailu, E.G. Effect of Lime Soaking and Cooking (Nixtamalization) on the Proximate, Functional and Some Anti-nutritional Properties of Millet Flour. *Aust. J. Technol.* **2010**, *14*, 131–138.
24. Ayinde, F.A.; Bolaji, O.T.; Abdus-Salaam, R.B.; Osidipe, O. Functional properties and quality evaluation of "kokoro" blended with beniseed cake (*Sesame indicum*). *Afr. J. Food Sci.* **2012**, *6*, 117–123. [CrossRef]
25. Olayiwola, I.O.; Folarnmi, F.; Adebowale, A.; Onabanjo, O.O.; Sanni, S.A.; Afolabi, W.A.O. Nutritional Composition and Sensory Qualities of Cocoyam-Based Recipes Enriched with Cowpea Flour. *J. Nutr. Food Sci.* **2012**, *2*, 1–6.
26. Prinyawiwatkul, W.; Beuchat, L.R.; McWatters, K.H.; Phillips, R.D. Fermented cowpea flour: Production and characterization of selected physico-chemical properties. *J. Food Process. Preserv.* **1996**, *20*, 265–284. [CrossRef]
27. Liezl, M.; Atienza, E.I. Proximate Properties, Functional Components, and Vitamin and Mineral Content of Nixtamalized Lagkitan Corn (*Zea mays* L.). *EC Nutr.* **2019**, *14*, 360–372.
28. Valderrama-Bravo, C.; Rojas-Molina, A.; Gutiérrez-Cortez, E.; Rojas-Molina, I.; Oaxaca-Luna, A.; De la Rosa-Rincón, E.; Rodríguez-García, M.E. Mechanism of calcium uptake in corn kernels during the traditional nixtamalization process: Diffusion, accumulation and percolation. *J. Food Eng.* **2010**, *98*, 126–132. [CrossRef]
29. Bressani, R.; Turcios, J.; De Ruiz, A. Nixtamalization Effects on the Contents of Phytic Acid, Calcium, Iron and Zinc in the Whole Grain, Endosperm and Germ of Maize. *Food Sci. Technol. Int.* **2002**, *8*, 81–86. [CrossRef]
30. Rosado, J.L.; Cassis, L.; Solano, L.; Duarte-Vazquez, M.A. Nutrient addition to corn masa flour: Effect on corn flour stability, nutrient loss and acceptability of fortified corn tortillas. *Food Nutr. Bull.* **2005**, *26*, 266–272. [CrossRef]
31. Pappa, M.R.; de Palomo, P.P.; Bressani, R. Effect of lime and wood ash on the nixtamalization of maize and tortilla chemical and nutritional characteristics. *Plant Foods Hum. Nutr.* **2010**, *65*, 130–135. [CrossRef]
32. Dejene, D. Study on the Effect of Traditional Processing on Proximate Composition and Bioavailability of Minerals in Chickpea (*Cicer arietinum*) Grown in Ethiopia. Master's Thesis, Addis Ababa University, Addis Ababa, Ethiopia, 2010.
33. Helen, W. Influence of Processing on Anti-Nutrients and Production of Some Value-Added Finger Millet Based Products. Master's Thesis, Addis Ababa University, Addis Ababa, Ethiopia, 2010.
34. Shemelis, T. Improvement of Energy and Nutrient Density of Sorghum Based Weaning Food Using Germination. Master's Thesis, Addis Ababa University, Addis Ababa, Ethiopia, 2009.
35. Karkle, E.N.L.; Beleia, A. Amino Acid Profile and Antinutritional Factors of Nixtamalized Maize Flour Supplemented with Sprouted Soybean Flour. *Ciênc. Tecnol. Aliment.* **2010**, *30*, 1056–1060. [CrossRef]
36. Zhu, L.J.; Liu, Q.Q.; Sang, Y.; Gu, M.H.; Shi, Y.C. Underlying reasons for waxy rice flours having different pasting properties. *Food Chem.* **2010**, *120*, 94–100. [CrossRef]
37. Oluwamukomi, M.; Eleyinmi, A.; Enujiugha, V. Effect of Soy Supplementation and Its Stage of Inclusion on the Quality of Ogi–A Fermented Maize Meal. *Food Chem.* **2005**, *91*, 651–657. [CrossRef]
38. Bolade, M.K.; Adeyemi, I.A.; Ogunsua, A.O. Influence of Particle Size Fractions on the Physicochemical Properties of Maize Flour and Textural Characteristics of a Maize-Based Non-fermented Food Gel. *Int. J. Food Sci. Technol.* **2009**, *44*, 646–655. [CrossRef]
39. Adedeji, O.; Nsakpuma, T. Functional properties of maize flour (*Zea mays*) and stability of its paste (towo) as influenced by processing methods and baobab (*Adansonia digitata*) pulp inclusion. *Ukr. J. Food Sci.* **2019**, *7*, 49–60. [CrossRef]
40. Hidalgo, A.; Brandolini, A.; Gazza, L. Influence of steaming treatment on chemical and technological characteristics of einkorn (*Triticum monococcum* L. ssp. *Monococcum*) wholemeal flour. *Food Chem.* **2008**, *111*, 549–555. [CrossRef]
41. Pongjaruvat, W.; Methacanon, P.; Seetapan, N.; Fuongfuchat, A. Influence of pregelatinized tapioca starch and transglutaminase on dough rheology and quality of gluten-free jasmine rice breads. *Food Hydrocoll.* **2014**, *36*, 143–150. [CrossRef]
42. Campus-Baypoli, O.N.; Rosas-Burgos, E.C.; Torres-Chávez, P.I.; Ramírez-Wong, B.; Serna-Saldívar, S.O. Physiochemical changes of starch during maize tortilla production. *Starch/Stärke* **1999**, *51*, 173–177. [CrossRef]
43. Villada, J.A.; Sánchez-Sinencio, F.; Zelaya-Ángel, O.; Gutiérrez-Cortez, E.; Rodríguez-García, M.E. Study of the morphological, structural, thermal, and pasting corn transformation during the traditional nixtamalization process: From corn to tortilla. *J. Food Eng.* **2017**, *212*, 242–251. [CrossRef]

44. Zaidul, I.S.M.; Norulaini, N.A.N.; Omar, A.K.M.; Yamauchi, H.; Noda, T. RVA analysis of mixtures of wheat flour and potato, sweet potato, yam and cassava starches. *Carbohydr. Polym.* **2007**, *69*, 784–791. [CrossRef]
45. Sandhu, K.S.; Singh, N. Some properties of corn starches II: Physicochemical, gelatinization, retrogradation, pasting and gel textural properties. *Food Chem.* **2007**, *101*, 1499–1507. [CrossRef]
46. Acevedo, B.A.; Thomphson, C.M.B.; Foutel, N.S.G.; Chaves, M.G.; Avanza, M.V. Effect of different treatments on the microstructure and functional and pasting properties of pigeon pea (*Cajanus cajan* L.), dolichos bean (*Dolichos lablab* L.) and jack bean (*Canavalia ensiformis*) flours from north-east Argentina. *Int. J. Food Sci. Technol.* **2017**, *52*, 222–230. [CrossRef]
47. Whistler, R.L.; Corbett, W.M. Oligosaccharides from partial acid hydrolysis of corn fiber hemicellulose. *J. Am. Chem. Soc.* **1955**, *77*, 6328–6330. [CrossRef]
48. Nche, P.; Nout, M.; Rombouts, F. The Effect of Cowpea Supplementation on the Quality of Kenkey, a Traditional Ghanaian Fermented Maize Food. *J. Cereal Sci.* **1994**, *19*, 191–197. [CrossRef]
49. Alvarado-Gil, J.J.; Vargas, H.; Sánchez-Sinencio, F.; González-Hernández, J.; Miranda, L.C.M. Effect of lime content on the processing conditions of cooked maize tortillas: Changes of thermal, structural, and rheological properties. *Opt. Eng.* **1997**, *36*, 348–357.
50. Santiago-Ramos, D.; Figueroa-Cárdenas, J.; Véles-Medina, J.J.; Reynoso-Camacho, R.; Ramos-Gómez, M.; Gaytán-Martínez, M.; Morales-Sánchez, E. Effects of annealing and concentration of calcium salts on thermal and rheological properties of maize starch during an ecological nixtamalization process. *Cereal Chem.* **2015**, *92*, 475–480. [CrossRef]
51. de los Ángeles Cornejo-Villegas, M.; Rincón-Londoño, A.; Del Real-López, A.; Rodríguez-García, M.E. The effect of Ca^{2+} ions on the pasting, morphological, structural, vibrational, and mechanical properties of corn starch–water system. *J. Cereal Sci.* **2018**, *79*, 174–182. [CrossRef]
52. Lin, M.J.; Humbert, E.S.; Sosulski, F.W. Certain functional properties of sunflower meal products. *J. Food Sci.* **1976**, *39*, 368–370. [CrossRef]
53. Kinsella, J.E.; Melachouris, N. Functional properties of proteins in foods: A survey. *Crit. Rev. Food Sci. Nutr.* **1976**, *7*, 219–280. [CrossRef]
54. Awuchi, C.G.; Igwe, V.S.; Echeta, C.K. The functional properties of foods and flours. *Int. J. Adv. Acad. Res.* **2019**, *5*, 139–160.

proceedings

Proceeding Paper

Enhancing Food Security and Nutrition Through Indigenous Agro-Product-Based Functional Foods: A Case Study on Composite Flour Development †

Chioma Bertha Ehis-Eriakha [1,*], Peace Omoikhudu Oleghe [1,2] and Fred Coolborn Akharaiyi [1,*]

1 Department of Microbiology, Faculty of Science, Edo State University, Iyamho 312102, Edo State, Nigeria; peaceoleghe@auchipoly.edu.ng
2 Department of Pharmaceutical Technology, School of Applied Sciences and Technology, Auchi Polytechnic, Auchi 312101, Edo State, Nigeria
* Correspondence: ekwuabu.bertha@edouniversity.edu.ng (C.B.E.-E.); akharaiyi.fred@edouniversity.edu.ng (F.C.A.); Tel.: +234-7038892250 (C.B.E.-E.)

† Presented at the CORAF's 2023 Symposium on Processing and Transformation of Agricultural Products in West and Central Africa: Achievements and Opportunities for Private Sector Engagement, Lome, Togo, 21–23 November 2023.

Abstract: The current rising food prices, influenced by importation costs, the global food crisis, as well as pre- and post-harvest losses, have contributed majorly to malnutrition and food insecurity. Therefore, utilizing technologies that harness our indigenous agro-products as composite flours to develop functional foods will address these issues. In this study, dry raw samples of perishable and healthy yellow potato, yellow maize and pigeon pea were obtained from the agricultural development program, Edo State, Nigeria, and authenticated and processed into gluten-free fermented composite flours. The flours were profiled physicochemically and nutritionally, providing valuable insight into their multiple benefits. An experimental design software (Design Expert 13.0.) was applied to achieve optimum blended flours regarding the ratio of sweet potato–pigeon pea–maize, and mix 5 (67.70:20.00:12.31) displayed more outstanding attributes than other blends for the production of biscuits, bread and cakes using creaming and mixing methods. Various standard tests for flours and products were appropriately carried out to evaluate the proximate, techno-functional, mineral, antioxidant, anti-nutrient, sensory and color values. Individual antioxidant parameters were improved across all products compared to wheat-based products (control) under the same production conditions, showing a statistical significance at $p < 0.05$. A similar trend was observed in the proximate, anti-nutritional and mineral contents, while all products had a desirable color outlook. A sensory evaluation revealed the general acceptability, while an in vivo animal experimental model revealed that all animals fed with the various product samples gained weight with improved general body organs and no evidence of disease. This research underscores the potential of harnessing agri-value chain approaches in developing functional foods and promoting food security.

Keywords: flour development; botanical value-chain; food fermentation; functional pastries

1. Introduction

Malnutrition and undernourishment remain an existential threat especially among vulnerable people globally [1,2]. In Sub-Saharan Africa, the large population, massive post-harvest food wastages and poor technological know-how in transforming botanicals

 https://doi.org/10.3390/proceedings2025118004

into highly nutritional, therapeutic and economic food products across the entire food value chain have caused economic crisis, poverty and, largely, food insecurity. For instance, using Nigeria as a case study, the 2021 preponderance of undernourished individuals was recorded as 15.9% [3]. By the end of 2024, the human population is estimated to rise to over 229.2 million, and with a yearly increase of over 2.41%, it is projected to rise to over 400 million by the year 2050 [4,5]. In 2022, the National Bureau of Statistics report on the Multidimensional Poverty Index (MPI) projected over 133 million Nigerians (63%) to be multidimensionally poor; food insecurity and nutrition were some of the parameters used in arriving at this conclusion [6]. Amalgamated nutritional support with supplementary and therapeutic diets is needed to address these challenges [7–10].

Although staples vary from place to place, cereals, tubers/root crops, legumes, grains and fruits/vegetables are generally inexpensive and are known to make up about 90% of the food calorie intake worldwide [2,11]. They are used alone or as composite flour combinations in preparing readily available functional foods. This diet therapy is necessary not only for survival but it can also help confer substantial health benefits providing some physiological effects and specific health and nutritional enhancements for the prevention and management of persistent ailments, such as polygenic diseases and hyperpiesia [12,13], because of their macronutrient's composition [14,15] and their various collections of ancillary metabolic substances having antioxidant capacities [16,17].

The current cultivation of wheat in many African nations is insufficient, and importation has become a massive drain on local economies. In 2021, Nigeria was ranked in the second position among the leading top five countries that imported the highest dollar value of wheat globally, with USD 3.32 billion [18]. With the threatening effects of the global climatic variations coupled with various armed conflicts affecting wheat production, there is a need to harness indigenous botanicals and develop them into agro-processed flour composite bio-resources as a more sustainable alternative in functional food product development [2,15]. This will aid in accomplishing Sustainable Development Goal (SDG) number two by mitigating post-harvest losses through the formulation of composite flours from indigenous raw botanicals and producing nutritionally enhanced baked products (bread, biscuits and cakes) from the developed raw composite flours. The adoption of this innovation would drastically reduce the amount spent on wheat importation [15].

Generally, composite food technology is an excellent way of creating distinctively novel food products from the combinations of conventional and unconventional botanicals. These composite flour developments will accelerate the further exploitation of native food crops for producing ready-to-eat, highly nutritious functional foods, such as bread, cakes and biscuits [10]. The nutritional quality and value and health derivatives obtained from using unseemly, underutilized plant materials as composite flour mixes in the preparation of ready-to-eat bakery and pastry staple food products cannot be overemphasized [2].

This study is aimed at developing gluten-free composite flours from a blend of perishable and healthy yellow potato, yellow corn and pigeon pea as a suitable alternative to wheat flour. These flour mixes were utilized in the production of ready-to-eat baked functional staples.

2. Methods

2.1. Sample Collection and Authentication

Dry tubers from the yellow-fleshed variety of sweet potato (*Ipomea batatas*), yellow-grain variety maize (*Zea mays*) and pigeon peas (*Cajanus cajan*) were bought from agricultural development program foodstuff sellers in Jattu, Etsako-West Local Government Area, Edo State, Nigeria and were confirmed taxonomically at the Plant Biology and Biotechnology Unit (Herbarium Curation Sub-division), Department of Biological Sciences, Edo

University Uzairue, Edo State, Nigeria. Wistar albino rats were obtained from the animal house, Federal University of Agriculture, Abeokuta, Ogun State.

2.2. Sample Processing and Fermentation

The sweet potato, maize and pigeon pea botanicals were each spontaneously fermented for 72 h at 28 ± 2 °C as reported by [19] to produce their respective fermented flours.

2.3. Experimental Design

The optimum blending used to produce the composite flours was obtained using the modified D-optimal mixture (Design Expert 13.0.). Experimental designs for mixtures in the food sector are described by [20]. Different blends were obtained by mixing the respective proportions of sweet potato, pigeon pea and maize flour as selected by the developed experimental design. The composite blends of these flours were thoroughly mixed using a mixer and packaged in different polyethylene bags for analysis.

2.4. Production of Baked Products

The baked products (biscuits, bread and cakes) were produced from blended composite flours, using traditional creaming and mixing methods by [21–23]. Additionally, wheat flour (100%) was used in preparing the respective control products.

2.5. Physicochemical Property Determination

2.5.1. Proximate Composition

The proximate compositions of the raw and composite flours and products were carried out according to the standard methods of [24]. Tests carried out include crude fiber, protein, ash, moisture, fat and carbohydrate.

2.5.2. Techno-Functional Composition of the Flours

The methods of [25] were used for the techno-functional analysis of the raw and composite flours for the following parameters: Bulk Density, Water Absorption Capacity (WAC), Oil Absorption Capacity (OAC), Foaming Capacity (FC), Swelling Capacity (SC), Least Gelation Concentration (LGC) and Emulsion Capacity (EC).

2.5.3. Mineral Analysis

The determination of mineral elements in the samples followed the dry ash methods outlined in [24]. The analyzed mineral elements of the samples are calcium and magnesium, using Atomic Absorption Spectroscopy (AAS) Model TAS-990, Yima, China, and sodium and potassium (flame photometer).

2.5.4. Antioxidant Activities

The methods of [26] were used in analyzing the antioxidant activities of the samples for the presence of 2,2-Diphenyl-1-picrylhydrazyl (DPPH) free radicals scavenging activity, iron (Fe^{2+}) chelation, hydroxyl (OH) radical scavenging activity, 2,2′–azinobis-(3-ethylbenzothiazoline-6-sulfonate) (ABTS) radical scavenging activity, Ferric-Reducing Antioxidant Property (FRAP) and the total phenol content.

2.5.5. Anti-Nutritional Composition

The anti-nutritional composition of the product samples was analyzed for tannin, oxalate, phytate and saponin contents according to the method of [27].

2.6. The Determination of the Color

The color of all the product samples was analyzed using the methods described by [28].

2.7. Sensory Evaluation

Sensory evaluation studies to determine customers' preferences were carried out for the baked bread, cake and biscuit products using both descriptive and discriminatory evaluation as described by [29] for the following characteristics: color, crispiness, texture, aroma, taste and overall acceptability.

2.8. In Vivo Animal Experiment

Fifty (50) healthy two-week-old Wistar albino weaning rats were used to determine the anthropometric parameters of the formulated products and the control products over 14 days following the method described by [30,31].

2.9. Analysis of Data

Data generated were subjected to statistical analyses such as ANOVA and the Duncan multiple range test. All data samples were obtained in triplicates and the results were presented as mean ± SD. The means were separated at $p < 0.05$ significant difference.

3. Results

The ratio of sweet potato–pigeon pea–maize (67.70:20.00:12.31) in mix 5 displayed more outstanding physicochemical and nutritional attributes than other blends for the production of baked products. The data presented in Table 1 compare the proximate properties of the composite flour (mixture 5) with the mean proximate value of the respective raw fermented botanical flours. It reveals that the composite flour (mixture 5) had all of its proximate parameters in moderate proportions, showing a statistical significance ($p < 0.05$).

Table 1. Proximate properties of raw and composite flours.

Parameter	Sweet Potato	Pigeon Pea	Maize	Composite Flour Mixture 5
Moisture content (%)	6.666 ± 0.05 ^{a}	4.105 ± 0.04 ^{c}	4.190 ± 0.02 ^{c}	5.869 ± 0.35 ^{b}
Fat (%)	2.535 ± 0.08 ^{d}	5.645 ± 0.02 ^{b}	6.095 ± 0.08 ^{a}	4.477 ± 0.16 ^{c}
Crude fiber (%)	4.675 ± 0.21 ^{c}	7.665 ± 0.06 ^{a}	7.225 ± 0.15 ^{b}	4.866 ± 0.63 ^{c}
Ash (%)	2.365 ± 0.06 ^{c}	3.230 ± 0.13 ^{a}	2.435 ± 0.05 ^{c}	2.553 ±0.13 ^{b}
Protein (%)	7.265 ± 0.22 ^{c}	11.665 ± 0.32 ^{a}	3.610 ± 0.20 ^{d}	9.000 ± 1.82 ^{b}
Carbohydrate (%)	76.494 ± 0.05 ^{a,b}	75.745 ± 0.41 ^{b,c}	77.805 ± 0.23 ^{a}	72.25 ± 1.60 ^{d}
pH	5.770 ± 0.72 ^{b}	5.770 ± 0.72 ^{b}	6.030 ± 0.01 ^{a}	5.253 ± 0.01 ^{c}

Data presented as mean ± SD; in each column different superscripts indicate diverse significant differences at $p \leq 0.05$.

The techno-functional properties of the fermented raw flours and the blended mixture 5 composite are shown in Table 2. These characteristics help in predicting and evaluating the precise behavior of each proximate property when designing and developing food. Most of the results for composite flour mixture 5 showed enhanced techno-functional properties with statistical significances ($p < 0.05$).

The proximate compositions of the bread, biscuit and cake samples made from mixture 5 of the fermented composite flour blends and the wheat flour used as the control are shown in Table 3. It can be observed that the ash, carbohydrate and crude fiber contents of the fermented mixture 5 samples had higher values than their respective controls. These values (apart from the crude fiber contents) were statistically significant at $p < 0.05$.

Table 2. Techno-functional properties of raw and composite flours.

Parameter	Sweet Potato	Pigeon Pea	Maize	Composite Flour Mixture 5
WAC (%)	21.01 ± 0.14 ^{d}	23.03 ± 0.04 ^{b}	22.16 ± 0.01 ^{c}	25.00 ± 0.00 ^{a}
OAC (%)	17.45 ± 0.07 ^{b}	16.06 ± 0.08 ^{c}	15.99 ± 0.07 ^{d}	19.00 ± 0.00 ^{a}
EMULSION CAPACITY (%)	40.50 ± 0.00 ^{c}	41.67 ± 0.01 ^{a}	41.85 ± 0.03 ^{a}	41.06 ± 0.01 ^{b}
SWELLING CAPACITY (%)	426.60 ± 0.01 ^{d}	427.48 ± 1.43 ^{c}	448.45 ± 0.06 ^{b}	471.93 ± 0.00 ^{a}
FOAM CAPACITY (%)	2.01 ± 0.07 ^{d}	4.0 ± 0.00 ^{c}	5.0 ± 0.01 ^{b}	8.05 ± 0.07 ^{a}
LEAST GEL CONC. (mg/cm^3)	0.2 ± 0.00 ^{b}	0.2 ± 0.00 ^{b}	0.25 ± 0.00 ^{a}	0.21 ± 0.01 ^{b}
BULK DENSITY (gm/ cm^3)	0.69 ± 0.00 ^{a}	0.69 ± 0.00 ^{a}	0.70 ± 0.00 ^{a}	0.73 ± 0.00 ^{a}

Data presented as mean ± SD; in each column different superscripts indicate diverse significant differences at $p \leq 0.05$.

Table 3. Proximate composition of bread, biscuit and cake.

Sample	Ash Content (%)	Moisture Content(%)	Fat Content (%)	Crude Fiber Content (%)	Protein Content (%)	Carbohydrate (By Difference) Content (%)
CB	1.011 ± 0.02 ^{c}	10.324 ± 0.05 ^{b}	18.934 ± 0.0 ^{b}	4.140 ± 0.02 ^{c}	13.222 ± 0.01 ^{a}	52.370 ± 0.13 ^{b}
5FB	1.495 ± 0.14 ^{b}	8.875 ± 0.68 ^{c}	17.592 ± 0.07 ^{c}	5.752 ± 0.04 ^{a}	10.450 ± 0.04 ^{b}	55.835 ± 0.75 ^{a}
CA	3.163 ± 0.03 ^{b}	5.982 ± 0.01 ^{b}	21.161 ± 0.08 ^{a}	1.271 ± 0.04 ^{b}	12.805 ± 0.25 ^{a}	55.622 ± 0.12 ^{c}
5FA	3.711 ± 0.11 ^{a}	3.639 ± 0.05 ^{c}	17.339 ± 0.03 ^{c}	3.485 ± 0.01 ^{b}	11.918 ± 0.08 ^{b}	58.491 ± 0.70 ^{b}
CC	1.808 ± 0.12 ^{c}	13.918 ± 0.38 ^{a}	25.508 ± 0.11 ^{c}	7.116 ± 0.91 ^{a}	14.586 ± 0.06 ^{a}	36.780 ± 0.67 ^{b}
5FC	3.197 ± 0.00 ^{b}	11.859 ± 0.19 ^{b}	24.179 ± 0.12 ^{b}	7.401 ± 0.12 ^{a}	12.813 ± 0.13 ^{b}	40.838 ± 0.58 ^{a}

Data presented as mean ± SD; in each column different superscripts indicate diverse significant differences at $p \leq 0.05$. **Key: CB = Control Bread; FB = Fermented Bread; CA = Control Biscuit; FA = Fermented Biscuit; CC = Control Cake; and FC = Fermented Cake.**

The mineral compositions of the bread, biscuit and cake samples made from mixture 5 of the fermented composite flours and the wheat flour used as the control are shown in Table 4. The result shows that all the baked samples had some level of minerals present. These are positive effects since food products are required to be mineral-rich. However, the lower the sodium content, the better. Out of all the formulated products (including the controls), the Fermented Cake (5FC) performed better because it has a very high potassium content (23.005 ppm) and the lowest sodium content (15.798 ppm). In many instances the controls obtained higher values than the products produced due to the more refined process of the wheat flour.

Table 4. Mineral composition of bread, biscuit and cake.

Samples	Sodium (Na) (ppm)	Potassium (K) (ppm)	Calcium (Ca) (ppm)	Magnesium (Mg) (ppm)
CB	18.701 ± 0.00 ^{d}	24.100 ± 0.00 ^{b}	12.702 ± 0.00 ^{a}	17.499 ± 0.00 ^{a}
5FB	18.011 ± 0.02 ^{d}	20.007 ± 0.01 ^{c}	10.706 ± 0.01 ^{c}	14.298 ± 0.00 ^{c}
CA	27.501 ± 0.00 ^{a}	20.601 ± 0.00 ^{d}	14.099 ± 0.00 ^{a}	9.200 ± 0.00 ^{e}
5FA	17.105 ± 0.01 ^{e}	19.803 ± 0.00 ^{e}	10.505 ± 0.01 ^{d}	15.160 ± 0.06 ^{a}
CC	20.096 ± 0.01 ^{d}	18.501 ± 0.00 ^{f}	15.001 ± 0.00 ^{a}	7.901 ± 0.00 ^{f}
5FC	15.798 ± 0.00 ^{f}	23.005 ± 0.01 ^{c}	13.901 ± 0.00 ^{b}	11.890 ± 0.01 ^{c}

Data presented asmean ± SD; in each column different superscripts indicate diverse significant differences at $p \leq 0.05$. **Key: CB = Control Bread; FB = Fermented Bread; CA = Control Biscuit; FA = Fermented Biscuit; CC = Control Cake; and FC = Fermented Cake.**

The antioxidant compositions of the breads, biscuit and cake samples made from mixture 5 of the fermented composite flour and the wheat flour used as the control are given in Table 5. The ranges observed among some antioxidant parameters such as OH and FRAP are quite wide, indicating significant differences at $p < 0.05$ between the various baked products in terms of their antioxidant potential, while for DPPH, ABTS and Fe^{2+} chelation, the antioxidant parameters have no statistical variations at $p < 0.05$. Across all three samples, the ABTS, DPPH and FRAP values obtained for the fermented composite flour were consistently higher than their control samples.

Table 5. Antioxidant properties of bread, biscuit and cake.

Sample	OH (%)	ABTS (mMol/g)	DPPH (%)	Fe^{2+} Chelation (%)	FRAP (mg/g)
CB	42.810 ± 0.61^{a}	0.010 ± 0.00^{c}	71.345 ± 0.23^{b}	7.360 ± 0.36^{b}	7.292 ± 0.13^{c}
5FB	36.960 ± 0.64^{b}	0.019 ± 0.00^{b}	72.547 ± 0.41^{b}	5.788 ± 0.93^{b}	10.500 ± 0.27^{b}
CA	42.143 ± 0.34^{d}	0.025 ± 0.00^{a}	52.502 ± 0.82^{c}	2.980 ± 0.23^{f}	5.842 ± 0.09^{f}
5FA	60.000 ± 0.45^{a}	0.026 ± 0.00^{a}	55.608 ± 0.44^{b}	12.848 ± 0.14^{d}	26.309 ± 0.76^{a}
CC	40.22 ± 0.76^{d}	0.007 ± 0.00^{d}	26.275 ± 0.04^{g}	12.747 ± 0.46^{e}	8.342 ± 0.10^{g}
5FC	43.157 ± 0.25^{c}	0.025 ± 0.00^{a}	72.657 ± 0.50^{c}	16.440 ± 0.21^{d}	18.314 ± 0.02^{d}

Data presented as mean ± SD; in each column different superscripts indicate diverse significant differences at $p \leq 0.05$. **Key: CB = Control Bread; FB = Fermented Bread; CA = Control Biscuit; FA = Fermented Biscuit; CC = Control Cake; FC = Fermented Cake; OH = Hydroxyl Radical; ABTS = 2,2′-azinobis-(3-ethylbenzothiazoline-6-sulfonate); DPPH = 2,2-Diphenyl-1-picrylhydrazyl; Fe^{2+} Chelation = iron chelation; FRAP = Ferric-Reducing Antioxidant Property.**

The anti-nutritional composition of the bread, biscuit and cake samples made from mixture 5 of the fermented composite flour (Table 6) showed the presence of anti-nutritional contents with a statistical significance at $p < 0.05$ for all the samples. Across all three samples, the fermented composite flour values obtained for saponins and oxalates were consistently higher than their control samples. The color analysis of the bread, biscuit and cake samples made from the fermented composite flour showed a statistical significance at $p < 0.05$, resulting in a desirable darker color outlook for all the fermented composite flour samples (except cake) when compared with their controls. For the bread with the fermented composite flour, the color value was 1.20 ± 0.01 vs. 0.11 ± 0.01 for the Control Bread, while the color of the biscuits made from fermented mixture 5 was 2.325 ± 0.01 vs. 0.803 ± 0.00 for the control. For the cake, 0.389 ± 0.00 was recorded for the fermented mixture 5 sample and 0.098 ± 0.00 for the control, indicating a higher absorbance at 560 nm.

Table 6. Anti-nutritional composition of bread, biscuit and cake.

Sample	Tannins (mg/g)	Oxalates (mg/g)	Phytates (mg/g)	Saponins (%)
CB	1.316 ± 0.00^{a}	0.315 ± 0.00^{c}	1.650 ± 0.00^{d}	14.494 ± 0.03^{c}
5FB	0.917 ± 0.01^{c}	3.731 ± 0.07^{b}	1.728 ± 0.18^{c}	21.424 ± 0.01^{a}
CA	0.720 ± 0.00^{d}	0.902 ± 0.00^{e}	1.858 ± 0.01^{a}	12.497 ± 0.00^{e}
5FA	1.461 ± 0.02^{b}	2.261 ± 0.01^{c}	1.742 ± 0.01^{b}	23.495 ± 0.03^{a}
CC	0.708 ± 0.02^{d}	0.241 ± 0.04^{g}	0.929 ± 0.00^{f}	13.711 ± 0.23^{f}
5FC	0.754 ± 0.02^{c}	1.791 ± 0.01^{d}	1.441 ± 0.00^{e}	20.654 ± 0.06^{a}

Data presented as mean ± SD; in each column different superscripts indicate significant differences at $p \leq 0.05$. **Key: CB = Control Bread; FB = Fermented Bread; CA = Control Biscuit; FA = Fermented Biscuit; CC = Control Cake; FC = Fermented Cake.**

The results from the sensory evaluation of the fermented composite baked products (bread, biscuits and cake) using the composite mixture 5 flour (Figure 1) show that the organoleptic characteristics of the biscuit were highest among the three baked products, while bread consistently had lower scores for all sensory parameters. The values obtained from each sensorial parameter for these baked products all showed a statistical significance at $p < 0.05$. All the products assessed were judged to be generally acceptable.

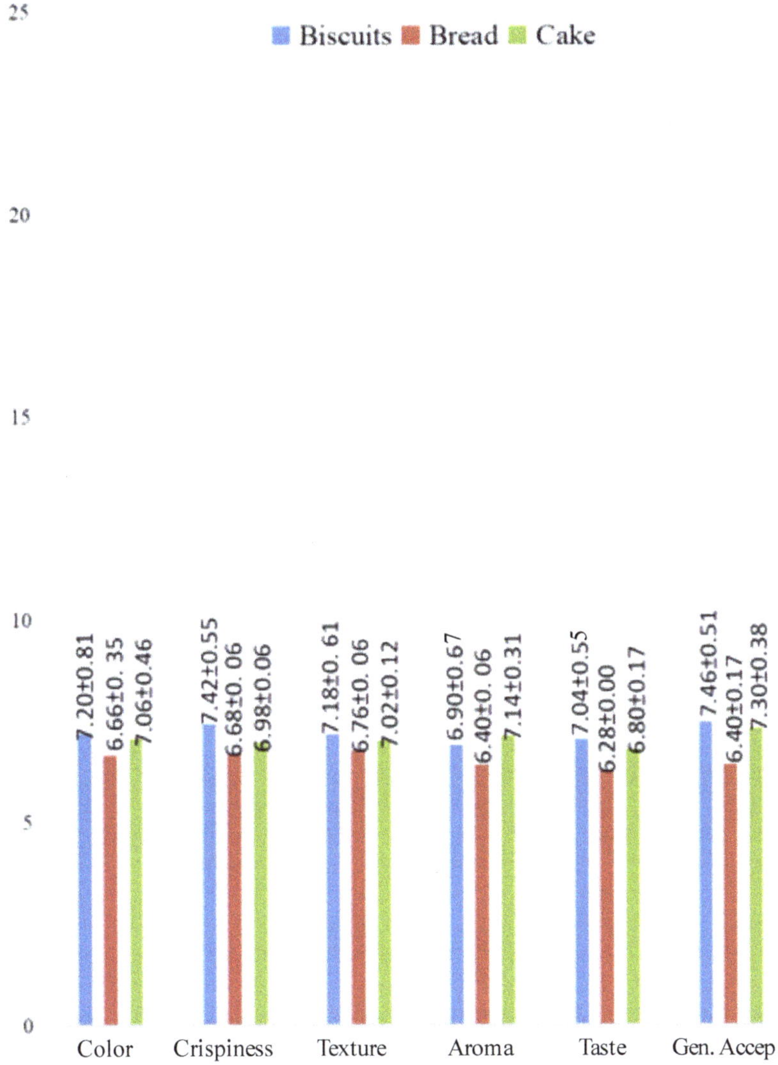

Figure 1. Sensory evaluation of fermented bread, biscuits and cake made from composite mixture 5 flour.

The results of the in vivo animal studies on the percentage weight gain of animals fed with the formulated bread, biscuit and cake products made from composite flour blends along with their respective control products show that there was a significant percentage weight gain for all the experimental animal models fed (Figure 2). The composite flour supports a similar weight gain as the control flour in biscuits and bread. By Day 14, animals fed with the composite biscuits (21.5%) and bread (15.2%) had comparable weight gains to those on the Control Biscuits (17.3%) and bread (10.5%). However, the Control Cake led to greater weight gain (33.8%) than the composite cake (23.2%), suggesting differences in their nutrient composition or digestibility.

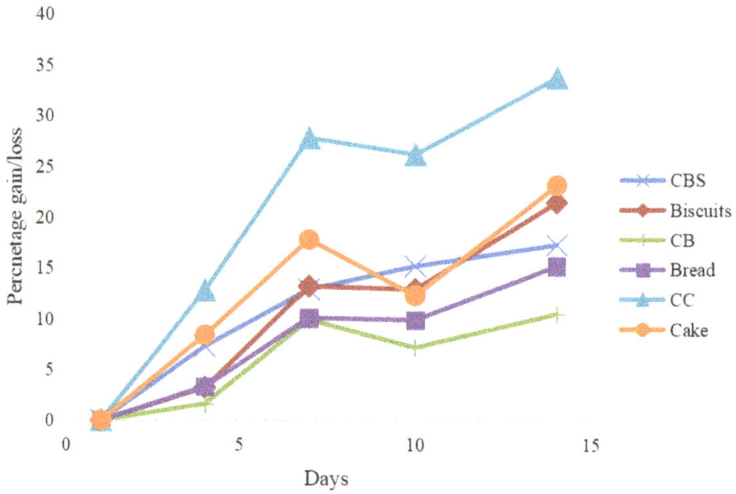

Figure 2. Percentage of weight gain of animals fed with formulated products made from composite mixture 5 flour along with their respective control products. Key: CBs: Control Biscuits, CB: Control Bread, and CC: Control Cake.

4. Discussion

The proximate composition of the raw and composite flours in Table 1 shows significant variations in moisture, fat, crude fiber, ash, protein, carbohydrates and pH, which influence the nutritional quality and functionality [15,32]. The moisture content varied among the samples, with the composite mixture 5 flour indicating a more moderate shelf stability than each botanical flour. The fat content was highest in the maize and lowest in the sweet potato, with the composite flour exceeding previously reported values for pigeon pea–sorghum blends [33]. The crude fiber content also varied, with the composite flour showing a similarity to earlier studies on pigeon pea–sorghum blends [33]. The ash content, indicative of mineral presence, differed among the samples, aligning with reported values of [33]. The protein content was highest in the pigeon pea and lowest in the maize, with the composite flour surpassing the range observed in cassava–pigeon pea blends [34]. The carbohydrate content was highest in the maize and lowest in composite flour 5, which is comparable to values in cassava–pigeon pea blends [34]. Overall, the composite flour presents a balanced nutritional profile, demonstrating how flour blending enhances the nutritional quality and functionality, making it suitable for diverse food applications. In agreement with the findings of other researchers, the fermentation in this study enhanced the functional parameters of both the flour and baked products.

The techno-functional properties of raw and composite flours in Table 2 show statistically significant variations ($p < 0.05$) in functional parameters such as the WAC, OAC, Emulsion Capacity, Swelling Capacity, foam capacity, Least Gelation Concentration and Bulk Density, which are critical for food processing and formulation. The composite flour mixture 5 exhibited the highest WAC (25.00%) and OAC (19.00%), aligning with studies by [35,36], which highlighted improved absorption properties due to ingredient synergy. The Emulsion Capacity varied slightly across samples and was influenced by the protein composition [36]. The Swelling Capacity, essential for texture and viscosity, was highest in the composite flour (471.93%), which is consistent with [37] who noted that ingredient proportions affect swelling indices. The foam capacity was also highest in the composite flour (8.05%), which is similar to [36] who observed enhanced foaming properties in certain blends. The Least Gelation Concentration varied slightly, with the maize flour requiring the highest value (0.25 mg/cm^3), while the Bulk Density remained comparable across samples (0.69–0.73 g/cm^3), supporting findings by [36]. These results confirm that composite flours can be optimized to enhance specific functional properties, making them suitable for various food applications.

The proximate analysis of the baked products shows that there are significant differences ($p < 0.05$) in the ash, moisture, fat, crude fiber, protein and carbohydrate contents between each baked product and their 100% wheat-based control products. The higher proximate parameter values obtained in the composite flour baked products showed that they are more nutritionally enhanced than their wheat-based control products, and this is consistent with the findings of [38–40].

The mineral content analysis of the baked products made from the fermented mixture 5 composite flour and their 100% wheat-based products showed statistically significant variations ($p < 0.05$). The results showed a significant presence of biofunctional minerals in the baked products. This could be attributed to the fact that apart from the fermentation process, the botanical flour blends are known to be good sources of minerals. These combinations help to enhance the bioavailability of minerals in the flour blends and their food products [41]. The higher mineral content observed in control products compared to their fermented counterparts can be attributed to several factors inherent from the cultivation of the botanicals to the processing technologies employed in their development into flour forms as well as differences in flour types. Also, some microorganisms involved in fermentation may assimilate free minerals from the substrate for their metabolic activities, leading to a measurable decrease in mineral concentrations in the final product [42,43]. Additionally, during the processing of the raw gluten-free botanicals into composite flours, some steps, such as soaking, can cause soluble minerals to leach into the fermentation medium, resulting in their loss during subsequent production stages [44]. However, it is important to note that fermentation can enhance the mineral bioavailability by degrading anti-nutritional factors like phytic acid, which binds minerals and inhibits their absorption [45,46]. Therefore, despite a decrease in the measurable mineral content, the minerals present in fermented products are often more bioavailable, offering improved nutritional benefits [43].

The antioxidant composition of the baked products made from the fermented mixture 5 composite flour showed higher OH and DPPH values and moderate Fe^{2+} chelation and FRAP values, indicating increased free radical scavenging activities. Overall, these results suggest that the fermentation process could improve the antioxidant properties of breads, biscuits and cakes made from the composite flour blends of sweet potato, pigeon pea and maize. These findings are consistent with previous studies that have shown fermentation to enhance the antioxidant properties of foods [47–49]. Antioxidants are vital in both food preservation and human health. In food products, they prevent oxidative deterioration, thereby maintaining the taste, color and nutritional value. This preservation extends the shelf life and ensures food safety [50]. In humans, antioxidants neutralize harmful free radicals, reducing oxidative stress linked to chronic diseases such as heart disease, cancer and neurodegenerative disorders [51]. The regular consumption of antioxidant-rich foods supports overall health by protecting cells from damage and bolstering the immune system [52].

The result of the anti-nutritional composition shows that the levels of anti-nutritionals (anti-nutrients/phytochemicals) vary depending on the baked product. In general, the fermentation process reduces the levels of anti-nutritionals, such as tannins, in food products by breaking down their complex molecules into simpler forms that are more easily absorbed by the body, while flavonoids are beneficial compounds with antioxidant properties, which may protect against cellular damage and reduce the risk of chronic diseases [53].

These findings are consistent with previous studies that have reported variations in the levels of anti-nutritional factors in composite flour and foods [54–56].

For the color analysis, the higher color intensity observed in the fermented baked products could be attributed to the Maillard reaction, which occurs during fermentation and baking, leading to the formation of brown pigments and enhancing the color of the final product [57]. This may have also influenced the sensory properties positively since color is a crucial factor that influences the visual appeal of food [58].

Based on the sensory evaluation results, all the baked products (cake, bread and biscuits) were generally acceptable by the panelists. The general acceptability is the conclusive assessment of the products' overall quality by consumers, and it is important in determining the success of such products in the marketplace. However, biscuits received higher scores for color, crispiness, texture, aroma, taste and general acceptability compared to the other two baked products. This suggests that biscuits may be the preferred choice. These findings are consistent with those of [59,60].

Animals (Wistar albino rats) were used to determine the anthropometric, hematological and histopathological parameters of the formulated bread, biscuit and cake products obtained from the composite flour. This becomes necessary in this research to ascertain the products' toxicity or their values as good food. From this study, all animals fed with the various baked product samples gained weight with no evidence of disease and with optimal organ functionality. Therefore, the consumption of the bread, biscuit and cake products made from fermented flour blends of sweet potato, maize and pigeon pea will be of good nutrition, and they are safe for human consumption.

5. Conclusions

This study demonstrates that blending sweet potato, pigeon pea and maize flours, followed by fermentation, enhances the nutritional profile and functional properties of baked biscuit, bread and cake products. Specifically, the composite flour exhibited an increased protein content, improved water and oil absorption capacities and elevated antioxidant activities, contributing to a better texture, shelf stability and potential health benefits. Fermentation also reduces anti-nutritional factors, like tannins, oxalates and phytates, thereby improving the mineral bioavailability. Sensory evaluations indicated the high acceptability of the baked products, mainly biscuits, suggesting their potential for consumer preference. Animal studies confirmed these products' safety and nutritional adequacy, supporting their suitability for human consumption. These findings underscore the potential of composite flours in addressing nutritional deficiencies and improving food security.

The possibility and feasibility of harnessing and further exploiting fermented indigenous agro-processed flour blends of sweet potato, pigeon pea and maize as ternary composite bio-resources in producing and developing enhanced baked functional foods cannot be overemphasized.

Author Contributions: Managed the analyses of data and the manuscript writing, contributed in designing the research, C.B.E.-E.; contributed in designing the work, the literature search and the writing of the manuscript, P.O.O.; designed the work and wrote the manuscript, F.C.A. All authors have read and agreed to the published version of the manuscript.

Funding: This research is funded by the Tertiary Education TrustFund (TETFUND) Batch Number 10 2022.

Institutional Review Board Statement: This study was conducted in accordance with the affirmation of Nigeria and approved with an ethical clearance certificate by the Auchi Polytechnic Ethical Research Committee (Ref.AP/SAST/23/06; 5 April 2023).

Informed Consent Statement: Not applicable.

Data Availability Statement: The data presented in this study can be made available on request from the corresponding author.

Conflicts of Interest: The authors declared no conflict of interest.

References

1. Food and Agriculture Organization (FAO). *The State of Food Security and Nutrition in the World 2022. Repurposing Food and Agricultural Policies to Make Healthy Diets More Affordable*; FAO: Rome, Italy, 2022; ISBN 978-92-5-136499-4, ISSN 2663-807X. [CrossRef]
2. Oleghe, P.O.; Akharaiyi, F.C.; Ehis-Eriakha, C.B. Harnessing indigenous agro-processed flour blends as composite bioresources in functional food development. *Food Res.* **2024**, *8*, 235–249. [CrossRef] [PubMed]
3. Dyvik, E.H. Number of Undernourished/Hungry People Worldwide 2000–2022. Available online: https://www.statista.com/statistics/264900/number. (accessed on 25 January 2024).
4. Countrymeters. Nigeria Population (2024) Live. Available online: https://countrymeters.info/en/Nigeria (accessed on 25 January 2024).
5. Sasu, D.D. Undernourishment and Food Insecurity in the Nigerian Population 2004–2022. Available online: https://www.statista.com/statistics/1262212/undernourishment (accessed on 18 January 2024).
6. National Bureau of Statistics. Multidimensional Poverty Index (MPI). Available online: www.nigerianstat.gov.ng (accessed on 25 September 2024).
7. Butnariu, M.; Sarac, I. Functional Food (Review). *Int. J. Nutr.* **2019**, *3*, 7–16. [CrossRef]
8. Galanakis, C.M.; Aldawoud, T.M.S.; Rizou, M.; Rowan, N.; Ibrahim, S. Food ingredients and active compounds against the coronavirus disease (COVID-19) pandemic: A comprehensive review. *Foods* **2020**, *9*, 1701. [CrossRef] [PubMed]
9. Chen, G.; Li, Y.; Li, X.; Zhou, D.; Wang, Y.; Wen, X.; Wang, C.; Liu, X.; Feng, Y.; Li, B.; et al. Functional foods and intestinal homeostasis: The perspective of in-vivo evidence. *Trends Food Sci. Technol.* **2021**, *111*, 475–482. [CrossRef]
10. Oleghe, P.O.; Akharaiyi, F.C.; Ehis-Eriakha, C.B. Compositions nutrient and antinutrients of biscuits prepared from fermentedand unfermented ternary mixture flours. *J. Food Qty. Haz. Control* **2024**, *11*, 13–25. [CrossRef]
11. Su, W.H.; He, H.J.; Sun, D.W. Non-Destructive and rapid evaluation of staple foods quality by using spectroscopic techniques: A review. *Crit. Rev. Food Sci. Nutr.* **2016**, *57*, 1039–1051. [CrossRef]
12. John, R.; Singla, A. Functional foods: Components, health benefits, challenges, and major projects. *DRC Sustain. Future J. Environ. Agric. Energy* **2021**, *2*, 61–72. [CrossRef]
13. Shah, A.M.; Tarfeen, N.; Mohamed, H.; Song, Y. Fermented foods: Their health-promoting components and potential effects on gut microbiota. *Fermentation* **2023**, *9*, 118. [CrossRef]
14. Cámara, M.; Sánchez-Mata, M.C.; Fernández-Ruiz, V.; Cámara, R.M.; Cebadera, E.; Domínguez, L. A review of the role of micronutrients and bioactive compounds on immune system supporting to fight against the COVID-19 disease. *Foods* **2021**, *10*, 1088. [CrossRef]
15. Oleghe, P.O.; Akharaiyi, F.C.; Ehis-Eriakha, C.B.; Oboh, E.J. Microbiological and Physiochemical Properties of Fermented and Unfermented Sweet Potato Flour. *Eur. Food Sci. Eng.* **2023**, *4*, 1–9. [CrossRef]
16. Olorunfemi, M.A.; Awolu, O.O.; Enujiugha, V.N. Evaluation of the chemical, antinutritional and antioxidant properties of composite flour comprising native and modified acha (*Digitaria exilis* stapf) flour supplemented with mango kernel seed and soy cake flours. *Food Sci. Technol. Int.* **2021**, *28*, 40–49. [CrossRef] [PubMed]
17. Adeyanju, A.A.; Bamidele, O.P. Nutritional composition, *in-vitro* starch digestibility and antioxidant activities of composite flour made from wheat and mature, unripe pawpaw (*Carica papaya*) fruit flour. *Nutrients* **2022**, *14*, 4821. [CrossRef] [PubMed]
18. Observatory Economic Complexity. Wheat in Nigeria. OEC Report. Available online: https://oec.world/en/profile/bilateral-product/wheat/reporter/nga (accessed on 15 April 2023).
19. Oleghe, P.O.; Akharaiyi, F.C.; Ehis-Eriakha, C.B. Phylogenetic identification of microbes from fermented botanicals used in gluten-free composite flour mixes. *Foods Raw. Mater.* **2025**, *13*, 82–93. [CrossRef]
20. Squeo, G.; De Angelis, D.; Leardi, R.; Summo, C.; Caponio, F. Background, Applications and Issues of the Experimental Designs for Mixture in the Food Sector. *Foods* **2021**, *10*, 1128. [CrossRef]
21. Chinma, C.E.; James, S.; Imam, H.; Ocheme, O.B.; Anuonye, J.C.; Yakubu, C.M. Physicochemical and sensory properties and in-vitro digestibility of biscuits made from blends of tigernut (*Cyperus esculentus*) and pigeon pea (*Cajanus cajan*). *Niger. J. Nutr. Sci.* **2011**, *32*, 55–62. [CrossRef]
22. Begum, R.; Uddin, M.J.; Rahman, M.A.; Islam, M.S. Comparative study on the development of maize flour based composite bread. *J. Bangladesh Agric. Univ.* **2013**, *11*, 133–139. [CrossRef]

23. Asimah, V.K.; Kpodo, F.M.; Adzinyo, O.A.; Dzah, C.S. Utilisation of brown rice flour and peanut paste in cake production. *Am J. Food Sci. Technol.* **2016**, *4*, 129–134. [CrossRef]
24. Association of Analytical Chemist (AOAC). *Official Methods of Analysis*, 19th ed.; Association of Analytical Chemist: Washington, DC, USA, 2012; Volume 2, pp. 121–130.
25. Sodipo, M.; Oluwagbenga, A.; Jolayemi, O.; Lawal, O. Development and nutritional evaluation of a complementary diet from fermented provitamin-a-biofortified maize (*Zea mays* L.) and germinated lentil seeds (*Lens culinaris*). *Croat. J. Food Sci. Technol.* **2020**, *12*, 90–100. [CrossRef]
26. Zhang, W.; Chen, H.; Wang, Z.; Lan, G.; Zhang, L. Comparative studies on antioxidant activities of extracts and fractions from the leaves and stem of *Epimedium koreanum* Nakai. *J. Food Sci. Technol.* **2013**, *50*, 1122–1129. [CrossRef]
27. Iombor, T.T.; Olaitan, I.N.; Ede, R.A. Proximate composition, antinutrient content and functional properties of soursop flour as influenced by oven and freeze drying methods. *Curr. Res. Nutr. Food Sci.* **2014**, *2*, 106–110. [CrossRef]
28. Bakker, J.; Arnold, G.M. Analysis of sensory and chemical data for a range of red port wines. *Am. J. Enol. Vitic.* **1993**, *44*, 27–34. [CrossRef]
29. Iwe, M.O. *Handbook of Sensory Method and Analysis*; Rejoint Communication Service Ltd.: Enugu, Nigeria, 2002; pp. 1–56.
30. Novelli, E.L.B.; Diniz, Y.S.; Galhardi, C.M.; Ebaid, G.M.X.; Rodrigues, H.G.; Mani, F.; Fernandes, A.A.H.; Cicogna, A.C.; Novelli-Filho, J.L.V.B. Anthropometrical parameters and markers of obesity in rats. *Lab. Anim.* **2007**, *41*, 111–119. [CrossRef] [PubMed]
31. Altunkaynak, B.Z.; Özbek, E. Overweight and structural alterations of the liver in female rats fed a high-fat diet: A serological and histological study. *Turk. J. Gastroenterol.* **2019**, *20*, 93–103.
32. Akubor, P.I. Proximate composition and selected functional properties of African breadfruit and sweet potato flour blends. *Plant Foods Hum. Nutr.* **1997**, *51*, 53–60. [CrossRef]
33. Nwakanma, C.I.; Obinwa, E.P.; Umeh-Idika, A.S. Proximate and functional properties of pigeon pea (*Cajanus cajan*) and sorghum (*Sorghum bicolor*) flour blends. *J. Home Econ. Res.* **2024**, *31*, 104–113.
34. Ihemeje, A.; Uneanya, G.C.; Odimegwu, N.E.; Echefu, E.O. Quality evaluation of cassava-pigeon pea composite flour and the cookies. *J. Agric. Food Sci.* **2017**, *15*, 43–53. [CrossRef]
35. Chiedu, U.C.; Malomo, S.A.; Ijarotimi, O.S.; Emojorho, E.E.; Arogundade, T.J. Physicochemical, nutritional and functional properties of composite flour blends from whole wheat, sweet potato, defatted peanut, and rice bran. *Eur. J. Nutr. Food Saf.* **2023**, *15*, 41–55. [CrossRef]
36. Peter-Ikechukwu, A.I.; Ogazi, C.G.; Uzoukwu, A.E.; Kabuo, N.O.; Chukwu, M.N. Proximate and functional properties of composite flour produced with date fruit pulp, toasted watermelon seed, and wheat. *J. Food Chem. Nanotechnol.* **2020**, *6*, 159–166. [CrossRef]
37. Bukuni, S.J.; Ikya, J.K.; Dinnah, A.; Bongjo, N.B. Chemical and functional properties of composite flours made from fermented yellow maize, Bambara groundnut, and mango fruit for 'ogi' production. *Asian Food Sci. J.* **2022**, *21*, 22–33. [CrossRef]
38. Onwuka, C.F.; Ameh, A.; Aliyu, M. Effect of baking time and composite flour blends on the physical, nutritional and sensory properties of bread. *J. Food Sci. Technol.* **2018**, *55*, 1973–1983.
39. Lei, V.; Amoaful, E.; Akonor, P.T. Changes in nutritional composition of fermented foods: An updated review. *Trends Food Sci. Technol.* **2019**, *88*, 1–10.
40. Verni, M.; Verardo, V.; Rizzello, C.G. How fermentation affects the antioxidant properties of cereals and legumes. *Foods* **2019**, *8*, 362. [CrossRef] [PubMed]
41. Oyarekua, M.A.; Adeoye, A.O.; Ojo, O. Comparable vitamin and mineral contents of co-fermented maize/carrot/pigeon pea and co-fermented maize/sweet potato/pigeon pea as infant complementary food. *Tropical. J. Nat. Prod. Res.* **2022**, *6*, 1331–1335. [CrossRef]
42. Kennedy, D.O.B. Vitamins and the Brain: Mechanisms, Dose and Efficacy—A Review. *Nutrients* **2016**, *8*, 68. [CrossRef]
43. Sharma, R.; Garg, P.; Kumar, P.; Bhatia, S.K.; Kulshrestha, S. Microbial fermentation and its role in quality improvement of fermented foods. *Fermentation* **2020**, *6*, 106. [CrossRef]
44. Rashwan, A.K.; Yones, H.A.; Karim, N.; Taha, E.M.; Chen, W. Potential processing technologies for developing sorghum-based food products: An update and comprehensive review. *Trends Food Sci. Technol.* **2021**, *110*, 168–182. [CrossRef]
45. Oghbaei, M.; Prakash, J. Effect of primary processing of cereals and legumes on its nutritional quality: A comprehensive review. *Cogent Food Agric.* **2016**, *2*, 1136015. [CrossRef]
46. Adebo, J.A.; Njobeh, P.B.; Gbashi, S.; Oyedeji, A.B.; Ogundele, O.M.; Oyeyinka, S.A.; Adebo, O.A. Fermentation of cereals and legumes: Impact on nutritional constituents and nutrient bioavailability. *Fermentation* **2022**, *8*, 63. [CrossRef]
47. Kim, J.; Yun, H.; Park, H.; Kim, M.; Chun, J. Effects of fermentation on the antioxidant activity of soy sauce. *J. Food Nutr. Res.* **2019**, *7*, 158–163.
48. Li, X.; Chen, H.; Liu, J.; Chen, X. Effect of fermentation on the antioxidant activity and phenolic compounds of tartary buckwheat flour. *J. Food Process Preserv.* **2019**, *43*, e13857.

49. Khalid, N.; Muhammad, A.I.; Siddiqui, M.T.H.; Hussain, M.A. Nutritional and antioxidant potential of maize flour fermented with *Saccharomyces cerevisiae* and *Lactobacillus plantarum*. *Food Sci. Nutr.* **2020**, *8*, 415–422.
50. Wilson, D.W.; Nash, P.; Buttar, H.S.; Griffiths, K.; Singh, R.; De Meester, F.; Horiuchi, R.; Takahashi, T. The role of food antioxidants, benefits of functional foods, and influence of feeding habits on the health of the older person: An overview. *Antioxidants* **2017**, *6*, 81. [CrossRef] [PubMed]
51. Pham-Huy, L.A.; He, H.; Pham-Huy, C. Free radicals, antioxidants in disease and health. *Int. J. Biomed. Sci. IJBS* **2008**, *4*, 89–96. [CrossRef] [PubMed]
52. Mueen-Ahmed, K.K. The Science of Antioxidants: Balancing the Pros and Cons for Our Health. *J. Pharmacol. Pharmacother.* **2023**, *14*, 160–162. [CrossRef]
53. Ullah, A.; Munir, S.; Badshah, S.L.; Khan, N.; Ghani, L.; Poulson, B.G.; Emwas, A.H.; Jaremko, M. Important flavonoids and their role as a therapeutic agent. *Molecules* **2020**, *25*, 5243. [CrossRef]
54. Şanlier, N.; Gökcen, B.B.; Sezgin, A.C. Health benefits of fermented foods. *Crit. Rev. Food Sci. Nutr.* **2019**, *59*, 506–527. [CrossRef]
55. Adiamo, O.Q.; Babajide, J.M.; Bamgbose, A.M.; Afolayan, A.O. Nutrient and anti-nutrient contents of maize/cassava/sweet potato/soybean flour blends extruded at different temperatures. *J. Food Sci. Technol.* **2016**, *53*, 3574–3582.
56. Ojokoh, A.O.; Onasanya, T.T. Effect of fermentation and extrusion on the nutrient and anti-nutrient composition of Soy beans (*Glycine max*, L) and Acha (*Digitaria exilis* Stapf). *Microbiol. Res. J. Int.* **2017**, *21*, 1–21. [CrossRef]
57. Kaur, A.; Singh, N.; Ezekiel, R. Effect of incorporation of legume flours on the quality characteristics of composite breads: A review. *J. Food Sci. Technol.* **2018**, *55*, 385–396.
58. Delgado-Pando, G.; Cofrades, S.; Ruiz-Capillas, C.; Jimenez-Colmenero, F. Healthier meat and meat products: Their role as functional foods. *Meat Sci.* **2013**, *93*, 28–36.
59. Sachanarula, S.; Chantarasinlapin, P.; Adisakwattana, S. Substituting whole wheat flour with Pigeon pea (*Cajanus cajan*) flour in Chapati: Effect on nutritional characteristics, color profiles, and In Vitro starch and protein digestion. *Foods* **2022**, *11*, 3157. [CrossRef]
60. Idowu, M.A.; Adebowale, A.A.; Adegunwa, M.O. Quality characteristics of biscuits produced from wheat, cocoyam and soybean flour blends. *J. Food Process Preserv.* **2018**, *42*, e13320. [CrossRef]

proceedings

MDPI

Proceeding Paper

Evaluation of Food Technologies and Farmers' Practices Related to Sorghum Cultivation in Central and Northern Benin †

Sylvain Megnonhou [1,*], David Montcho [1], Essegbemon Akpo [1,2], Judicaël Dandjlessa [3] and Adjaho Olatondji Eustache Kévin Kpocheme [3]

1 Laboratory of Plant, Horticultural and Forest Sciences, National University of Agriculture, Porto-Novo 01 BP 55, Benin; dmontchohambada@gouv.bj (D.M.); akpo.essegbemon@gmail.com (E.A.)
2 International Maize and Wheat Improvement Center, México-Veracruz, El Batán Km. 45, Mexico City 56237, Mexico
3 Laboratory of Phytotechnics, Physiology and Genetic Improvement of Plant Species (PAGEV), Faculty of Agricultural Sciences, University of Abomey-Calavi, Cotonou 01 BP 526, Benin; dandjlessa94@yahoo.com (J.D.); eveneustache2@gmail.com (A.O.E.K.K.)
* Correspondence: sylvainmegnonhou@gmail.com
† Presented at the CORAF's 2023 Symposium on Processing and Transformation of Agricultural Products in West and Central Africa: Achievements and Opportunities for Private Sector Engagement, Lome, Togo, 21–23 November 2023.

Abstract: This study explored food habits and traditional practices related to sorghum cultivation in Benin. Based on data collected from 289 producers, statistical analysis revealed that sorghum is mainly consumed as porridge (97.92%), paste (91.69%), akassa (40.48%), and used as fodder (30.45%). Additionally, its leaves are used in the treatment of anemia and malaria (40.48%), while its stems serve in shelter construction (55.36%). Factorial analysis shows that these uses vary across sociocultural groups: the Kotokoli, Lokpa, Yom, Mokolé, and Idatcha favor food consumption, the Bariba use it for medicinal purposes, and the Fulani transform it into a local beverage. These findings highlight the need to tailor sorghum breeding programs to sociocultural needs and current climate challenges.

Keywords: food habits; category of use; traditional practices; Benin

1. Introduction

Sorghum is one of the main staple crops in West Africa. It is a subsistence crop for hundreds of millions of people living in rural areas [1]. This cereal, capable to grow where other cereals are too demanding, is part of the basic diet of low-income populations and a forage crop for livestock [2]. However, this crop faces the challenges of climate change, which are expected to reduce its yield by 13% in West Africa [3]. Furthermore, projected climate change in the dry savannah of Northern Benin will likely reduce the efficiency of water and nitrogen use, and grain yields of maize and sorghum [4]. These threats will not go unanswered by farmers and scientific communities, who seek to use or develop new, more resilient and productive varieties. However, in peasant communities, producers adopt a wide diversity of local species and varieties that have culturally important advantages [5,6]. Indeed, they possess traditional knowledge about the value and properties of many plant species [7,8]. Thus, genetic resource improvement programs can only be sustainable if they integrate the social, cultural, and economic values associated with them by local communities. For this reason, knowledge and consideration of peasant perceptions are necessary and decisive in identifying measures to adapt local agriculture [9]. In this sense, ethnobotanical studies appear to be a good approach to understand, in a given region, the uses and sociocultural perceptions of genetic resources by local populations [10]. Endogenous knowledge, an essential component of the adoption and conservation of technologies, is important both in improving the daily lives of populations and in decision-making for the sustainable management of natural resources [11]. All this knowledge is indispensable in the development of new inclusive technologies aimed at providing effective solutions to threats at the community level [12]. It is therefore imperative to explore farmers' knowledge regarding threatened crops in order to enable the scientific community to develop new varieties that take into account the social context of their use [13,14]. It is in this perspective that the present study aimed to evaluate food technologies and farmers' practices related to the use of sorghum by different sociocultural groups in Central and Northern Benin.

2. Materials and Methods

2.1. Study Area

This study was conducted in three agroecological zones out of the eight that exist in Benin (Figure 1). The classification of these zones is based on the definition of relatively homogeneous zones using agro-pedological parameters, crop systems, spatial distribution of the population, and vegetation [15]. These zones were chosen based on their level of production and diversity in sorghum, and the interest that their communities attach to its use in Benin [16].

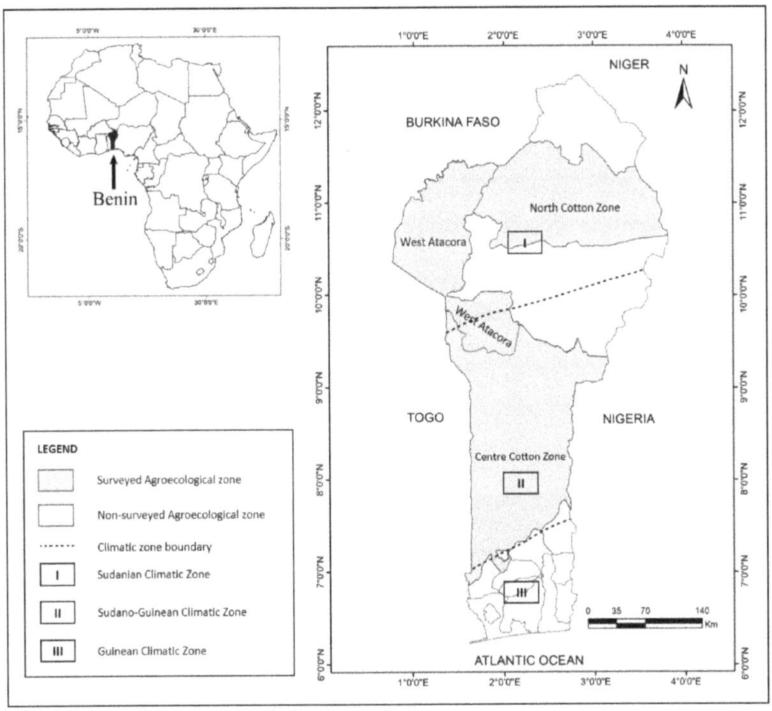

Figure 1. Study area location map.

2.2. Sampling Method

The sample size (n) was obtained using the binomial distribution proposed by Dagnélie (1998) [17]:

$$n = [(U_1 - \alpha/2)^2 \times P_i \times (1 - P_i)]/d^2$$

where Pi (75%) represents the proportion of producers who are aware of and use sorghum, and $U1 - \alpha/2 = 1.96$ indicates the value of the normal random variable for a risk α equal to 0.05. The estimated margin of error for any survey parameter is 5% (d). For the Pi values obtained, 289 producers were identified and randomly sampled within the different sociocultural groups (Table 1).

Table 1. Sampling groups.

Regions of Benin	Sociocultural Groups	Absolute Frequencies	Relative Frequencies (%)
Central	Bariba	13	4.50
	Idacha	38	13.15
	Mahi	19	6.57
	Peulh	10	3.46
Total	Total	80	27.68
Northern	Kotokoli	12	4.15
	Lokpa	11	3.81
	Bariba	75	25.95
	Mokolé	10	3.46
	Fulani	34	11.76
	Peulh	20	6.92
	Yom	47	16.26
Total	-	209	72.32
Total surveyed	-	289	100

2.3. Data Collection

The data collection process involved both individual structured interviews and focus group discussions. Using a well-designed questionnaire and an interview guide, data were first gathered through one-on-one interviews with the sampled producers. To complement and validate the information collected individually, focus groups were then conducted, allowing for collective discussions and data confirmation. The key data collected included the demographic and socio-economic characteristics of producers (such as gender, age, primary and secondary activities, education level, and sorghum cultivation area) and the various food, medicinal, artisanal, and cultural uses of sorghum across different sociocultural groups in Central and Northern Benin.

2.4. Data Analysis

For the reported forms of use, the relative citation frequencies (RCFs) were used to assess the relative importance of each use. In order to determine the distribution of knowledge in the population, the interviewed individuals were grouped according to their sociocultural affiliation. Contingency tables indicating the number of surveyed individuals mentioning a form of use and the plant part used by the sociocultural group were established. Correspondence factor analysis (CFA) was performed using R software version 3.4.1 to describe the relationship between sociocultural groups and usage categories, and the relationship between sociocultural groups and the plant parts used.

3. Results

3.1. Socio-Economic and Demographic Characteristics of the Respondents

Table 2 presents the socio-economic and demographic characteristics of the surveyed producers. The analysis of this table shows that the surveyed producers were mostly men (84.43%). They were also mostly adults between the ages of 31 and 60 (81.31%), and a small proportion (7.61%) were elderly. In terms of education level, the surveyed individuals were mostly uneducated (84.43%), with a few having completed primary (10.73%) and secondary education (4.84%). Furthermore, they were mostly engaged in agriculture (98.96%) and livestock farming (86.85%) as their main and secondary activities, respectively. Regarding sorghum production, 66.78% of the surveyed producers cultivated areas equal to or greater than one hectare (≥1 ha).

Table 2. Socio-economic and demographic characteristics of the respondents.

Sociodemographic Characteristics		Number	%
Gender	Male	244	84.43
	Female	45	15.57
Ages	18 to 30 years (young)	32	11.07
	31 to 60 years (adults)	235	81.31
	Over 60 years old (elderly)	22	7.61
Education level	Primary	31	10.73
	Secondary	14	4.84
	None	244	84.43
Main activity	Agriculture	286	98.96
	Livestock	3	1.04
	Trade	0	0.00
Secondary activity	Livestock	251	86.85
	Agriculture	26	9.00
	Trade	12	4.15
Sorghum cultivation areas	<1 ha	96	33.22
	≥1 ha	193	66.78

3.2. Categories and Forms of Sorghum Use in Central and Northern Benin

Table 3 reveals seven categories of sorghum uses by producers, including food, beverage, agronomic, artisanal, cultural, medicinal, and cosmetic. Food use is the most mentioned, especially in the form of porridge (97.92%) and dough (91.69%). Furthermore, the grains of this crop are used in the preparation of various traditional drinks such as tchakpalo (74.74%), tchoukoutou (62.98%), and sodabi (25.95%). These different drinks are produced using traditional methods with red sorghum grains. The same table shows that sorghum is adopted by farmers as a leading crop. However, 79.24%, 26.30%, 25.95%, and 24.57% of the respondents, respectively, indicated the use of sorghum leaves and stems as livestock feed, organic fertilizer, support, and cushion for yam mounds. These uses of sorghum facilitate livestock rearing and yam production in the surveyed communities. Additionally, the stems are valued in artisanal constructions by the surveyed farmers. In fact, 83.74%, 58.13%, and 55.36% of the respondents, respectively, use sorghum stems to make fences (Figure 2), mats, and traditional houses (Figure 3). On a cultural level, a minority (15.92%) of the respondents mentioned the adoption of sorghum stems in funeral ceremonies as a widow's stick. This practice was mentioned among the Yom people in Northern and Central Benin and involved the widow holding the sorghum stem for four months to protect her from ghost spirits. In terms of medicinal use, 40.48% of the respondents mentioned the use of red sorghum leaves in decoction to treat malaria and anemia.

Table 3. Categories and forms of sorghum use in Central and Northern Benin.

Categories of Use	Parts Used	Forms of Use	Number of Citations	FRC (%)
Food	Grains, leaves	Porridge	283	97.92
		Paste	265	91.69
		Fodder	229	79.24
		Akassa	117	40.48
		Feed	88	30.45
Beverage (traditional drinks)	Grains	Tchakpalo	216	74.74
		Tchoukoutou	182	62.98
		Sodabi	75	25.95
Agronomic	Stems, leaves	Organic fertilizer	76	26.30
		Yam support	71	24.57
		Cushions (yam)	75	25.95
Artisanal	Stems, leaves	Fence	242	83.74
		Traditional houses	160	55.36
		Mats	168	58.13
Cultural	Stems	Widow's stick	46	15.92
Medicinal	Leaves	Malaria and anemia	117	40.48
Cosmetic	Stems, leaves	Soap	86	29.76
		Packaging	65	22.49

FRC: Frequency relative of citation.

Figure 2. Fence made from sorghum stems in Copargo.

Figure 3. Traditional shelter (**left**) and mats (**right**) made from sorghum stems in Kandi.

3.3. Relationship Between the Different Categories and Forms of Sorghum Use and the Different Sociocultural Groups of Northern and Central Benin

The correspondences resulting from the projection of the different usage categories and sociocultural groups onto the axis systems of the AFC (Figure 4) show that the Kotokoli, Lokpa, Yom, Mokolé, and Idatcha prefer the food usage of the species, while its medicinal usage is favored by the Bariba. On the other hand, the Peulhs use it more for artisanal construction purposes (traditional houses) and traditional drinks (tchoukoutou and tchakpalo). Its usage as a cultic object (widow's stick) is more specific to the Lokpa. The other sociocultural groups, not circled in Figure 4, are not associated with a specific usage category.

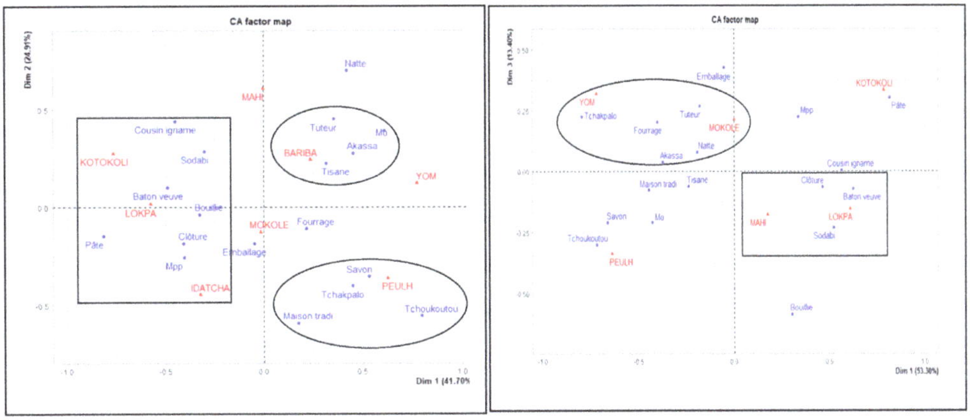

Figure 4. Relationship between usage form and the different sociocultural groups of Central and Northern Benin.

3.4. Relationship Between the Parts of the Plant Used and the Sociocultural Groups

The correspondences resulting from the projection of the different parts used and sociocultural groups onto the axis systems of the AFC (Figure 5) show that the Peulh, Yom, Lokpa, Kotokoli, and Mahi attach more importance to the stems and grains, while the Bariba and Idatcha rely more on the use of the leaf in addition to the grains. The other sociocultural groups not circled are not associated with the specific use of a plant part.

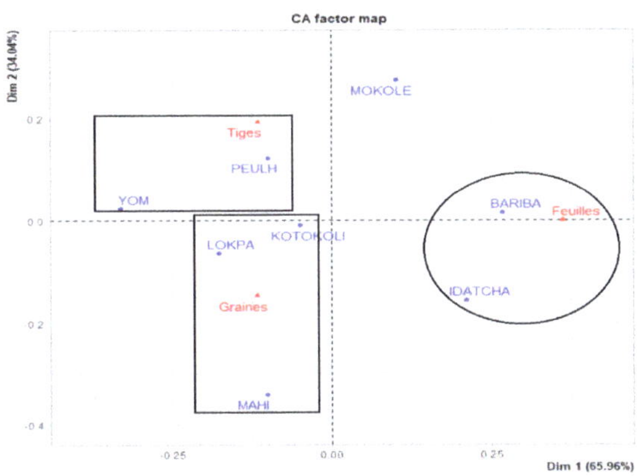

Figure 5. Relationship between plant part used and different sociocultural groups.

4. Discussion

This study highlighted the various uses of sorghum by different sociocultural groups in Central and Northern Benin. Thus, food use is the most mentioned in all the surveyed sociocultural groups. This reveals the importance of this crop in the diet of the Benin population. This importance was described in the studies by Chantereau et al. (2013) [18], who stated that sorghum is one of the cereals used as a staple food in many populations of dry tropical regions in Africa, Asia, and Central America, where production is mainly for self-consumption. However, the surveyed producers used sorghum grains to make porridge, akassa, paste, and traditional drinks (tchoukoutou, tchakpalo, and sodabi). These forms of sorghum use have also been reported in several previous studies in Benin and elsewhere in Africa [19,20]. In addition, the use of sorghum residues as fodder and the use of grain by-products as feed for animals, particularly ruminants, were mentioned by 79.24% and 30.45% of the surveyed individuals, respectively. This form of sorghum use demonstrates its importance in livestock farming among the communities in Central and Northern Benin, which have significant livestock herds. This contributes to reducing the difficulties associated with transhumance, which is currently a major problem in agriculture in developing countries. Djohy et al. (2023) [21] confirmed this usefulness of sorghum when they reported that the residues can cover up to 77% of the livestock's feed requirements during the dry season. Other authors, such as McCuistion et al. (2019) [22], found that when properly processed and balanced with other ingredients, sorghum can serve as a primary source of cereals in animal diets. Agronomically, sorghum leaves and stems are used by farmers to facilitate soil fertilization (26.30%) and support (stakes) yam vines in cultivation (24.57%). These results differ from those mentioned by other authors regarding the uses of sorghum. Indeed, the studies by Dahlberg (2019) [23] and Rao et al. (2019) [24] revealed that sweet sorghum is similar to grain sorghum but has the ability to accumulate sugars in the stems without significantly reducing grain production. It is therefore used as a raw material for first-generation biofuel, with the sugars from the grains and stems being used to produce bioenergy, while energy sorghum or biomass sorghum is increasingly considered as a potential raw material for lignocellulosic biofuel production. This knowledge is still new and rarely disseminated to rural communities [25]. This could explain the non-mention of this use by the surveyed producers. Furthermore, the use of sorghum in traditional medicine was mentioned by some respondents. Thus, according to 40.48% of the producers, the use of red sorghum leaves in decoction helps fight against malaria and anemia. This recipe complements the one related to the antihyperglycemic activity of sorghum grains demonstrated by Ofosu et al. (2020) [26]. The medicinal potential of sorghum has also been mentioned by Stefoska-Needham et al. (2015) [25], who added that sorghum-based diets could play a role in the prevention of chronic diseases such as diabetes, obesity, and heart disease. In addition to these food and medicinal uses, sorghum provides many other services to the community. Thus, culturally, sorghum stems were used as "widow's stick" among the Yom people in North and Central Benin. This practice involved having the sorghum stem held by the new widow for four months to protect her from the spirits of ghosts. Bricas (1998) [27] addressed this in the same way by demonstrating that sorghum remains an identity symbol and has a reputation for being a healthy food. Associated with rural tradition, individualism, and urban socialization, the consumption of sorghum reflects a certain urban food modernity. Furthermore, the use of sorghum stems in artisanal constructions, such as mats, enclosures, shelters, etc., motivates the Fulani and Mokolé people to cultivate sorghum more in Northern Benin. This could explain the presence of long-stem sorghum varieties (>200 cm) reported by Kayodé et al. (2006) [28] in their study on sorghum diversity in Northern Benin. Speaking of the relationships between the category and forms of use and sociocultural groups, the uses of sorghum are sometimes dependent on sociocultural groups. However, the Kotokoli, Lokpa, Yom, Mokolé, and Idatcha prefer the food use of the species, while its medicinal use is more mentioned by the Bariba. On the other hand, the Fulani people use it more for artisanal constructions (traditional houses) and traditional drinks (tchoukoutou and tchakpalo). Sop et al. (2012) [29] mentioned similar observations regarding woody species in Burkina Faso.

The diversity of uses of a species can be explained by the local microclimate that governs its distribution and availability [29]. It can also be due to the diversity of the needs of local communities [30,31].

5. Conclusions

This study investigated food technologies and peasant practices related to the use of sorghum by different sociocultural groups in Central and Northern Benin. This investigation identified the various forms in which producers use sorghum and its derivatives. Thus, the interviewed producers adopted sorghum in various ways for food, agronomic, artisanal, cultural, and medicinal purposes. However, this study identified food use as the most cited among all the different sociocultural groups, especially in the form of porridge, paste, akassa, and traditional beverages (tchoukoutou and tchakpalo). In addition to these food needs, this study highlighted numerous other uses that communities derive from sorghum cultivation. For example, the leaves of red sorghum are used in the treatment of anemia and malaria. Besides these uses, the leaves are also used to feed livestock and fertilize fields. As for the stems, they are of great importance in the lives of the surveyed communities as they are used for building shelters and various other useful artisanal constructions for rural populations. In terms of the relationship between categories and forms of use and the different sociocultural groups surveyed, the uses of sorghum are sometimes dependent on sociocultural groups. However, the Kotokoli, Lokpa, Yom, Mokolé, and Idatcha prefer the food use of the species, while its medicinal use is more cited by the Bariba. On the other hand, the Fulani people use it more for artisanal construction purposes (traditional houses) and traditional beverages (tchoukoutou and tchakpalo). These different specificities in the uses of sorghum by communities should be considered in research programs in order to propose new varieties that not only address current climate challenges but also sociocultural needs.

Author Contributions: Conceptualization; methodology, S.M., D.M. and E.A.; methodology, S.M., D.M. and E.A.; validation, D.M., E.A., J.D. and A.O.E.K.K.; investigation, S.M.; data curation, S.M.; writing—original draft, S.M.; writing—review and editing, S.M., D.M., E.A., J.D. and A.O.E.K.K.; supervision, D.M. and E.A. All authors have read and agreed to the published version of the manuscript.

Funding: This study was financed under funding agreement N°81067102 between the Swiss Agency for Development and Cooperation (SDC) and the West and Central African Council for Agricultural Research and Development (CORAF).

Institutional Review Board Statement: Not applicable.

Informed Consent Statement: Not applicable.

Data Availability Statement: All the data used to support the findings of this study are available within the research article.

Acknowledgments: Special thanks to the producers who made themselves available to answer my questionnaire and participated in the focus groups during this study.

Conflicts of Interest: The authors declare no conflicts of interest.

References

1. Kangama, C.O. Importance of Sorghum Bicolor in African's Cultures. *J. Agric. Environ. Sci.* **2017**, *6*, 134–137. [CrossRef]
2. Proietti, I.; Frazzoli, C.; Mantovani, A. Exploiting Nutritional Value of Staple Foods in the World's Semi-Arid Areas: Risks, Benefits, Challenges and Opportunities of Sorghum. *Healthcare* **2015**, *3*, 172–193. [CrossRef] [PubMed]
3. Sultan, B.; Guan, K.; Kouressy, M.; Biasutti, M.; Piani, C.; Hammer, G.L.; McLean, G.; Lobell, D.B. Robust Features of Future Climate Change Impacts on Sorghum Yields in West Africa. *Environ. Res. Lett.* **2014**, *9*, 104006. [CrossRef]

4. Amouzou, K.A.; Lamers, J.P.A.; Naab, J.B.; Borgemeister, C.; Vlek, P.L.G.; Becker, M. Climate Change Impact on Water- and Nitrogen-Use Efficiencies and Yields of Maize and Sorghum in the Northern Benin Dry Savanna, West Africa. *Field Crops Res.* **2019**, *235*, 104–117. [CrossRef]
5. Helida, A.; Zuhud, E.A.; Hardjanto, H.; Purwanto, Y.; Hikmat, A. Index of Cultural Significance as a Potential Tool for Conservation of Plants Diversity by Communities in The Kerinci Seblat National Park. *J. Manaj. Hutan Trop. J. Trop. For. Manag.* **2015**, *21*, 192–201. [CrossRef]
6. Supiandi, M.I.; Mahanal, S.; Zubaidah, S.; Julung, H.; Ege, B. Ethnobotany of Traditional Medicinal Plants Used by Dayak Desa Community in Sintang, West Kalimantan, Indonesia. *Biodiversitas J. Biol. Divers.* **2019**, *20*, 1264–1270. [CrossRef]
7. Masengo, C.A.; Bongo, G.N.; Robijaona, B.; Ilumbe, G.B.; Koto-Te-Nyiwa, J.-P.N.; Mpiana, P.T. Étude ethnobotanique quantitative et valeur socioculturelle de Lippia multiflora Moldenke (Verbenaceae) à Kinshasa, République Démocratique du Congo. *Rev. Mar. Sci. Agron. Vét.* **2021**, *9*, 93–101.
8. Sourabie, S.; Zerbo, P.; Yonli, D.; Boussim, J.I. Connaissances Traditionnelles Des Plantes Locales Utilisées Contre Les Bio-Agresseurs Des Cultures et Produits Agricoles Chez Le Peuple Turka Au Burkina Faso. *Int. J. Biol. Chem. Sci.* **2020**, *14*, 1390–1404. [CrossRef]
9. Oloukoi, J.; Yabi, I.; Houssou, C.S. Perceptions et Stratégies Paysannes d'adaptation à La Variabilité Pluviométrique Au Centre Du Bénin. *Int. J. Biol. Chem. Sci.* **2019**, *13*, 1366. [CrossRef]
10. Hu, R.; Lin, C.; Xu, W.; Liu, Y.; Long, C. Ethnobotanical Study on Medicinal Plants Used by Mulam People in Guangxi, China. *J. Ethnobiol. Ethnomedicine* **2020**, *16*, 40. [CrossRef]
11. Adjiba, S.T.C.; Adegbola, Y.P.; Yabi, J.A. Genre, Diffusion et Adoption Des Technologies de Gestion Durable Des Terres Dans Les Petites Exploitations Familiales Des Pays En Voie de Développement: Une Revue: Gender, Diffusion and Adoption of Sustainable Land Management Technologies in Small Family Farms in Developing Countries: A Review. *Int. J. Biol. Chem. Sci.* **2022**, *15*, 2118–2140. [CrossRef]
12. Shikuku, K.M. Information Exchange Links, Knowledge Exposure, and Adoption of Agricultural Technologies in Northern Uganda. *World Dev.* **2019**, *115*, 94–106. [CrossRef]
13. Acheampong, P.P.; Addison, M.; Wongnaa, C.A. Assessment of Impact of Adoption of Improved Cassava Varieties on Yields in Ghana: An Endogenous Switching Approach. Cogent Econ. *Finance* **2022**, *10*, 2008587. [CrossRef]
14. Martiarena, M.L.; Temudo, M.P. Endogenous Learning and Innovation in African Smallholder Agriculture: Lessons from Guinea-Bissau. *J. Agric. Educ. Ext.* **2023**, *30*, 161–179. [CrossRef]
15. Issa, M. *Sanni Cadre de Gestion Environnementale et Sociale*; Ministère de l'Agriculture de l'Elevage et de la Pêche: Cotonou, Bénin, 2009.
16. Dossou-Aminon, I.; Dansi, A.; Ahissou, H.; Cissé, N.; Vodouhè, R.; Sanni, A. Climate Variability and Status of the Production and Diversity of Sorghum (*Sorghum bicolor* (L.) Moench) in the Arid Zone of Northwest Benin. *Genet. Resour. Crop Evol.* **2016**, *63*, 1181–1201. [CrossRef]
17. Dagnélie, P. *Statistiques Théoriques et Appliquées*; De Boeck: Bruxelles, Belgique, 1998.
18. Chantereau, J.; Cruz, J.-F.; Ratnadass, A.; Trouche, G.; Fliedel, G. *Le Sorgho*; éditions Quae: Versailles, France, 2013; ISBN 978-2-7592-2062-5.
19. Dabija, A.; Ciocan, M.E.; Chetrariu, A.; Codină, G.G. Maize and Sorghum as Raw Materials for Brewing, a Review. *Appl. Sci.* **2021**, *11*, 3139. [CrossRef]
20. Adebo, O.A. African Sorghum-Based Fermented Foods: Past, Current and Future Prospects. *Nutrients* **2020**, *12*, 1111. [CrossRef]
21. Djohy, G.L.; Bouko, B.S.; Djohy, G.; Dossou, P.J.; Yabi, J.A. Contribution des résidus de culture à la réduction du déficit alimentaire des troupeaux de ruminants dans l'Ouémé Supérieur au Bénin. *Cah. Agric.* **2023**, *32*, 13. [CrossRef]
22. McCuistion, K.C.; Selle, P.H.; Liu, S.Y.; Goodband, R.D. *Sorghum as a Feed Grain for Animal Production. Sorghum and Millets*; Elsevier: Amsterdam, The Netherlands, 2019; pp. 355–391. ISBN 978-0-12-811527-5.
23. Dahlberg, J. *The Role of Sorghum in Renewables and Biofuels. Sorghum: Methods and Protocols*; Zhao, Z.-Y., Dahlberg, J., Eds.; Methods in Molecular Biology; Springer: New York, NY, USA, 2019; pp. 269–277. ISBN 978-1-4939-9039-9.
24. Rao, P.S.; Vinutha, K.S.; Kumar, G.S.A.; Chiranjeevi, T.; Uma, A.; Lal, P.; Prakasham, R.S.; Singh, H.P.; Rao, R.S.; Chopra, S.; et al. Sorghum: A Multipurpose Bioenergy Crop. In *Agronomy Monographs*; Ciampitti, I.A., Vara Prasad, P.V., Eds.; Soil Science Society of America: Madison, WI, USA, 2019; pp. 399–424. ISBN 978-0-89118-628-1.
25. Stefoska-Needham, A.; Beck, E.J.; Johnson, S.K.; Tapsell, L.C. Sorghum: An Underutilized Cereal Whole Grain with the Potential to Assist in the Prevention of Chronic Disease. *Food Rev. Int.* **2015**, *31*, 401–437. [CrossRef]
26. Ofosu, F.K.; Elahi, F.; Daliri, E.B.-M.; Yeon, S.-J.; Ham, H.J.; Kim, J.-H.; Han, S.-I.; Oh, D.-H. Flavonoids in Decorticated Sorghum Grains Exert Antioxidant, Antidiabetic and Antiobesity Activities. *Molecules* **2020**, *25*, 2854. [CrossRef]
27. Bricas, N. *Cadre Conceptuel et Méthodologique Pour L'analyse de la Consommation Alimentaire Urbaine en Afrique*; CIRAD: Montpellier, France, 1998; 48p.

28. Kayodé, A.P.; Linnemann, A.R.; Nout, M.R.; Hounhouigan, J.D.; Stomph, T.J.; Smulders, M.J. Diversity and Food Quality Properties of Farmers' Varieties of Sorghum from Bénin. *J. Sci. Food Agric.* **2006**, *86*, 1032–1039. [CrossRef]
29. Sop, T.K.; Oldeland, J.; Bognounou, F.; Schmiedel, U.; Thiombiano, A. Ethnobotanical Knowledge and Valuation of Woody Plants Species: A Comparative Analysis of Three Ethnic Groups from the Sub-Sahel of Burkina Faso. *Environ. Dev. Sustain.* **2012**, *14*, 627–649. [CrossRef]
30. Assogba, G.A.; Fandohan, A.B.; Salako, V.K.; Assogbadjo, A.E. Usages de Bombax Costatum (Malvaceae) Dans Les Terroirs Riverains de La Réserve de Biosphère de La Pendjari, République Du Bénin. *Bois Forets Trop.* **2017**, *333*, 17. [CrossRef]
31. Houètchégnon, T.; Gbèmavo, D.S.J.C.; Ouinsavi, C.A.I.N.; Sokpon, N. Structural Characterization of Prosopis Africana Populations (Guill., Perrott., and Rich.) Taub in Benin. *Int. J. For. Res.* **2015**, *2015*, 101373. [CrossRef]

proceedings

MDPI

Proceeding Paper

Influence of Cultivation Practices on Yield and Spoilage of *Kponan* Yam (*Dioscorea cayenensis-rotundata*) During Storage †

Yapo Hypolithe Kouadio [1,*], Kouakou Nestor Kouassi [2,3], Gbè Aya Jacqueline Konan [2], Kouame Aristide Kouakou [2], Bomo Justine Assanvo [1] and Yao Denis N'dri [2]

1 Laboratory of Biochemistry Biotechnology and Food Sciences, UFR Science and Technology, Université Alassane OUATTARA, Bouaké 01 P.O. Box V 18 , Côte d'Ivoire; bomoj@yahoo.com

2 Laboratory of Food Biochemistry and Technology of Tropical Products, Department of Food Science and Technology, Université Nangui ABROGOUA, Abidjan 02 P.O. Box 801, Côte d'Ivoire; nestorkksi@yahoo.fr (K.N.K.); gbekonan18@gmail.com (G.A.J.K.); aristidekouame1981@gmail.com (K.A.K.); ndri_denis@yahoo.fr (Y.D.N.)

3 Centre Suisse de Recherches Scientifiques en Côte d'Ivoire, Abidjan 01 P.O. Box 1303, Côte d'Ivoire

* Correspondence: yapohypolithe@gmail.com

† Presented at the CORAF's 2023 Symposium on Processing and Transformation of Agricultural Products in West and Central Africa: Achievements and Opportunities for Private Sector Engagement, Lome, Togo, 21–23 November 2023.

Abstract: The *Kponan* yam (*Dioscorea cayenensis-rotundata*) is a culturally and economically important crop in Côte d'Ivoire but faces significant post-harvest losses, reducing farmers' and traders' incomes. This study aimed to identify strategies to minimize these losses during storage. To this end, *Kponan* yams from Bondoukou, Bouna, and Kouassi-Kouassikro, key production areas, were grown in experimental plots and monitored over three months in both field and warehouse storage. The results showed that yams harvested in Kouassi-Kouassikro (2.53 ± 0.06 kg) and Bondoukou (2.37 ± 0.09 kg) were heavier than those from Bouna (2.01 ± 0.11 kg). Storage conditions influenced spoilage: yams stored in pits had lower alteration rates (40.00% to 48.57%) compared to those stored under straw huts or trees (100%). In warehouses, alteration rates ranged from 72.29% (Bondoukou) to 100% (Kouassi-Kouassikro), with rot rates varying from 47.05% to 70.00%, respectively. Weight losses varied from 15.15% to 36.67% in warehouses versus 8.47% to 42.86% in field storage. These results underline the importance of storage methods, with pit storage significantly reducing deterioration.

Keywords: *Kponan* yam; post-harvest losses; cultivation practices; conservation practices; food security; Côte d'Ivoire

1. Introduction

The genus *Dioscorea*, part of the Dioscoreaceae family, encompasses around 600 species primarily found in the subtropical and temperate regions of Africa, India, Southeast Asia, Australia, and the Americas [1]. Yam serves as a vital food crop for over 13 tropical countries, sustaining more than 500 million people [2,3]. Yam is a widely consumed staple food in West Africa [4,5]. Its value chain supports the livelihoods of approximately 5 million people, including rural farming households, traders, transporters, and processors [6]. Yam is a vital source of carbohydrates, vitamins, essential minerals, and fiber, and it has a low glycemic index [7,8]. Yam cultivation is used both for self-consumption and for generating income through the sale of surplus produce [9]. As a result, yam has become a cash crop that supplies urban markets [10,11]. Nigeria stands as the world's leading yam producer, contributing more than 50 million tons, which represents over 67% of the global production, which was estimated at 75.14 million tons in 2021 [3].

In Côte d'Ivoire, approximately 60% of the population relies on yam as a staple food [12]. Accounting for over 10% of the global production (7.85 million tons in 2021), Côte d'Ivoire stands as the third-largest yam producer, following Nigeria (67%) and Ghana (11%) [3]. Yam takes the lead in food crop production [13], positioning Côte d'Ivoire at the forefront of consumption among major producing countries, with an average of 502 kcal per capita per day [3,14].

Although yam tubers play a significant role in the nutritional well-being and economy of populations, they are perishable and seasonal products. Notably, the losses are higher in early-harvested tubers [15]. These post-harvest losses are attributed to storage-related difficulties, resulting in weight losses ranging from 65% to 85% [16]. Moreover, losses of up to 50% of the harvest are attributable to factors such as dehydration, harvest injuries, pests, rats, and the use of chemicals [17].

Within the yam varieties cultivated in Côte d'Ivoire, those belonging to the *Dioscorea cayenensis-rotundata* complex exhibit a higher susceptibility to rot [18]. Regrettably, the *Kponan* variety, highly favored by consumers in the city of Abidjan [18–20], is particularly susceptible to rot during storage [18]. This particular yam variety, available in wholesale markets in Abidjan, originates from six different geographical areas, with the primary sources being Bondoukou (60%), Bouna (33.3%), and Kouassi-Kouassikro (6.7%) [20].

Several authors have emphasized that the cultivation, harvesting, and storage techniques used for *Kponan* yams, both in the field and in urban warehouses, play a crucial role in their shelf life [11,17,21,22]. This study aims to examine the impact of cultivation practices on the yield of *Kponan* yams, as well as the spoilage rate during storage both in warehouses and in the field.

2. Material and Methods

2.1. Material

The biological material used in this study consists of the *Kponan* yam variety (*Dioscorea cayenensis-rotundata*), cultivated and harvested in experimental fields located in Bondoukou, Bouna, and Kouassi-Kouassikro [21].

2.2. Methods

2.2.1. Implementation of the Experimental Design

The experimental field used a completely randomized block design with three blocks, each subdivided into three plots (one per yam origin: Bouna, Bondoukou, Kouassi-Kouassikro) as described in [21] (Figure 1). Within each zone, *Kponan* yams from Bouna, Bondoukou, and Kouassi-Kouassikro were cultivated in the same field, employing the cultivation techniques of selected local producers [21]. Land clearing in the fields was carried out using herbicides in Bouna, Bondoukou, and Kouassi-Kouassikro, as described by Kouadio et al. [17]. The height of the ridges varied between 70 and 100 cm across the three areas. The distance between the ridges varied between 30 and 50 cm in Kouassi-Kouassikro, and between 50 and 100 cm in Bondoukou and Bouna. Regarding staking, in Bouna and Kouassi-Kouassikro, the stakes were made up of shrubs left in the field. In Bondoukou, in addition to shrubs, corn stalks were used as stakes. Field maintenance, mainly consisting of weeding, was carried out manually with a daba in Bouna (2 weeding sessions: May and July) and in Bondoukou (4 weeding sessions: May to August). In Kouassi-Kouassikro, weeding was first carried out with herbicide (in May) and then a second time with a hoe just before the first harvest [17]. Regarding crop association, yams were grown alongside cashew trees in Bouna and Kouassi-Kouassikro, whereas in Bondoukou, they were intercropped with both cashew trees and maize. The first yam harvest was conducted when the tubers reached six months of physiological maturity. Tubers were harvested using dabas in Bondoukou and Bouna but a hoe in Kouassi-Kouassikro [17,22].

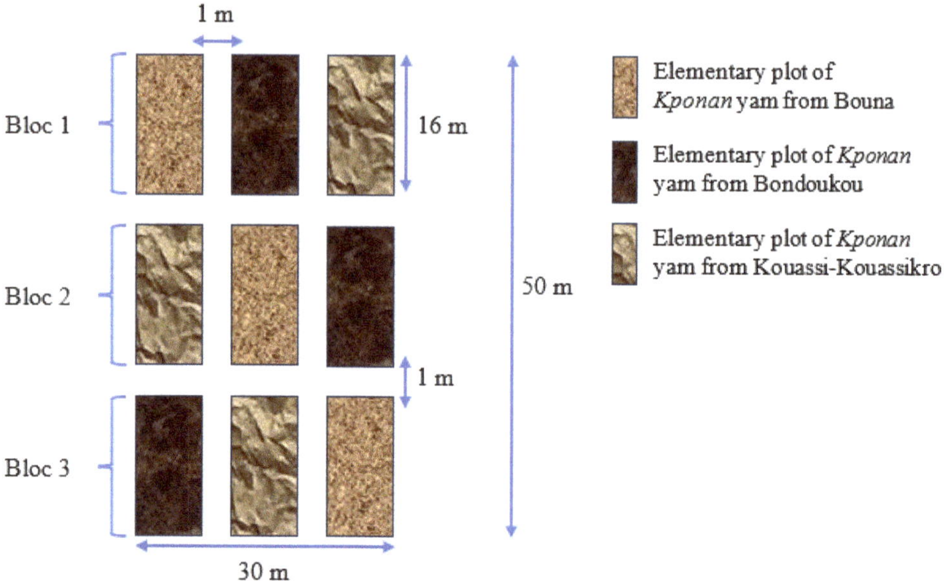

Figure 1. Experimental layout of a yam cultivation plot [21].

2.2.2. Tuber Weight

Tuber weight was determined using a pre-calibrated hand scale which displays the weight directly on screen. The tubers were weighed individually and their average weight (kg) was calculated [21].

2.2.3. Storage Conditions for *Kponan* Yams

Healthy yam tubers from each experimental plot, free from bruises and cuts, were sorted and divided into two lots. One of these lots was kept on-site in the experimental fields. The second lot was carefully transported to Abidjan to be stored in a warehouse.

In Bouna, the storage technique used was pit storage. Three pits were dug per block—one for each yam origin (Bouna, Bondoukou, Kouassi-Kouassikro)—resulting in a total of nine pits across the three blocks. Each pit contained the following: 38 yam tubers from Bondoukou, 38 yam tubers from Bouna, and 40 yam tubers from Kouassi-Kouassikro. Thus, in total, 114 tubers from Bondoukou, 114 tubers from Bouna, and 120 tubers from Kouassi-Kouassikro were stored in the pits.

In Bondoukou, the storage method employed was straw hut storage. One straw hut was constructed per block, totaling three straw huts. Each straw hut was divided into three sections corresponding to the different yam origins. The quantities stored per straw hut were as follows: 41 yam tubers from Bondoukou, 38 yam tubers from Bouna, and 44 yam tubers from Kouassi-Kouassikro. This amounted to 123 tubers from Bondoukou, 114 tubers from Bouna, and 132 tubers from Kouassi-Kouassikro stored in barns.

Finally, in Kouassi-Kouassikro, the yams were stored in open-air piles under the shade of a large tree, arranged into nine piles (one per origin and sub-plot). The quantities per sub-plot were as follows: 37 yam tubers from Bondoukou, 36 yam tubers from Bouna, and

35 yam tubers from Kouassi-Kouassikro. In total, 111 tubers from Bondoukou, 108 tubers from Bouna, and 105 tubers from Kouassi-Kouassikro were stored in this way.

The yams transported to Abidjan were stored in a facility set up at Nangui ABROGOUA University for a storage period of three months. The *Kponan* yam tubers were placed on stacked platforms and classified according to their origin and cultivation area. The storage facility had wall openings to facilitate air circulation. The average storage temperature was 27.98 °C, with an average relative humidity of 85.61% [17].

The number of healthy tubers stored was 115 for yams from Bondoukou cultivated in Bouna, 114 for those from Bouna cultivated in Bouna, 118 for those from Kouassi-Kouassikro cultivated in Bouna, 122 for those from Bondoukou cultivated in Bondoukou, 114 for those from Bouna cultivated in Bondoukou, 133 for those from Kouassi-Kouassikro cultivated in Bondoukou, 108 for those from Bondoukou cultivated in Kouassi-Kouassikro, 105 for those from Bouna cultivated in Kouassi-Kouassikro, and 100 for those from Kouassi-Kouassikro cultivated in Kouassi-Kouassikro.

Temperature and relative humidity were recorded three times a week and measured four times a day (at 8 a.m., 12 p.m., 4 p.m., and 8 p.m.) [23].

2.2.4. Determination of Tuber Damage During Storage

Each month, yams are checked for mold, wilt, tuber desiccation, and rot. The damage rate was calculated as a ratio of the number of damaged yams to the total number of yams in stock [22].

$$\%\ \text{Damage} = \frac{\sum \text{Damaged yams}}{\text{Total yams}} \times 100 \quad (1)$$

The rate of weight loss (due to wilting or dehydration) was determined according to the following equation [21]:

$$\text{Weight loss}\ (\%) = \frac{M_i - \sum M_f}{M_i} \times 100 \quad (2)$$

M_i: Initial weight of tubers at harvest;
M_t: Final weight of tubers after storage.

2.2.5. Data Analysis

Analyses of variance (ANOVA) were performed with Stastica 7.1 software to compare means within the same production zone during storage. Differences were highlighted using Tukey's test at the 5% significance level.

3. Results

3.1. Distribution of Tuber Weight at Harvest

Figure 2 illustrates the distribution of yam tuber weight at harvest according to the cultivation practice in Bouna, Bondoukou, and Kouassi-Kouassikro. The average weight of freshly harvested tubers varied significantly (ANOVA, $p < 0.05$), ranging from 2.01 ± 0.11 (Bouna) to 2.53 ± 0.06 kg (Kouassi-Kouassikro) (Figure 2a). Following the Gaussian distribution (Figure 2b), yam tubers with greater weight (weight > 2.30 kg) were harvested in Kouassi-Kouassikro (2.53 ± 0.06 kg) and in Bondoukou (2.37 ± 0.09 kg), unlike those in Bouna, which recorded a lower weight (2.01 ± 0.11 kg).

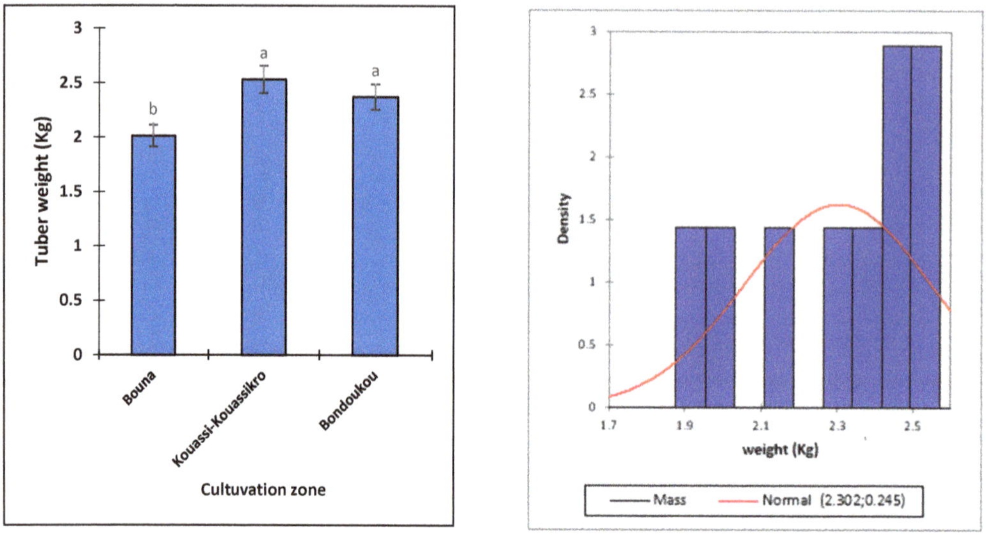

(a) (b)

Figure 2. Tuber weight at harvest. (**a**) Average tuber weight as influenced by cultivation practices. (**b**) Distribution of tuber weight (kg) showing a normal (Gaussian) pattern, indicating that weight values measured at harvest follow a normal distribution.

3.2. Damage Recorded During Field Storage

During the storage of yam tubers in the field (Table 1), those kept in pits encountered fewer damages (40.00 to 48.57%) compared to those stored under thatched roofs and those stored in the open air under the shade of a tree (100%). The storage duration was one month in the open air, two months under straw huts, and three months in pits (Figure 3). Open-air-stored yams from Bouna had the highest rot incidence (60.00%), followed by Bondoukou and Kouassi-Kouassikro. Yams from Bondoukou stored under thatched roofs (Bondoukou Thatch) had a rot rate of 57.14%. Meanwhile, yams from Bouna, Bondoukou, and Kouassi-Kouassikro stored in pits exhibited the lowest rates of rot (22.86 to 31.43%). Regarding weight loss, yams stored under thatched roofs recorded the highest rates of tubers losing weight (31.43 to 42.86%), while yams stored in pits showed the lowest rate of tubers losing weight. As for internal tuber cracks, the rates ranged from 5.71 to 8.58% for yams stored in pits, from 8.57 to 11.43% for those stored under straw huts, and from 11.23 to 14.29% for those stored in the open air.

Table 1. Impacts of cultural practices and the conservation method on *Kponan* deterioration.

	Rot (%)	Weight Loss	Crack (%)	Total Losses
Bouna*Open air	60.00 ^{a}	25.71 ^{e}	14.29 ^{a}	100.00 ^{a}
Bondoukou*Open air	57.14 ^{ab}	28.57 ^{d}	14.29 ^{a}	100.00 ^{a}
Kouassi-Kouassikro*Open air	57.14 ^{ab}	31.43 ^{c}	11.43 ^{b}	100.00 ^{a}
Bouna*Straw hut	48.58 ^{c}	42.86 ^{a}	8.57 ^{c}	100.00 ^{a}
Bondoukou*Straw hut	57.14 ^{ab}	31.43 ^{c}	11.43 ^{b}	100.00 ^{a}
Kouassi-Kouassikro*Straw hut	54.29 ^{b}	34.29 ^{b}	11.43 ^{b}	100.00 ^{a}
Bouna*Pit	22.86 ^{e}	8.57 ^{g}	8.58 ^{c}	40.00 ^{d}
Bondoukou*Pit	25.72 ^{e}	11.42 ^{f}	5.71 ^{d}	42.85 ^{c}
Kouassi-Kouassikro*Pit	31.43 ^{d}	11.43 ^{f}	5.71 ^{d}	48.57 ^{b}

Data are presented as mean ± standard deviation. Values in the same column that have different superscript letters differ significantly at the 5% level, based on Tukey's test for variance analysis. (*) indicates an interaction between the factors.

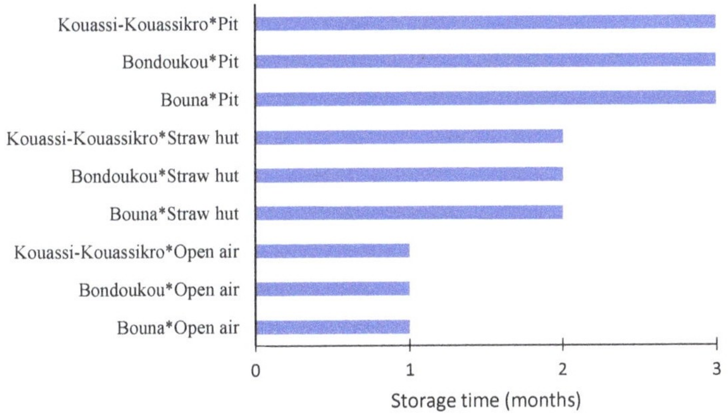

Figure 3. Impact of farming and conservation practices on tuber storage duration. (*) indicates an interaction between the factors.

3.3. Damages Recorded During Storage in the Warehouse

During storage in the warehouse, the rates of deterioration of yams varied according to their origin. For yams cultivated in Bondoukou, the rates of deterioration ranged between 72.29% and 77.35%, respectively, for those from Kouassi-Kouassikro and from Bouna. For yams cultivated in Bouna, the rates of deterioration fluctuated between 79.40% and 84.81%. Regarding yams cultivated in Kouassi-Kouassikro, all were deteriorated by 100% after 2 months of storage in the warehouse. Yams cultivated in Bondoukou and Bouna could be preserved for 3 months (Figure 4). The storage duration was two months for yams cultivated in Kouassi-Kouassikro, and three months for those cultivated in Bouna and Bondoukou (Figure 5).

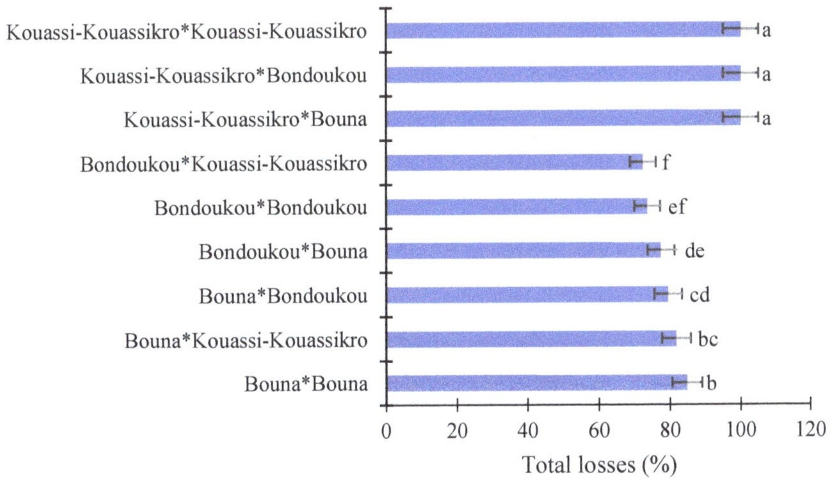

Figure 4. Impacts of cultivation zone and cutting origin on *Kponan* deterioration in the warehouse. Different superscript letters indicate significant differences at the 5% level ($p < 0.05$, Tukey's test). (*) indicates an interaction between the factors.

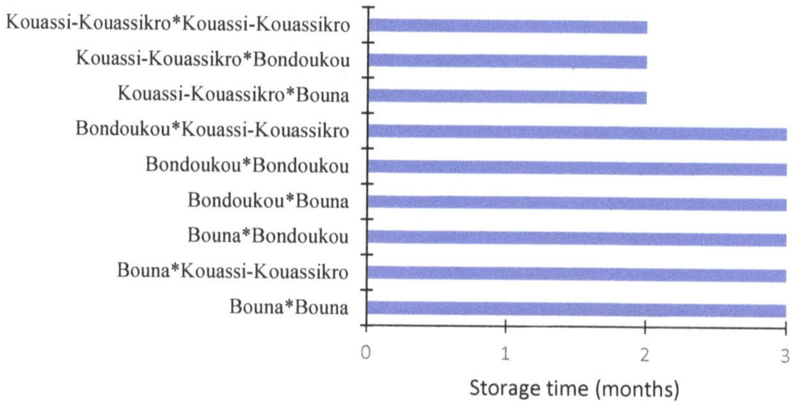

Figure 5. Duration of tuber conservation in the warehouse based on origin and cultivation zone. (*) indicates an interaction between the factors.

4. Discussion

The impacts of cultivation and conservation practices on the physical parameters of *Kponan* yam were assessed. At harvest, a significant variability in the physical parameters of yams was observed. In terms of weight, the results revealed a dispersion of the average weight of freshly harvested tubers, ranging from 2.01 kg to 2.53 kg. These findings corroborate those reported by Ola et al. [24], who also noted a similar average weight of *Dioscorea cayenensis-rotundata* tubers. According to the Gaussian distribution, tubers of a lower weight (weight < 2.30 kg) were mainly harvested in Bouna, while those of a greater weight (weight > 2.30 kg) were harvested in Kouassi-Kouassikro and Bondoukou.

These findings highlight that *Kponan* yam yield depends more on cultivation zone conditions (such as rainfall and soil) than cutting origin. Meteorological data from SODEXAM for the year 2021 underscored the significant impact of rainfall on yam yield, particularly in the Kouassi-Kouassikro region, which recorded higher precipitation compared to Bondoukou and Bouna. This impact of rainfall on the yield of *Kponan* yam in the Zanzan region (Bouna and Bondoukou) was also observed by Kouakou and Anoh [11], who found that although *Kponan* yam is not highly demanding in terms of precipitation, its yield is negatively affected by rainfall.

Regarding conservation, rot appeared as the main alteration of yams, whether in the field or in storage, regardless of the origin of the cuttings or the conservation method. These results are consistent with previous studies [22]. *Kponan* yam tubers stored in pits in Bouna showed lower rates of alteration compared to those stored under straw huts in Bondoukou or exposed to open air in Kouassi-Kouassikro. The conservation duration seems to be influenced by the location and method of storage, with more effective conservation in pits and in areas characterized by a dry season. These findings are consistent with previous research, emphasizing the importance of pit storage for optimal conservation duration [25]. According to previous studies [11,22], the use of chemical fertilizers and pesticides is inversely correlated with the shelf life of *Kponan* yam.

Furthermore, the use of herbicides was identified as a key factor affecting conservation duration, with an excessive use of herbicides in Kouassi-Kouassikro, which could explain the observed differences in yam conservation compared to Bondoukou and Bouna.

According to Kouadio et al. [17], herbicides are not directly responsible for tuber rot. However, excessive or improper use of herbicides, combined with other inadequate agricultural practices, can create conditions conducive to tuber deterioration [17]. Certain components of herbicides, such as nitrogen, can influence tuber quality. For instance, an excess of nitrogen may alter their chemical composition, making them more vulnerable to diseases and rot. Therefore, balanced fertilization is essential [26].

Furthermore, environmental conditions, particularly temperature (27.97–28.87 °C) and relative humidity (85.6–90.15%) which were recorded during the storage of the yam tubers, can accelerate their rotting. High humidity likely promotes fungal proliferation, explaining the accelerated tuber degradation observed in warehouse storage. Similar findings were reported by Osunde and Orhevba [23], who observed that tubers stored under ventilated conditions had a lower rot rate (1.85%) compared to those stored without ventilation (12.03%).

Other factors, such as the presence of internal cracks in the tubers or the shocks sustained during their transport from the field to the city, can also contribute to the decay of *Kponan* yam. Aidoo et al. [27] demonstrated that the rate of decay was higher in tubers visibly pierced by rhizomes compared to healthy, unpierced tubers. Furthermore, Assiri et al. [28] suggested that decay is caused by fungal agents that invade the tissues of tubers damaged by insects, nematodes, or mechanical actions during and after harvest.

Bosrotsi [29] highlighted that physical damage sustained during harvesting operations is a major contributor to post-harvest losses in roots and tubers. It is crucial to handle tubers carefully to prevent damage, as injured tubers have poor storage potential and deteriorate quickly. *Kponan* yam rot is mainly caused by improper storage practices, adverse environmental conditions, and physical damage during harvest and transportation. The proper management of herbicide use, effective pit storage, and sustainable agricultural practices are essential to reduce post-harvest losses and improve the shelf life of tubers.

5. Conclusions

The study demonstrates that Kponan yam deterioration is driven by storage conditions and cultivation practices. Tubers maintain better physical quality when grown with moderate herbicide use and stored in pits. Tubers can be stored for two to three months in storage and up to three months in pits. However, this storage period is reduced to one month when yams are cultivated and stored in the open air, as in Kouassi-Kouassikro.

Author Contributions: Designed the research, Y.H.K. and K.N.K.; conducted the statistical analysis, Y.H.K.; drafted the article with input from all other authors, G.A.J.K., K.A.K., B.J.A. and Y.D.N. All authors have read and agreed to the published version of the manuscript.

Funding: This research received no external funding.

Institutional Review Board Statement: Not applicable.

Informed Consent Statement: Not applicable.

Data Availability Statement: All relevant data are presented in the paper; additional information can be obtained by contacting the corresponding author.

Acknowledgments: The authors express their sincere gratitude to the producers for their invaluable assistance.

Conflicts of Interest: The authors declare that there are no conflicts of interest.

References

1. Wang, S.; Gao, W.; Liu, H.; Chen, H.; Yu, J.; Xiao, P. Studies on the physicochemical, morphological, thermal and crystalline properties of starches separated from different *Dioscorea opposita* cultivars. *Food Chem.* **2006**, *99*, 38–44. [CrossRef]
2. Onyeka, T.; Petro, D.; Ano, G.; Etienne, S.; Rubens, S. Resistance in water yam (*Dioscorea alata*) cultivar in the French West Indies to anthracnose disease based on tissue culture-derived whole-plant assay. *Plant Pathol.* **2006**, *55*, 671–678. [CrossRef]
3. FAOSTAT. 2022. Available online: https://www.fao.org/faostat/en/#data/QCL (accessed on 5 August 2022).
4. Asiedu, R.; Sartie, A. Crops that feed the World 1. Yams Yams for income and food security. *Food Sec.* **2010**, *2*, 305–315. [CrossRef]

5. Darkwa, K.; Olasanmi, B.; Asiedu, R.; Asfaw, A. Review of empirical and emerging breeding methods and tools for yam (*Dioscorea* spp.) improvement: Status and prospects. *Plant Breed.* **2020**, *139*, 474–497. [CrossRef]
6. Mignouna, D.B.; Akinola, A.A.; Abdoulaye, T.; Alene, A.D.; Manyong, V.; Maroya, N.G.; Aighewi, B.A.; Kumar, L.P.; Balogun, M.; Lopez-Montes, A.; et al. Potential returns to yam research investment in sub-Saharan Africa and beyond. *Outlook Agric.* **2020**, *49*, 215–224. [CrossRef]
7. Akinola, A.A.; Ezeorah, S.N.; Nwoko, E.P. Modelling the rehydration characteristics of White Yam. *West Indian J. Eng.* **2019**, *41*, 70–76.
8. Dappah, K.D.; Kouassi, K.N.; Kouadio, Y.H.; N'Dri, Y.D.; Amani, N.G. Glycemic index and glycemic load of yam (*Dioscorea cayenensis-rotundata*) porridge with Talinum triangulare sauce tested in human subjects. *GSC Biol. Pharm. Sci.* **2023**, *22*, 056–063. [CrossRef]
9. Maliki, R.; Sinsin, B.; Floquet, A. Evaluating yam-based cropping systems using herbaceous legumes in the savannah transitional agro-ecological zone of Benin. *J. Sustain. Agric.* **2012**, *36*, 440–460. [CrossRef]
10. Adifon, F.H.; Yabi, I.; Vissoh, P.; Balogoun, I.; Dossou, J.; Saïdou, A. Écologie, systèmes de culture et utilisations alimentaires des ignames en Afrique tropicale: Synthèse bibliographique. *Cah. Agric.* **2019**, *28*, 22. [CrossRef]
11. Kouakou, K.P.; Anoh, K. Geotraçabilité de l'igname *Kponan* de Bondoukou. *Rev. Trimest. Des Sci. Soc. RSS-PASRES* **2019**, *23*, 2311–5890.
12. INS. *Recensement Général de la Population et de l'Habitat (RGPH): Technical Report*; Ministère de l'Economie, du Plan et du Développement Résultats Globaux: Abidjan, Côte d'Ivoire, 2021; p. 37.
13. Girardin, O.; Nindjin, C. Amélioration de la conservation de l'igname en milieu villageois. *Sempervira* **2015**, *5*, 1025–8167. [CrossRef]
14. Maroya, N.; Balogun, M.; Aighewi, B.; Mignouna, D.B.; Kumar, P.L.; Asiedu, R. Transforming Yam Seed Systems in West Africa. In *Root, Tuber and Banana Food System Innovations*; Springer International Publishing: Cham, Switzerland, 2022; pp. 421–451. [CrossRef]
15. Kouamé, J.; Gnangui, S.N.; Koné, F.M.T.; Kouamé, L.P. Use of Kinetic and Thermodynamic Parameters for the Prevention of Enzymatic Browning of Edible Yam *Dioscorea cayenensis-rotundata* cv. "Zrèzrou" Department of Food Science and Technology, University Nangui Abrogoua. *Int. J. Curr. Microbiol. App.Sci.* **2017**, *6*, 4176–4187. [CrossRef]
16. Corneille, A.; Agbidinoukoun, A.; Agbangla, C.; Adjanohoun, A.; Kumulungui, B.; Ondo, P.; Nzang, R.; Souza, P. Capacité morphogénétique in Vitro de quelques accessions d'ignames du Complexe *Dioscorea cayenensis/D. rotundata* cultivées au Bénin et évaluation de la quantité d'ADN Matériel régénéré. *Rev. Ivoir. Sci. Technol.* **2012**, *20*, 68–86.
17. Kouadio, Y.H.; Kouassi, K.N.; Kouassi, K.B.; Konan, G.A.J.; Kouakou, K.A.; Dappah, K.D.; N'dri, Y.D.; Amani, N.G. Effect of warehouse storage on the alteration, cooking and organoleptic characteristics of *Kponan* yam (*Dioscorea cayenensis-rotundata*) of Côte d'Ivoire. *Heliyon* **2024**, *10*, e40014. [CrossRef]
18. Kouadio, Y.H.; Kouassi, K.N.; Konan, Y.; N'Dri, Y.D.; Amani, N.G. Practices and damages encountered by wholesale traders during '*Kponan*' yams (*Dioscorea cayenensis-rotundata*) storage in Abidjan, Côte d'Ivoire. *Int. J. Biol. Chem. Sci.* **2022**, *16*, 1570–1579. [CrossRef]
19. Bagal, M.; Vittori, M. Les Indications Géographiques en Côte d'Ivoire, Produits Potentiels et Cadre Juridique Pertinent. 2010. Available online: https://www.origin-gi.com/wp-content/uploads/2010/11/papier-pays_cte-divoire_origin1.pdf (accessed on 16 May 2025).
20. Kouakou, K.P.; Kouassi, C.A.; Anoh, K.P. Le marche de gros de l'igname *Kponan* à Abidjan (Côte D'ivoire). *Eur. Sci. J.* **2019**, *15*, 218–238. [CrossRef]
21. Kouadio, Y.H.; Kouassi, K.N.; Nimaga, D.; N'Dri, Y.D.; Amani, N.G. Impacts of Field Harvesting and Conservation Practices on the Physical Quality of Yam "*Kponan*" (*Dioscorea cayenensis-rotundata*) of Côte D'Ivoire. *Open J. Appl. Sci.* **2022**, *12*, 1979–1994. [CrossRef]
22. Kouassi, K.N.; Kouadio, Y.H.; Kouassi, K.B.; N'Dri, Y.D.; Amani, N.G. Impacts of storage practices on the physical, culinary and sensory quality of *Kponan* yam (*Dioscorea cayenensis-rotundata*) from Côte d'Ivoire during storage. *J. Sci. Food Agric.* **2023**. [CrossRef] [PubMed]
23. Osunde, Z.D.; Orhevba, B.A. Effects of storage conditions and storage period on nutritional and other qualities of stored yam (*Dioscorea* spp.) tubers. *Afr. J. Food Agric. Nutr. Dev.* **2009**, *9*, 1684–5374. [CrossRef]
24. Ola, F.A.; Akande, F.B.; Fadeyibi, A.; Busari, R.A. Physical parameters of some varieties of yam tubers relevant in the design of mechanical yam harvester. *Tech. Sci. J. Community Dev. Afr.* **2021**, *2*, 39–45.
25. FIRCA. *Répertoire de Technologies de Conservation et de Transformation de l'igname et de la Banane Plantain*; Fonds Interprofessionnel pour la Recherche et le Conseil Agricoles. Groupe le Prestige: Abidjan, Côte d'Ivoire, 2010; 155p.
26. Ngue, B.T.; Mbairanodji, A.; Njualem, D. Guide des techniques de production et de conservation d'igname (*Dioscorea* spp.). 2007. Available online: http://www.mangalani-consult.org/fichiers/ressources/FA08_Igname.pdf (accessed on 16 May 2025).

27. Aidoo, A.K.; Arthur, S.; Bolfrey-arku, G.; Osei, K.; Lamptey, J.N.L.; Mochiah, M.B.; Ohene-mensah, G.; Frimpong, F.; Appiah-kubz, I.; Owusu danquah, E. Storage rot of seed yam resulting from speargrass injuries. *Afr. Crop Sci. J.* **2020**, *28*, 203–211. [CrossRef]
28. Assiri, K.P.; Koutoua, S.; Assi, S.T. Champignons responsables de pourritures de l'igname Dioscorea cayenensis-rotundata variété "Kponan" en post récolte. *J. Appl. Biosci.* **2017**, *111*, 10957–10968. [CrossRef]
29. Bosrotsi, C.K.; Addo, A.; Dzisi, K.A.; Agodzo, S.K. Development of a Yam Harvester using Finite Element Method. *Int. J. Mod. Stud. Mech. Eng.* **2017**, *3*, 35–44.

proceedings

Proceeding Paper

The Design, Development and Evaluation of a Biomass-Fueled Cabinet Dryer for High-Quality Cassava Flour Production †

Suraju Adegbite [1,*], Wahabi Asiru [1], Oyewole Okewole [2], Wasiu Awoyale [2], Peter Kolawole [2], Thiano Diallo [2] and Adebayo Abass [2]

1 Federal Institute of Industrial Research Oshodi (FIIRO), P.M.B.21023, Ikeja Lagos 100261, Nigeria
2 International Institute of Tropical Agriculture (IITA), P.M.B.5320, Oyo Road, Ibadan 200001, Nigeria; oyewoleokewole@yahoo.com (O.O.); p.kolawole@cgiar.org (P.K.); t.diallo@cgiar.org (T.D.); a.abass@cgiar.org (A.A.)
* Correspondence: adeyemi40@gmail.com; Tel.: +234-7063011356
† Presented at the CORAF's 2023 Symposium on Processing and Transformation of Agricultural Products in West and Central Africa: Achievements and Opportunities for Private Sector Engagement, Lome, Togo, 21–23 November 2023.

Abstract: Cassava is an essential staple crop in developing nations, ensuring food security and generating income for smallholder farmers. The lack of modern drying technology in rural regions requires effective strategies to reduce post-harvest losses and improve product quality. This study introduces a biomass-powered cabinet dryer designed to produce high-quality cassava flour (HQCF), featuring a heat exchanger, blower, and drying chamber constructed from locally sourced materials. The drying chamber contains two racks, each equipped with 10 trays, accommodating 40 to 80 kg of wet cassava mash per batch. The dryer, operating at temperatures between 50 and 115 °C for a duration of 10–120 min, decreased moisture content from 38.21% to around 11%. Drying efficiencies varied from 59.22% to 88.24%, whereas biomass efficiencies ranged from 3.0% to 4.0%. The results demonstrate the dryer's efficiency in drying, while the study emphasizes the necessity of customized technical solutions to assist rural populations in cassava-producing areas.

Keywords: cassava; design; biomass cabinet dryer; high-quality cassava flour; drying efficiency; heating energy

1. Introduction

Cassava is a crucial staple crop in many developing regions, essential for food security and income for smallholder farmers [1]. It serves multiple purposes, such as a famine-reserve crop, rural staple, urban convenience food, industrial raw material, cash crop, and foreign exchange source. Processing cassava into various products boosts the economies of tropical nations, with Nigeria seeing increased output due to such processing. However, the lack of modern drying technology in rural areas leads to inconsistent drying, lower product quality, and higher post-harvest losses, hindering the production of value-added cassava products [2].

Cassava roots are versatile, used for human consumption, livestock feed, and industrial products such as starch and flour. Cassava faces post-harvest challenges, despite low production costs; hence, efficient post-harvest processing is crucial for converting raw cassava into consumable or industrial products such as flour, starch, and chips [3]. Reliable technologies are essential for making cassava processing profitable, reducing losses and enhancing food security and economic sustainability. The processing phase includes peeling, washing, cutting, drying, and milling, all of which impact the final product's quality [3]. High-quality cassava flour (HQCF) can stimulate rural industrialization, elevate market value, and improve farmers' earnings and livelihoods. HQCF is versatile, suitable for household use, pastries, urban convenience foods, and industrial products like textiles, plywood, paper, and packaged foods. It can also be blended with wheat flour for making bread, biscuits, cookies, and pies [4].

Drying cassava roots is crucial for producing high-quality products for industrial and human use, but small-scale farmers often struggle with the costs of fossil fuel-powered and solar dryers [5]. Traditional drying methods are labor intensive, time consuming, and susceptible to weather, crop damage, and quality loss due to insect infestation. There is a critical need for affordable, eco-friendly drying technology accessible to rural farmers. Biomass-based drying presents a viable solution, reducing post-harvest losses and improving cassava product quality through controlled heat processes.

Biomass plays a crucial role in rural areas, providing around 90% of energy needs, primarily for cooking, and biomass waste can also generate electricity and heat in the food industry [6]. Various biomass sources like wood, charcoal, sawdust, and agricultural residues such as maize cobs, coconut shells, and rice husks are suitable for biomass gasification [7]. Mishra [8] developed a prototype drying system with energy storage using solar and biomass energy for drying potatoes and chayote. Muhammad et al. [9] designed a biomass dryer with coconut shells and an IoT-based monitoring system to optimize drying processes for various materials, including cassava chips. Biomass has proven effective as a heat source in drying trials for plantain, cashew nuts, tomatoes, okra, peppers, copra, and peanuts in laboratory or prototype settings [10,11].

Large-scale biomass dryers for cassava drying are primarily manufactured in countries such as China and Brazil. Companies like Zhengzhou Taida Drying Equipment Co., Ltd., and Shanghai Yongxian Machinery Manufacturing Company specialize in commercial biomass dryers, including the Osa cassava dreg dryer [12]. However, research on locally designed biomass heat-driven cabinet dryers for large-scale cassava drying to produce high-quality cassava flour (HQCF) is very limited, especially in developing countries like Nigeria. Locally developed biomass heat-generated cabinet dryers offer a cost-effective, culturally relevant, and sustainable solution. They empower local communities, promote economic growth, minimize environmental impact, and are easier to maintain compared to imported dryers.

The biomass-fueled dryer for cassava drying offers a promising solution to rural challenges, combining innovative design with locally available biomass as a renewable energy source. This research aims to explore the feasibility and practicality of this dryer. It focuses on designing, developing, and evaluating the performance of a biomass-fueled dryer tailored for cassava drying. This is a sustainable solution which addresses environmental issues while providing efficient drying capabilities in rural areas.

2. Materials and Methods

2.1. Description of the Biomass-Fueled Cabinet Dryer

The biomass-fueled cabinet dryer comprises a heat exchanger, main blower, and drying chamber (Figure 1). The heat exchanger is where biomass combustion occurs to generate heat for the drying process. The drying chamber contains two racks, each with 10 drying trays, totaling 20 trays. The main blower draws hot air from the heat exchanger and channels it through a diffuser, ensuring even heat distribution in the drying chamber. The drying process employs conduction, convection, and radiation for effective heat transfer. Conductive heat transfer occurs through the walls of the heat exchanger, while convective heat transfer facilitates heat distribution through hot gases and forced airflow, thereby ensuring uniform moisture removal. Radiative heat exchange enhances evaporation, while effective thermal management—via airflow control, combustion efficiency, and temperature distribution—facilitates the production of high-quality cassava flour with reduced energy waste. The dryer takes about 30 min to reach a steady state, with the drying chamber temperature ranging from 90 to 115 °C. This dryer was installed at the cassava processing center at the Open-Door International System in Akileye, located 12 km off Owode–Idiroko road via Ota, Ogun State, Nigeria. The biomass-fueled cabinet dryer is as shown in Figure 1.

Figure 1. The biomass-fueled cabinet dryer.

2.2. Materials

The materials and components for the dryer's fabrication were sourced locally to minimize environmental impact, maximize economic benefits, and ensure accessibility for smallholder farmers. Materials were chosen based on their ability to withstand heat, vibration, humid air fatigue, and stress during operation without failure. Aluminum and stainless steel were selected for sections in direct contact with cassava mash to prevent corrosion, contamination, and to resist the toxic nature of cassava. Mild steel and medium carbon steel, known for their robustness and ductility, were used for the dryer's construction. The materials used for the dryer evaluation included palm kernel shells, wet cassava mash, plastic containers, a weighing scale, a moisture meter, a digital hygrometer, a digital anemometer, temperature probe, a stopwatch, and fuel (diesel) to power the generator.

2.3. Methods

2.3.1. Design Analysis of the Biomass-Fueled Cabinet Dryer

Drying Chamber: The volume of the drying chamber was determined using the relationship described by Adzimah and Seckley [13], as shown in Equation (1).

$$Volume = lbh. \tag{1}$$

where l, b, and h were the length (2.5 m), breadth (1.4 m), and height (2.7 m) of the drying chamber. Therefore, the volume of the drying chamber was 9.45 m^3. The dimension of the drying tray were 1.20 m, 1.00 m, and 0.03 m in terms of length, breadth, and height, respectively.

Hopper: The hopper is the device through which biomass materials are fed into the combustion chamber for heat generation in the heat exchanger. The volume (V_h) of the hopper was determined as described by Imanogor [14] using Equation (2).

$$V_h = \frac{1}{3}[(A + B)HL] \tag{2}$$

where A is the hopper's top side, B its bottom, H its height, and L its length.

Screw auger: Beneath the hopper is the cylindrical trough which house the screw auger for conveyance of the biomass materials into the combustion chamber. The pitch (Pa), diameter ($\varnothing_{xa}$), and capacity of the screw auger (Q_m) were computed using the relationship given by David and Micheal [15] as shown in Equations (3)–(5).

$$P_a = \frac{4v_a \varnothing_{xa} x_a}{\frac{\pi}{4}\omega_a\left({\varnothing_{xa}}^2 - {\varnothing_{ia}}^2\right)} \quad (3)$$

$$\varnothing_{xa} = \varnothing_t - 2c \quad (4)$$

$$Q_m = \frac{\pi \times (\varnothing_{xa} - \varnothing_{ia})^2}{4} \times P_a \times \omega_a \times s \times i \times 60 \quad (5)$$

where $\varnothing_{ia}$ and $\varnothing_{xa}$ were the auger's internal diameter (m) and external diameter (m), v_a was the screw auger's downward flow velocity of material (ms^{-1}), x_a was the distance along the length of the auger (m), ω_a was the auger's angular velocity (rpm), c was the clearance between the trough and the screw auger, s was the specific mass of the palm kernel shell (Nm^{-3}), and i was the degree of screw's trough filling (%).

Selection of the Blower: The blower horsepower was 3.86 Hp, which was determined using the relationship as described by Adzimah and Seckley [13], as shown in Equation (6)

$$Fan\ horse\ power\ (P) = \frac{volume\ air\ flow\ rates \times total\ static\ pressure}{6320 \times fan\ efficiency} \quad (6)$$

where an air flow rate of 3843.44 cfm, a static pressure of 5.4 inches of water were determined based on the dimension of the drying chamber, and a fan efficiency of 80% was selected because most industrial blowers have an efficiency range of 70–85%, as reported by Adzimah and Seckley [13]. Therefore, a centrifugal fan with electric motor of 5 Hp, which draws heat from the heat exchanger for effective heat distribution within the drying chamber, was selected.

Design of Shaft

Selection of pulleys: Equation (7) expresses the relationship between the blower shaft with the shaft of electric motor based on speed ratio in accordance with the procedures of [16].

$$N_1\varnothing_1 = N_2\varnothing_2 \quad (7)$$

where N_1 and N_2 are the speeds of the electric motor (driver) and the blower shaft (rpm) (driven), $\varnothing_1$ and $\varnothing_2$ are the diameters of the driver pulley (mm) and the pulley driving the blower shaft (mm). A similar procedure was followed to determine the pulley specifications for the screw auger.

Selection of the Machine Belts: Using the dryer power rating and standard V-belts according to IS: 2494–1974, Khurmi, R. S. and Gupta [16], a V-belt of class A type and class B type were chosen for the screw auger and the blower shaft, respectively. Equations (8) and (9) were used to determine the center distance, C, of the pulleys and the length of the belt (L_b) according to [16].

$$C = \frac{\varnothing_1 + \varnothing_2}{2} + \varnothing_1 \quad (8)$$

$$L_b = 2C + \frac{\pi}{2}(\varnothing_1 + \varnothing_2) + \frac{1}{4C}(\varnothing_2 - \varnothing_1)^2 \quad (9)$$

Belt tensions: the belt tensions on the tight and slack sides of the belt drive (T_1 and T_2) for the screw auger and that of the main blower were determined as described by Harikumar [17] using Equations (10)–(13).

$$\alpha_1 = 180 - 2sin^{-1}\left(\frac{\varnothing_2 - \varnothing_1}{2C}\right) \& \alpha_2 = 180 + 2sin^{-1}\left(\frac{\varnothing_2 - \varnothing_1}{2C}\right) \tag{10}$$

$$\frac{T_1 - M_b {v_b}^2}{T_2 - M_b {v_b}^2} = e^{\mu\alpha cosec(\partial/2)} \tag{11}$$

$$v_b = \frac{2\pi r_b N_b}{60} \tag{12}$$

$$P_1 = (T_1 - T_2)v_b \tag{13}$$

where μ is the frictional coefficient of belt running on a pulley, ∂ is the belt groove angle in radian, V_b is the belt velocity, and P_1 is the power transmission of the belt drive which is a function of the power rating of the electric motor. According to Khurmi and Gupta [16], the shaft's design was based on the shaft's fluctuating torque, bending moment, and shearing force. Equation (14) was used to calculate the shaft's diameter (d_s).

$${d_s}^3 = \frac{16}{\pi S_s}\sqrt{(K_b M_{b1})^2 + (K_t M_{t1})^2} \tag{14}$$

where S_s is the shaft constant, K_b is the combined shock and fatigue factor applied to the bending moment, K_t is the combined shock and fatigue factor applied to the torsional moment, M_{b1} is the screw auger shaft's resultant bending moment (Nm), and M_{t1} is the screw auger shaft's resultant torsional moment.

2.4. Dryer Fabrication

The isometric drawing of the dryer, which served as a reference during its fabrication, is presented in Figure 2. Shafts were machined from rods, while angle iron and pipes were used as a framework. Material specifications were used to procure the necessary components. Fasteners, including bolts, nuts, and sturdy welds, were used. There were no crevices or fractures to be found because the welding and finishing were carried out properly to keep bacteria from growing in the crevices of the walls. Arc welding was used to assemble the components, and welds were filled, ground, and polished with the correct filler materials. Bolts and nuts were used to join various components while studs were strategically placed to provide proper tension and torque. Electric motors were selected and installed. The dryer was insulated with 100 mm thick fiber glass. Stainless steel was used to fabricate the drying trays and racks. All necessary adjustments and changes were made before the dryer was put to test.

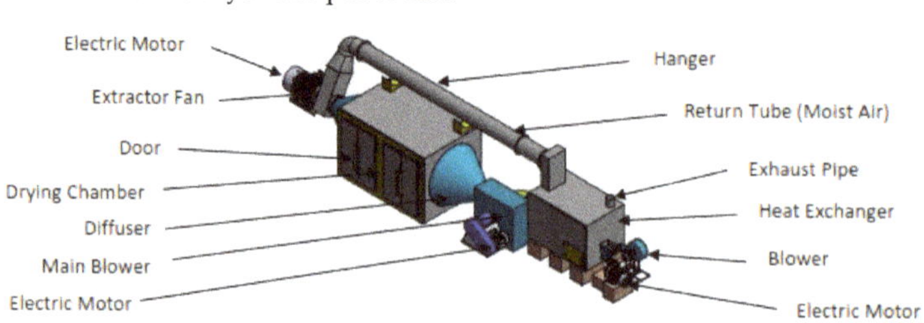

Figure 2. An isometric view of the cabinet dryer.

2.5. Energy Required for Drying

The energy required for drying was determined using the procedures of Famurewa et al. [18], as shown in Equation (15)

$$\mathrm{H} = \mathrm{MC}\Delta\theta + M_c L \quad (15)$$

where H is the energy required for drying, M is the mass of cassava mash per batch (150 kg), C is the specific heat capacity of cassava mash (1.445 kJ/kg k), $\Delta\theta$ is the change in temperature (80 °C), M_c is the mass of moisture to be removed (65 kg), t is the time of drying (2 h), and L is the latent heat of vaporization of water (2256 kJ/kg).

Quantity of Palm Kernels That Will Produce the Required Energy

Equation (16), in accordance with Famurewa et al. [18], was used to calculate the amount of palm kernel shell required to produce 163,980 kJ of heat required for drying 150 kg of cassava mash. Palm kernel shells have a calorific value of 23,604.71 kJ/kg. As a result, 6.95 kg of palm kernel was needed.

$$Amount\ of\ palm\ kernel\ used = \frac{Energy\ required\ for\ drying}{Calorific\ value} \quad (16)$$

2.6. Evaluation of the Cabinet Dryer

Cassava processing: Cassava roots were harvested, peeled, and grated using a mechanical cassava grater (500 kg/h, Nigerian made). The grated mash was pressed to form a cake, which was then pulverized. The weight and moisture content of the granulated mash were recorded. The initial weights of palm kernel shells, ambient conditions (humidity and temperature), and the time for loading the dryer were noted. The experiment was conducted in triplicates using a three-point loading mass per tray (2 kg, 3 kg, and 4 kg) with the average value determined. The cabinet dryer operated at a blower motor speed of 800 rpm with a varying mass of cassava mash (40 kg, 60 kg, and 80 kg) per batch. Air velocity, humidity, and temperature were monitored every 10 min. Each experiment was replicated three times. The dried flour was milled, weighed, and packaged.

Tools and equipment: The equipment used included a digital electronic weighing scale (Model A-12_300 kg capacity, manufactured in Shenzhen, China) for weighing wet cassava mash and dried flour, a digital anemometer (Model MS6252B, PEAKMETER, manufactured in Shenzhen, China) for air velocity measurement, a thermocouple thermometer (Model SD-947, REED instruments; manufactured in Taipei, Taiwan) for drying air and outlet air temperature measurement, and a digital hygrometer (Model TP49, ThermoPro; manufactured in Hongkong, SAR) for air humidity measurement. The moisture content of wet cassava mash was assessed using a wood moisture meter (Model MD-2G, OMICRON; manufactured in Ambala Cantt, Haryana, India) and confirmed via oven drying method. The process duration was tracked with a stopwatch. The dryer stabilized at 50 °C, over 30 min before loading the wet cake.

2.7. Thermal Analysis of the Dryer

Drying Rate: The drying rate was computed using the relationship given by Saravanan et al. [6], as shown in Equation (17).

$$R_d = \frac{M_i - M_d}{t} \quad (17)$$

where $\mathrm{R_d}$ is drying rate (kg/s); M_i is the initial mass of the sample before drying (kg); M_d is the final mass of the sample after drying; and t is the drying time (s).

Drying Chamber Efficiency: The drying chamber's efficiency was calculated according to Saravanan et al. [6], as shown in Equation (18).

$$\eta_d = \frac{T_1 - T_2}{T_1 - T_a} \tag{18}$$

where η_d is drying chamber efficiency; T_1 is the drying chamber inlet temperature (°C); T_2 is the drying chamber outlet temperature (°C); and T_a is the ambient temperature (°C).

Heating Energy: The drying was assessed as a two-step process to estimate how much energy was utilized to evaporate water from the mash. Thus, the sensible heating and evaporation process were evaluated as described by Rajvir et al. [19] using Equations (19) to (21), respectively.

$$Q_s = M_w C_p (T_1 - T_a) \tag{19}$$

where C_p is the specific heat capacity of water (kJ/kg K); M_w is the mass of water removed after drying (kg); T_1 is the drying chamber inlet temperature (°C); the T_a is the ambient temperature (°C); and Q_s is the sensible heating energy (kJ).

The second stage is the evaporation of moisture from the product and the heat required is given as follows:

$$Q_s = \left(h_g - h_f\right) \tag{20}$$

where h_g is the enthalpy of water as a vapor (kJ/kg) and h_f is the enthalpy of water as a liquid (kJ/kg) at the drying temperature obtained from the steam table; and Q_l is the evaporating heating energy (kJ). Therefore, the total heat requirement (Q_t) was determined using Equation (21)

$$Q_t = Q_s + Q_l \tag{21}$$

Biomass System/Overall Efficiency: This is the combustion efficiency of the heat source which was determined according to Rajvir et al. [19] using Equation (22).

$$\eta_s = \frac{L(TM_i - M_d)}{(FC - CV)} \tag{22}$$

where η_s is the overall efficiency; L is the latent heat of vaporization (2260 kJ/kg); Mi is the initial mass of sample before drying (kg); M_d is the final mass of sample after drying (kg); FC is the fuel consumption per batch; CV is the calorific value of palm kernel shells (23,605 kJ/kg).

2.8. Statistical Analysis

The data were analyzed using SPSS version 6.0 software. A one-way ANOVA with a completely randomized design was utilized to assess the impact of loading mass on parameters including the drying efficiency, heating energy, and moisture content of cassava flour.

3. Results

3.1. Moisture Content as a Function of Time

The drying profile of cassava flour at different loading masses is as shown in Figure 3. The moisture content of cassava mash decreases from 38% to equilibrium after 110 min. Efficient heat transfer is noted with a wider temperature difference between the drying air and cassava mash, leading to faster initial moisture removal. The final moisture content was 10.6% for 2 kg, 11.0% for 3 kg, and 11.8% for 4 kg loading masses.

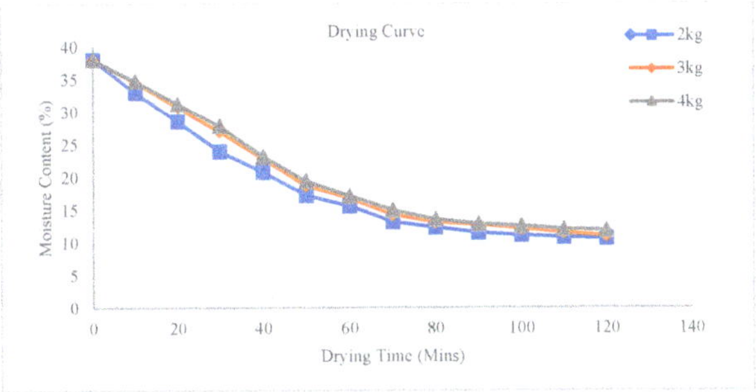

Figure 3. Drying curve showing the moisture content against the drying time at the loading masses of 2 kg, 3 kg, and 4 kg.

3.2. Drying Efficiency and Heating Energy

Figure 4a–c illustrates the drying efficiency and heating energy profiles during cassava flour drying across the loading masses of 2 kg, 3 kg, and 4 kg. Drying efficiency was maximum initially for all the loading masses, which gradually decreased with the application of heat and as the cassava mash approached equilibrium moisture content, reflecting reduced moisture evaporation. Across all loading masses, there was an inverse correlation between drying efficiency and heating energy. Figure 4a shows an average drying efficiency of 74.77% for the 2 kg batch. In Figure 4b, drying efficiency improves steadily, reaching a peak of approximately 69% at 100 min for the 3 kg batch, coinciding with the lowest heating energy recorded. Figure 4c indicates the maximum heating energy at 100 min for the 4 kg batch, with a consistently high drying efficiency of 88.24% throughout the process, the highest among the three batches.

Table 1 presents ANOVA results for the performance indices (drying efficiency, heating energy, and moisture content). The result shows that significant difference was observed at a probability level of ($p < 0.05$) for the drying efficiency and heating energy as affected by the loading mass (2 kg, 3 kg, and 4 kg), while the moisture content of the samples was unaffected by the loading mass (Table 1). Post hoc analysis, as presented in Table 2, shows that drying efficiency was substantially better on average with a loading mass of 4 kg (M = 0.88, SD = 0.06) than with a loading mass of 2 kg (M= 0.75, SD = 0.11) or 3 kg (M= 0.60, SD = 0.14). However, the drying efficiency was much higher at a loading mass of 2 kg than that of a loading mass of 3 kg (Table 2). Moreover, Table 2 shows that the heating energy recorded at the 4 kg loading mass (M = 78,894.14, SD = 2028.18) was likewise considerably higher than the heating energy recorded at the loading mass of 3 kg (M = 57,386.17, SD = 1683.56) and 2 kg (M = 42,593.30, SD = 921.05). In addition, Table 2 shows that the heating energy required for a 3 kg loading mass was much higher than that of 2 kg loading mass.

Table 1. ANOVA of the effect of loading mass on the drying performance of the cabinet dryer.

Variables	Loading Mass	Mean	Std. Deviation	F	Sig.
Drying Efficiency	2 kg	0.75	0.11	19.522	0.000
	3 kg	0.60	0.14		
	4 kg	0.88	0.06		
Heating Energy	2 kg	42,593.30	921.05	1538.574	0.000
	3 kg	57,386.17	1683.56		
	4 kg	78,894.14	2028.18		
Moisture Content	2 kg	18.94	9.23	0.122	0.885
	3 kg	20.25	9.44		
	4 kg	20.68	9.34		

Data presented as mean of triplicates ($n = 3$) at 95% confidence interval ($p < 0.05$).

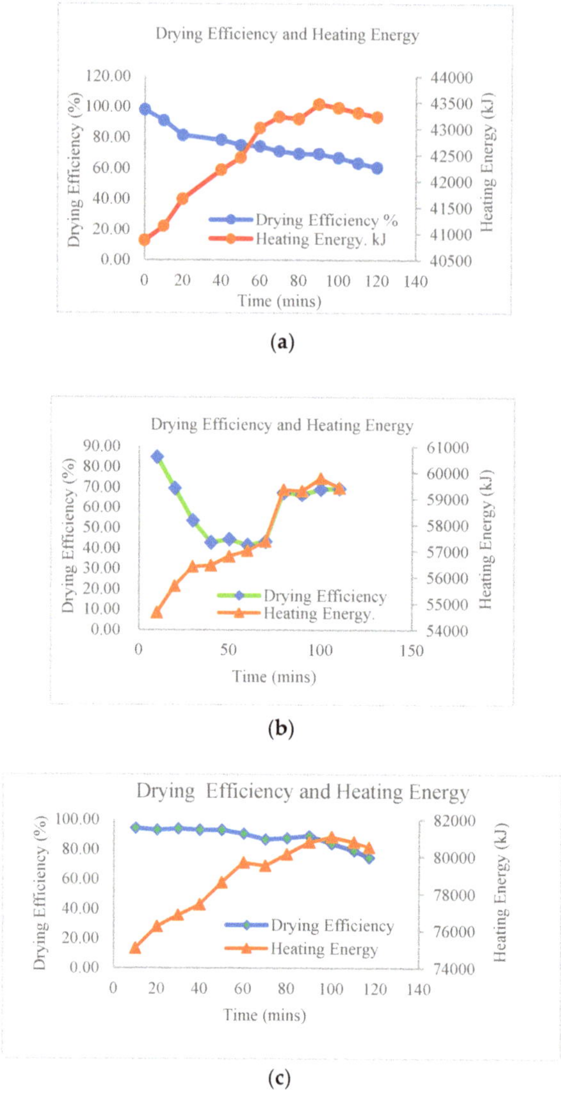

Figure 4. Graphs of drying efficiency against the heating energy at the drying time for different loading mass: (**a**) 2 kg; (**b**) 3 kg; and (**c**) 4 kg.

Table 2. Post hoc test on the loading mass difference.

Variables	Loading Group	Mean Difference	Sig.
Drying Efficiency	2 kg–4 kg	−0.13467	0.005
	2 kg–3 kg	0.14717	0.007
	3 kg–4 kg	−0.28183	0.000
Heating Energy	2 kg–3 kg	−14,792.87	0.000
	2 kg–4 kg	36,300.84	0.000
	3 kg–4 kg	−21,507.97	0.000

3.3. Drying Rate and Biomass Efficiency

Table 3 shows biomass efficiency ranging from 3.0% to 3.5% across various loading masses. The drying rates for flour samples were 0.014 kg/min, 0.20 kg/min, and 0.25 kg/min, while palm-kernel-shell feeding rates were 0.43 kg/min, 0.54 kg/min, and 0.59 kg/min for loading masses of 2 kg, 3 kg, and 4 kg, respectively. With an initial moisture content of 18% in the palm kernel shells, drying rates increased with higher loading masses due to a greater supply of heating energy. Optimal feeding rates of palm kernel shells are essential for generating and delivering heat to the drying chamber, as temperature control relies on these feeding rates.

Table 3. Overall biomass efficiency (%) at various loading masses and drying rates.

Loading Mass (kg)	Feed Rate of Palm Kernel Shell (kg/min.)	Total Mass of Palm Kernel Used (kg)	Initial Mass of Wet Mash (kg)	Final Mass of Flour (kg)	Drying Rate (kg/min)	Biomass/Overall Efficiency (%)
2.00	0.43	51.60	40.00	23.70	0.14	3.00
3.00	0.54	59.40	60.00	38.10	0.20	3.50
4.00	0.59	69.00	80.00	50.30	0.25	4.00

4. Discussion

The moisture content steadily decreases throughout the drying process until reaching equilibrium moisture content across all the loading masses. This drying behavior is consistent with findings from previous study on plantain and potato [18]. The final moisture content of the dried samples under varying loading masses meets the Nigerian standards for high-quality cassava flour (HQCF). This highlights the cabinet dryer's effectiveness in producing high-quality cassava flour with a suitable shelf life. Similar drying results have been reported in various studies focusing on biological materials processed in biomass heat-driven dryers. For instance, Godwin et al. [11] observed moisture content reductions from 87.6% to 17.3% (wb) for okra and from 90% to 15.3% (wb) for pepper. Likewise, Saravanan et al. [6] noted a reduction in cashew nut moisture content from 9% to 4% (wb).

Furthermore, the study evaluates the drying efficiency across different material capacities, recording efficiencies of 74.77%, 59.22%, and 88.24% for material capacities of 40 kg, 60 kg, and 80 kg, respectively. This highlights the dryer's capability to handle varying loads effectively, showcasing its versatility. The performance of this dryer compares favorably with drying results reported in previous studies on biomass heat-driven dryers and conventional dryers. For instance, a biomass cabinet dryer achieved an 89.67% drying efficiency for plantain in Famurewa et al.'s study [18], an industrial fruit and vegetable dryer attained an 84% drying efficiency for tomato in Ehiem et al.'s study [10], and a biomaterial cabinet dryer recorded an average drying efficiency of 79% for okra, pepper, and groundnut [11]. Also, the drying efficiency of 50.4% was recorded in a natural convection grain dryer for paddy and other cereals as reported by Mohapatra et al. [20], while Bukola and Ayoola [21] reported a drying efficiency of 85.4% in a mixed-mode solar dryer for yam chips. In addition, Adamade and Olaoye [22] reported drying efficiency of an electrically operated dryer, which ranged from 65.71 to 84.60% for Iru Woro and 65.60–84.87% for Iru Pete, respectively,

Moreover, Saravanan et al. [6] reported drying efficiencies ranging from 25% to 35%, averaging 30%, for a biomass dryer used for cashew nuts, attributing the lower efficiency to the inherent nature of the product. Sawant et al. [23] reported a drying efficiency of 25.20% for copra (a coconut derivative) in a waste-fired dryer, acknowledging its relatively lower efficiency. Drying efficiency is influenced by factors such as drying time, local environmental conditions, and the design of the drying system. The reduction in moisture content and the drying efficiencies recorded in the developed dryer suggest that the designed biomass-fueled cabinet dryer could serve as a practical solution for cassava flour production in rural communities.

The conventional drying of cassava using electrical energy or fossil fuel as input is a highly energy-intensive operation; hence, biomass drying could be one of the viable options for reducing the energy cost for cassava drying. Therefore, the biomass dryer is one of the suitable drying systems in rural areas where biomass is abundantly available. The comparative advantage of biomass dryer is that it can be effectively used during nights and even on rainy days without additional energy from other sources.

Furthermore, the study also highlights biomass efficiencies of 3.0%, 3.5%, and 4.0%. These values are significantly lower than those found in similar study on biomass dryers for cashew nut drying, which range from 4% to 16%, with the average value of 9.5% [6]. The biomass efficiencies recorded in the dryer indicate that the combustion efficiency within the drying chamber is limited. Hence, there is a need to expand the drying chamber to increase capacity and enhance heat transfer from the heat exchanger. In addition, a dehumidifier could be incorporated for the recirculation of the drying air to enhance the dryer efficiency. Despite these relatively modest biomass efficiencies, the overall drying efficiency underscores the dryer's effectiveness in achieving its primary objective.

This research focuses on designing, assessing, and analyzing the thermodynamic performance of a biomass-powered cabinet dryer for cassava processing. It evaluates moisture reduction, energy efficiency, and product quality, comparing the dryer to solar and electric alternatives to highlight its advantages and disadvantages. Although the study is limited to small-scale cassava drying, it provides useful insights for farmers, agro-processors, and policymakers looking for sustainable drying methods with possible applicability in other agricultural products. The study offers valuable insights into the performance of the biomass heat-generated cabinet dryer while emphasizing the importance of tailored technological solutions for addressing rural challenges. This approach underscores a holistic perspective that integrates socio-economic and environmental considerations into technology implementation in rural settings. By leveraging locally available materials and addressing the specific needs of the target community, the biomass heat-generated cabinet dryer proves to be a practical and efficient solution for cassava flour production in rural areas. This approach aligns with the overarching principles of sustainable and context-specific technology development.

The study provides useful insights; nonetheless, it has limitations, including the effects of biomass fuel variability, combustion efficiency, and environmental conditions on drying uniformity. It may also neglect long-term maintenance expenses, emissions, and difficulties in scaling for industrial application, as the emphasis is predominantly on small- to medium-scale drying. Moreover, elements such as air recirculation remain inadequately examined, potentially augmenting drying efficiency.

5. Conclusions

This study illustrates how a biomass-fueled dryer enhances cassava-drying processes compared to traditional methods, emphasizing its efficiency, controllability, and adaptability to varying conditions and loads. By utilizing locally available biomass as a renewable energy source, the dryer promotes sustainability and resilience in agricultural ecosystems. Despite its lower biomass efficiency, the dryer effectively reduces moisture levels, making it viable for small- and medium-scale farmers and enhancing economic opportunities through increased cassava processing for income generation. The use of locally sourced materials meets engineering standards, facilitating technology transfer and adoption in resource-limited settings. The dryer supports economic growth, environmental sustainability, and improved livelihoods for smallholder farmers, contributing to the comprehensive development of rural communities. Future research should prioritize improving biomass efficiency, the incorporation of dehumidifier for air recirculation, scalability, and operational ease to ensure widespread adoption across diverse rural contexts.

Author Contributions: Conceptualization, A.A. and W.A. (Wahabi Asiru); methodology, S.A., O.O., W.A. (Wasiu Awoyale), T.D. and P.K.; resources, A.A.; data collection, O.O.; data analysis, S.A.; writing—original draft, O.O. and S.A.; preparation, O.O. and S.A.; writing—review and editing, O.O. and S.A.; supervision, S.A. and W.A. (Wahabi Asiru); project administration, A.A. All authors have read and agreed to the published version of the manuscript.

Funding: Technologies for African Agricultural Transformation. Available online: https://taat-africa.org/ (accessed on 16 May 2025) & CGIAR Research Program on Roots, Tubers, and Bananas (RTB).

Institutional Review Board Statement: Not applicable.

Informed Consent Statement: Not applicable.

Data Availability Statement: The datasets generated and/or analyzed during the current study are included in the manuscript.

Acknowledgments: The study received financial support from the CGIAR Research Program on Roots, Tubers, and Bananas (RTB), the Cassava Compact of TAAT Program led by International Institute of Tropical Agriculture (IITA). In addition, Federal Institute of Industrial Research, Oshodi (FIIRO) Lagos-Nigeria also supported the study with human resources. All these efforts are appreciated.

Conflicts of Interest: The authors declare that there are no conflicts of interest.

References

1. Chukwuneke, J.L.; Achebe, C.H.; Okolie, P.C.; Okafor, E.A. The Effect of Variety and Drying on the Engineering Properties of Fermented Ground Cassava. *Int. J. Sci. Knowl. (Comput. Inf. Technol.)* **2013**, *1*, 13–27.
2. Adegbite, S.A.; Abass, A.; Olukunle, O.J.; Olalusi, A.P.; Asiru, W.B.; Awoyale, W. Mass and Energy Balance Analysis of Pneumatic Dryers for Cassava and Development of Optimization Models to Increase Competitiveness in Nigeria. *Curr. J. Appl. Sci. Technol.* **2019**, *32*, 1–11. [CrossRef]
3. Tran, T.; Andrade, L.A.T.; Chapius AAbass, A.; Sarta, M.; Adegbite, S.A.; Ojide, M.; Lukombo, S. *Root, Tuber and Banana Food System Innovations: Value Creation for Inclusive Outcomes*; Springer: Cham, Switzerland, 2022; ISBN 978-3-030-92021-0/978-3-030-92022-7. [CrossRef]
4. Adebayo, K.; Abayomi, L.; Abass, A.; Dziedzoave, N.T.; Forsythe, L.; Hillocks, R.J.; Kleih, U.K. Sustainable inclusion of smallholders in the emerging high-quality cassava flour value chains in Africa: Challenges for agricultural extension services. *J. Agric. Ext.* **2010**, *14*, 1–10. [CrossRef]
5. Zvinavashe, E.; Elbersen, H.W.; Slingerland, M.; Kolijn, S.; Sanders, J.P.M. Cassava for food and energy: Exploring potential benefits of processing of cassava into cassava flour and bioenergy at farmstead and community levels in rural Mozambique. *Biofuels Biorefining* **2011**, *5*, 151–164. [CrossRef]
6. Saravanan, D.; Wilson, V.H.; Sudhakar, K. Design and performance evaluation of biomass dryer for cashew nut Processing. *Adv. Appl. Sci. Res.* **2015**, *6*, 101–111.
7. Yank, A.; Ngadi, M.; Kok, R. Physical properties of rice husk and bran briquettes under low pressure densification for rural applications. *Biomass Bioenergy* **2016**, *84*, 22–30. [CrossRef]
8. Mishra, L.; Hauchhum, L.; Gupta, R. Development, and performance investigation of a novel solar-biomass hybrid dryer. *Appl. Therm. Eng.* **2022**, *211*, 118492. [CrossRef]
9. Kattimani, M.A.; Ahmed, M.N.; Akram, A. Design, and fabrication of centrifugal air blower test rig. *Int. J. Therm. Eng. (IJTE)* **2022**, *10*, 11–18. [CrossRef]
10. Ehiem, J.C.; Irtwange, S.V.; Obetta, S.E. Design and development of an industrial fruit and vegetable dryer. *Res. J. Appl. Sci. Eng. Technol.* **2009**, *1*, 44–53.
11. Godwin, E.A.; David, N.O.; Olugbenga, A.F.; Ubong, D.O. Design and Development of an Agricultural and Bio-materials Cabinet Tray Dryer. *Sci. Res.* **2016**, *4*, 174–182.

12. Zhengzhou Taida Drying Equipment Co., Ltd.; Shanghai Yongxian Machinery Manufacturing Company. Osa Cassava Dreg Dryer. 2024. Available online: https://www.hntaida.com (accessed on 25 January 2024).
13. Adzimah, K.S.; Seckley, E. 2009 Improvement on the design of a cabinet grain dryer. *Am. J. Eng. Appl. Sci.* **2009**, *2*, 217–228.
14. Imanogor, P.A.; Olawepo, B.B.; Itsisor, D.O. Development of a motorized maize shelling machine. *Indian J. Eng.* **2018**, *15*, 127–133.
15. Jones, D.D.; Kocher, M.F. Auger design for uniform loading of granular material: I. Rectangular cross–section containers. *Biol. Syst. Eng. Pap. Publ.* **1995**, *38*, 157–1162.
16. Khurmi, R.S.; Gupta, J.K. *A Textbook of Machine Design (S.I Units)*; Eurasia Publishing House (PVT) Limited: New Delhi, India, 2015; pp. 16–52.
17. Harikumar, A. *Design of Machine Elements 1. Module 13-Introduction to Belt Drives. Version 2 ME*; IIT Kharagpur: Kharagpur, India, 2014.
18. Famurewa, J.A.V.; Adejumo, P.; Adejumo, A. Performance Evaluation of a Biomass Fueled Cabinet Dryer. *Open Acad. J. Adv. Sci. Technol.* **2017**, *1*, 12–19.
19. Rajvir, S.; Maheshwari, R.C.; Ojha, T.P. Development of husk fired furnace. *J. Agric. Eng. Res.* **1980**, *25*, 109–120.
20. Mohapatra, S.S.; Lakshmi, D.V.N.; Mahanta, P. Performance Evaluation of Natural Convection Grain Dryer using Phase Change Material for Quality Drying of Paddy. *Int. J. Agric. Food Sci. Technol.* **2013**, *4*, 523–530.
21. Bukola, O.B.; Ayoola, P.O. Performance Evaluation of a Mixed-Mode Solar Dryer. *AU J. Technol.* **2008**, *11*, 225–231.
22. Adamade, C.A.; Olaoye, J.O. Performance Evaluation of a Dryer for Processed Locust Bean Condiments. *Agrosearch* **2014**, *14*, 103–112. [CrossRef]
23. Sawant, A.A.; Sonawane, S.P.; Sane, S.A. Design, Development and Performance Evaluation of Waste-Fired Copra Dryer. *Int. J. of Agric. Eng.* **2008**, *1*, 110–112.

proceedings

MDPI

Proceeding Paper

Germination and Drying as a Traditional Processing Method for Soybean Incorporation in Fish Feed in Togo †

Toï N'feide [1,2,*], Soudah Boma [1,3], Abdoul-Mountholib Kondi [4], Ekanao Tedihou [5], N'pagyendou Lare [5], Kossi T. A. Apaloo [6], Mouhamadou Amadou Ly [2], Yao Lombo [7] and Essodina Talaki [4]

1 Station de Recherche Halieutique d'Agbodrafo, Institut Togolais de Recherche Agronomique (ITRA-Togo), Lomé BP 1163, Togo; bomasoudah@gmail.com
2 Unité de Formation et de Recherche Sciences Agronomiques de l'Aquaculture et des Technologies Alimentaires, Université Gaston Berger de Saint Louis (UFR S2ATA/UGB), Saint Louis BP 234, Senegal; mouhamadou.ly@ugb.edu.sn
3 Centre International de Recherche-Développement sur l'Elevage en Zone Subhumide (CIRDES), Unité de Recherche «Maladies à Vecteurs et Biodiversité» (UMaVeB), Bobo-Dioulasso BP 559, Burkina Faso
4 Ecole Supérieure d'Agronomie, Université de Lomé (ESA/UL), Lomé BP 1515, Togo; isaackondi@gmail.com (A.-M.K.); talakiessodina@yahoo.fr (E.T.)
5 Laboratoire de Défense des Cultures et Biosécurité, Institut Togolais de Recherche Agronomique (ITRA-Togo), Lomé BP 1163, Togo; tekanao@gmail.com (E.T.); lare.eric@yahoo.fr (N.L.)
6 Laboratoire de Contrôle Qualité des Aliments, Institut Togolais de Recherche Agronomique (ITRA-Togo), Lomé BP 1163, Togo; aaparodrig@gmail.com
7 Direction Générale, Institut Togolais de Recherche Agronomique (ITRA-Togo), Lomé BP 1163, Togo; yaolombo@yahoo.fr
* Correspondence: sergenfeide@gmail.com or nfeide.toi@ugb.edu.sn
† Presented at the CORAF's 2023 Symposium on Processing and Transformation of Agricultural Products in West and Central Africa: Achievements and Opportunities for Private Sector Engagement, Lomé, Togo, 21–23 November 2023.

Abstract: The effects of germination (fresh germinated soybean [SG1F]) and post-germination drying (dried germinated soybean [SG1S]) on the crude protein (CP) levels, aflatoxin content, survival, and growth performance of Nile tilapia fry (*Oreochromis niloticus*) were evaluated. Raw soybean (SC) and roasted soybean (ST) served as controls. Laboratory analyses revealed CP contents of 11.74%, 38.72%, 38.42%, and 34.72% for SG1F, SG1S, RT, and RS, respectively. Aflatoxin levels were 2.3 µg/kg for SG1S, 1.66 µg/kg for RS, 1.60 µg/kg for RT, and 0.00 µg/kg for SG1F. Experimental diets, formulated to be isoproteic (35.44%) and isolipidic (8.73%), were prepared using flour from these soybean treatments and tested on tilapia fry with an average initial weight of 11.86 ± 2.15 g. The study was conducted in a completely randomized design with three replicates in 1 m^3 tanks stocked at 35 fry per tank over 56 days. Weight gains were 23.40 ± 10.21 g, 18.93 ± 8.67 g, 16.30 ± 9.92 g, and 16.07 ± 5.55 g for RT, RS, SG1S, and SG1F, respectively. Survival rates were 100%, 90.67%, 89.33%, and 88% for SG1F, RS, RT, and SG1S, respectively. Daily growth rates showed a consistent upward trend for all diets from the start to the end of the experiment.

Keywords: germinated and dried soybean meal; aflatoxins; crude protein content; antinutritional factors; growth performance; tilapia fry; Togo

1. Introduction

Aquaculture is arguably the fastest-growing food production sector, accounting for nearly 52% of global fish consumption and seen as a key solution to meeting the rising demand for aquatic products. It plays a significant role in food security and nutrition in the 21st century [1]. In Togo, aquaculture contributes only 5.08% of national fish production. However, total fish production—both from aquaculture and capture fisheries—meets just 30% of the country's domestic output [2], despite Togo's natural potential for aquaculture development [3]. Tilapia farming (genus *Oreochromis*) dominates, primarily in semi-intensive systems [4,5]. Thus, improving local aquaculture production is critical to reducing dependence on declining marine fisheries and imports.

Accelerating the adoption of sustainable and equitable aquaculture practices requires advancing policies, management, investments, and innovation [1]. Feed, a critical link accounting for 60–70% of production costs, is a priority area for technological innovation [6], as it significantly influences fish farm productivity. Fishmeal is the primary protein source in aquafeed [7,8], but its rising cost drives the need for alternative protein sources.

Scientific efforts to improve tilapia feed have focused on valorizing agricultural by-products [3–5,9,10]. However, alternative soybean (*Glycine max*) processing methods, such as dormancy breaking through water soaking, remain underexplored. Soybean, rich in digestible proteins, has superior nutritional value compared to cereal proteins [11,12], making it the second-most important protein source in fish feed after fishmeal. However, soybeans contain antinutritional compounds like tannins, saponins, phytic acid, protease inhibitors, gossypol, and aflatoxins [13,14].

Dormancy-breaking techniques aim to enhance protein bioavailability while reducing antinutritional factors, offering an alternative to roasting, which is cost-intensive and can degrade nutritional quality if poorly managed.

This study aims to improve aquaculture productivity in Togo through alternative soybean processing methods for sustainable fish feed. Specifically, it evaluates the effects of germination and post-germination drying on crude protein content and aflatoxin levels in soybeans. Additionally, it conducts a preliminary assessment of diets incorporating germinated and dried soybeans on the growth performance of Nile tilapia fry (*Oreochromis niloticus*).

2. Methodology

2.1. Study Area

Soybean samples were sourced from local markets and feed mills in Lomé. Bromatological analyses were conducted at the Food Quality Control Laboratory of ITRA, and feeding experiments were performed at the Agbodrafo Fisheries Research Station (SRHA) from 2 December 2020 to 15 May 2021. SRHA is situated in the Maritime Region under the Togolese Institute for Agronomic Research (ITRA), affiliated with the Ministry of Agriculture, Livestock, and Rural Development (MAEDR).

Located in a tropical Guinean climate, Agbodrafo experiences four seasons: a major dry season (November–March), a major rainy season (March–July), a minor dry season (August–September), and a minor rainy season (September–November). Average annual temperatures are 26.9 °C, with March being the warmest month (28.5 °C) and August the coolest (25 °C). Average annual rainfall is 888 mm, with the region's hydrographic network dominated by three rivers (Zio, Haho, and Boko), draining into Lake Togo [15].

2.2. Study Materials

The biological material primarily consisted of mixed-sex juveniles of Nile Tilapia (*Oreochromis niloticus*) from the Akosombo strain, with an average weight of 11.86 ± 2.15 g, sourced from the Fisheries Research Station in Agbodrafo.

2.3. Methods

2.3.1. Soybean Sample Collection

For this study, ten soybean samples were collected, comprising five samples of roasted soybeans and five samples of raw soybeans. These were sourced from the five largest sales outlets in Lomé and soybean collectors from major production areas. Composite samples were then created by combining equal quantities of the collected soybeans to mitigate regional variability in soybean composition due to environmental and agricultural practices [16].

2.3.2. Soaking and Dormancy Break in Soybean Sample Processing

The germination process for sprouted soybean treatments was scheduled to ensure all samples were analyzed on the same day (day 8, Table 1).

Table 1. Chronogram of the soybean germination process.

Day 1	Day 2	Day 3	Day 4	Day 5	Day 6	Day 7	Day 8
Soaking Sample 1	Incubation Sample 1 Soaking Sample 2	Germination Sample 1 Incubation Sample 2	Drying Sample 1 Germination Sample 2	Drying Sample 1 Stop Sample 2	Drying Sample 1	Drying Sample 1	Stop Sample 1

Two raw soybean samples (200 g each) were soaked in separate trays containing 1.232 L of water at an average temperature of 31 °C for 24 h. After soaking, the soybeans were drained and incubated for dormancy breaking. Over three days, the seeds were spread on moistened cloth and covered with similar material. Monitoring involved temperature checks and periodic water sprays to replicate traditional methods.

Upon tegument emergence, the first sample was removed on day 4 and air-dried at an average temperature of 31.68 °C to produce dried sprouted soybeans by day 8 (Figure 1). A second batch was soaked on day 5 to ensure fresh sprouted soybeans on day 8.

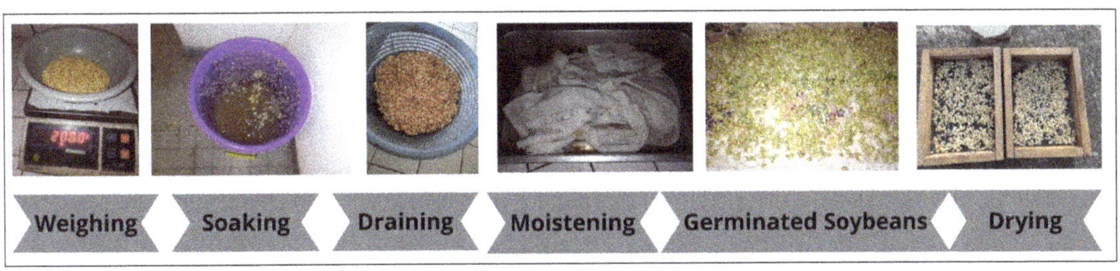

Figure 1. Soybean germination process.

To achieve this, two (02) raw soybean samples, each weighing 200 g, were soaked in two containers, each containing 1.232 L of water, for 24 h at an average temperature of 31 °C. After 24 h, the seeds were removed from the water for draining and incubation. The dormancy-breaking process lasted three (03) days, during which the soybean seeds were spread on a water-moistened cloth and covered with the same type of fabric. Monitoring involved measuring the temperature until the complete emergence of the seed coats, while regularly sprinkling water to simulate the traditional dormancy-breaking process.

After the seed coats had fully emerged, the first sample was removed on Day 4 for drying at an average ambient temperature of 31.68 °C, resulting, on Day 8, in soybean sprouts dried one (01) day after germination (Figure 1).

A second batch of raw soybeans was started on Day 5 for germination, ensuring that the complete emergence of seed coats from this batch coincided with the eighth day of the first batch. This process yielded fresh soybean sprouts one day after germination, under an average ambient temperature of 31 °C.

2.3.3. Laboratory Analyses

Sample Preparation

Soybeans subjected to various treatments and fishmeal were analyzed in the laboratory. A total of nine (09) samples were prepared, labeled, and transported to the ITRA quality control laboratory for bromatological analysis and aflatoxin detection (Table 2).

Table 2. Samples prepared for analysis.

Number	Samples	Bromatological Analyses	Aflatoxin Detection
1	Fresh germinated soybean (1 day)	1	1
2	Dried germinated soybean (1 day)	1	1
3	Raw soybean	1	1
4	Roasted soybean	1	1
5	Fishmeal	1	-

Determination of the Chemical Composition of Soybean Samples and Aflatoxin Detection

The analyses focused on determining the moisture content, organic components (proteins, lipids, and carbohydrates), and mineral content (ash) of soybean samples and fishmeal. The detection and quantification of aflatoxins in soybean samples were conducted using the standard AOAC (Association of Official Analytical Chemists, VICAM kit, number 991.31, September 2010 version) method for all soybean samples. The methods employed are summarized in Table 3.

Table 3. Analytical methods used in the laboratory.

Type of Analysis	Methods
Physico-chemical characteristics (%*w*/*w*)	
Moisture content	NF ISO 11294
Protein content	NF V 03-050
Ash content	NF ISO 7514
Lipid content	NF V 03 924
Carbohydrate content	CAC/vol. IX ed. 1
Aflatoxin detection (μg/kg)	
Aflatoxin quantification	AOAC

2.3.4. Feed Formulation

Selection of Ingredients

A list of 30 ingredients derived from databases on aquaculture feed formulation [17–19] was used. Based on this list, the ingredients were selected based on their availability in Togo [3,4]. The composition of crude protein, lipids, and metabolizable energy of the raw materials used is presented in Table 4.

According to Ranirinaharilala [20], feed formulation involves calculating ingredient proportions to meet the nutritional needs of the target species while optimizing cost. In this study, an Excel-based programming model using the "Solver" add-in, developed by the research team at the Agbodrafo Fisheries Research Station [9], was employed to formulate iso-protein diets. This model optimizes feed cost and quality by utilizing agro-industrial products and by-products. The soybean treatments were incorporated uniformly at a rate of 30%. However, for germinated soybean, the fresh weight obtained after germination was used to prevent excess moisture, which complicates pellet molding, as 100 g of raw soybean nearly doubles in weight after germination. The cassava mash was fixed at 10% to meet gelatinization needs.

Table 4. Raw material incorporation rates.

Ingredients	Incorporation Rate (%)			
	SG1F	**SG1S**	**SC**	**ST**
Maize	5	5	5	5
Broken rice	10	10	10	10
Cubed bran	7.5	7.5	7.5	7.5
Soybean cake	19.5	19.5	19.5	19.5
Groundnut oil	3	3	3	3
Growth concentrate 5%	2	2	2	2
DL-Methionine (99%)	0.5	0.5	0.5	0.5
Soybean	30	30	30	30
Fishmeal	22.5	22.5	22.5	22.5
Total	100	100	100	100
Theoretical Chemical Composition				
Crude protein (PB, %)	35.44	35.44	35.44	35.44
Crude lipids (LB, %)	8.73	8.73	8.73	8.73
Digestible energy (ED, kcal/kg)	**2931.52**	2931.52	2931.52	2931.52

SG1F: Diet containing fresh 1-day germinated soybean; **SG1S**: diet containing dried 1-day germinated soybean; **SC**: diet containing raw soybean; **ST**: diet containing roasted soybean; **PB**: crude protein; **LB**: crude lipids; **ED**: digestible energy.

2.3.5. Production of Experimental Feeds

Preparation of Ingredient Batches for Feed Formulation

A total of four (04) experimental diets were formulated for fish, incorporating different soybean treatments: roasted soybean composite (ST), fresh germinated soybean composite (1 day after sprouting) (SG1F), dried germinated soybean composite (1 day after sprouting) (SG1S), and raw soybean composite (SC). The control diet contained roasted soybean composite.

Feed Processing

Ingredients were sequentially ground using a mill, with cleaning between batches to prevent cross-contamination. Soybean samples were ground just before feed production to ensure immediate use. Macro-ingredients (maize, soybean, roasted soybean, wheat bran, fishmeal) were weighed with a 1 g precision digital scale, while micro-ingredients (growth concentrate, methionine, and shell powder) were measured with a 0.01 g precision scale, following a predetermined formulation. The mixing process involved five steps: (i) blending macro-ingredients, (ii) mixing micro-ingredients, (iii) combining both, (iv) incorporating heated cassava mash for granulation, and (v) adding groundnut oil. A manual machine with a 3 mm mesh shaped the feed into strands, which were sun-dried at ~35 °C for two days, followed by shade-drying at ~30 °C. The dried feed was crushed into pellets for ease of feeding. To prevent storage-related degradation, feed was produced weekly based on estimated requirements.

Packaging

The feed was packaged in sacks and stored on laboratory workbenches. The sacks were exposed to sunlight every two days to prevent mold development.

2.3.6. Testing of Experimental Feeds

Experimental Setup

The experimental design used was a triplicate system comprising 12 above-ground tanks of identical volume (1 m^3 each), arranged in two rows on a concrete platform separated by a drainage channel. Each tank was equipped with two pipes: an inlet pipe located laterally at the top with perforations at the base, and an outlet pipe situated at the bottom, externally elevated. Water was supplied through an open system based on the principle of communicating vessels, where used water was evacuated from the base while fresh water was introduced at the top. The entire system was housed in a well-ventilated building (Figure 2). The allocation of different feed regimes to the tanks was undertaken randomly.

Figure 2. Tanks used for the experiment.

Stocking

Stocking was carried out at a density of 35 juveniles per m^3 per tank, with juveniles averaging 11.86 $\pm$ 2.15 g in weight, totaling 420 juveniles across the 12 tanks.

Fish Feeding

Fish feeding commenced three (03) days after stocking to allow the fish to completely empty their stomachs and acclimate to their new environment. The fish were hand-fed to satiation with the formulated feed three times daily (08:00, 12:00, and 16:00) for 56 days. Water inflow was interrupted during feeding.

Tank Maintenance

Tank maintenance was carried out twice weekly by lowering the outlet pipes to remove waste and decomposed feed residues. On sampling days, tanks were completely emptied and cleaned.

Sampling

Four (04) samplings were conducted during the experiment, the first 14 days after stocking, and subsequently every 14 days to assess juvenile growth performance (weight). Fish were starved the day before sampling, and individual weights were systematically recorded.

2.3.7. Data Collection

Data collection involved the regular monitoring of water quality parameters and growth performance of juveniles.

Measurement of Physico-Chemical Parameters of Water

Water quality parameters were measured twice weekly, on Mondays and Thursdays/Wednesdays before feeding, over eight weeks. Parameters assessed included pH, dissolved oxygen, salinity, turbidity, and temperature to evaluate water quality.

Measurement of Juvenile Growth Performance

Juvenile growth performance (weight) was assessed every 14 days during sampling.

Growth and survival parameters

The following parameters were calculated:

Weight Gain (WG): The difference between final weight (Wf) and initial weight (Wi):

$$WG = Wf - Wi$$

Survival Rate (SR): The ratio of final (Nf) to initial (Ni) fish numbers, expressed as a percentage:

$$SR = 100 \times (Nf/Ni)$$

Specific Growth Rate (SGR, %/day): The daily weight increase as a percentage of body weight, independent of fish size: $SGR = 100 \times [(lnWf - lnWi)/t]$ ($t = duration\ of feeding$)

Daily Growth Rate (DGR): Evaluates growth speed, calculated as daily weight gain:

$$DGR = (Wf - Wi)/\Delta t$$

$$(\Delta t = duration\ in\ days)$$

Feed utilization parameters

Feed Conversion Ratio (FCR): The ratio of feed intake to biomass gain over a given period:

$$FCR = Qa/(Bf - Bi)$$

Feed Efficiency Ratio (FER): The reciprocal of FCR, representing weight gain per unit of dry feed:

$$FER = 1/FCR \text{ or } (Bf - Bi)/Qa$$

2.3.8. Statistical Analysis of Collected Data

Data were entered into an Excel spreadsheet and analyzed using SPSS (version 22) for mean and standard deviation calculations. A one-way ANOVA ($p \leq 0.05$) was conducted to evaluate treatment significance, followed by Duncan's test to identify significant differences between group means.

3. Results

3.1. Chemical Composition of Soybean Samples

The results of the chemical characteristics of the soybean samples, including moisture content, protein, lipid, ash, and carbohydrate levels, are presented in Table 5. The recorded values range from 5.44% (ST) to 62.94% (SG1F) for moisture content; 11.74% (SG1F) to 38.50% (SG1S) for protein content; 1.64% (SG5F) to 5.93% (ST) for ash content; 4.71% (SG1F) to 13.13% (ST) for lipid content; and 18.97% (SG1F) to 39.87% (SC) for carbohydrate content.

Table 5. Physicochemical characteristics of the different soybean treatments.

Physicochemical Characteristics (%m/m)	Ingredients				
	SG1F	SG1S	SC	ST	FP
Moisture content	62.94	7.87	8.07	5.44	3.26
Protein content	11.74	38.50	34.72	38.42	55.21
Ash content	1.64	4.50	4.65	5.93	6.47
Lipid content	4.71	12.90	12.69	13.13	11.93
Carbohydrate content	18.97	36.23	39.87	37.08	23.13

SG1F: Freshly germinated soybeans, one day post-germination; SG1S: dried germinated soybeans, one day post-germination; SC: raw soybeans; ST: roasted soybeans.

3.2. *Aflatoxin Levels in Treated Soybean Samples*

Aflatoxin levels in treated soybean samples were below the European Union's maximum permissible limit of 4 μg/kg [21]. The highest value was recorded for SG1S (2.30 μg/kg), while SG1F exhibited the lowest (0.00 μg/kg) (Table 6).

Table 6. Total aflatoxin quantification.

Sample Type	Results (μg/kg)
ST	1.6
SC	1.66
SG1S	2.30
SG1F	0.00

SG1F: Freshly germinated soybeans, one day post-germination; SG1S: dried germinated soybeans, one day post-germination; SC: raw soybeans; ST: roasted soybeans.

3.3. *Physico-Chemical Parameters of Water*

Table 7 summarizes the average values of water's physico-chemical parameters, including temperature, dissolved oxygen, pH, and salinity, across the different diets. Analysis of variance indicated no significant differences ($p \leq 0.05$) in the effect of these diets on the evaluated parameters. All values fell within the optimal range for tilapia farming, except for dissolved oxygen levels.

Table 7. Average values of water physico-chemical parameters.

Diets	Parameters			
	T °C	O_2 (mg/L)	pH	Salinity (mg/L)
SG1F	29.44 ± 0.80 ᵃ	1.13 ± 0.38 ᵃ	7.91 ± 1.67 ᵃ	0.31 ± 0.03 ᵃ
SG1S	29.26 ± 0.79 ᵃ	1.34 ± 0.79 ᵃ	7.97 ± 1.28 ᵃ	0.31 ± 0.03 ᵃ
SC	29.15 ± 0.74 ᵃ	1.17 ± 0.45 ᵃ	7.92 ± 1.04 ᵃ	0.31 ± 0.03 ᵃ
ST	29.16 ± 0.81 ᵃ	1.38 ± 0.57 ᵃ	8.61 ± 1.98 ᵃ	0.32 ± 0.04 ᵃ

SG1F: Diet containing fresh one-day germinated soybean; **SG1S:** diet containing dried one-day germinated soybean; **SC:** diet containing raw soybean; **ST:** diet containing roasted soybean; **T °C:** temperature; **O_2:** dissolved oxygen; **pH:** hydrogen potential; **Salinity:** water salinity. The values in the table are expressed as means ± standard deviation; (a) statistically identical.

3.4. *Growth Performance of Fingerlings by Diet*

Figure 3 illustrates the growth performance of fish across the experimental period (from day 0 to day 56) under the various dietary treatments. The Figure shows that fingerlings exhibited growth under all treatments during each monitoring session. However, the highest growth was observed with the control feed (ST), followed by SC, SG1S, and SG1F. Variance analysis revealed a statistically significant difference ($p \leq 0.05$) among the dietary treatments.

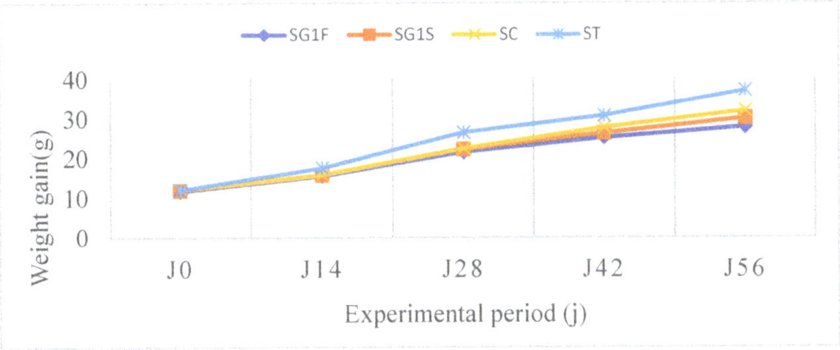

Figure 3. Evolution of the average weight of fish under different dietary treatments. **SG1F:** Diet incorporating freshly germinated soybeans (one day); **SG1S:** diet incorporating dried germinated soybeans (one day); **SC:** diet incorporating raw soybeans; **ST:** diet incorporating roasted soybeans; **J0:** day zero; **J14:** first monitoring session; **J28:** second monitoring session; **J42:** third monitoring session; **J56:** fourth monitoring session.

3.5. Growth Parameters

Table 8 presents the mean values of growth parameters (WG, SGR, and FCR) for the different diets. Variance analysis revealed a significant effect ($p \leq 0.05$) of the diets on the growth parameters. Specifically, Duncan's test indicated that the ST diet had a greater impact on WG, SGR, and FCR than the other diets. Conversely, SG1F exhibited the lowest values, indicating a less positive impact on growth parameters compared to the others. The effects of SG1F and SG1S diets were statistically identical for WG and FCR.

Table 8. Mean values of growth parameters for different dietary treatments.

Diets	Parameters		
	WG (g)	**SGR (%/d)**	**DGR (g/d)**
SG1F	16.07 ± 5.55 [c]	1.54 ± 0.45 [d]	0.29 ± 0.1 [c]
SG1S	16.30 ± 9.92 [c]	1.64 ± 0.52 [c]	0.29 ± 0.2 [c]
SC	18.93 ± 8.67 [b]	1.74 ± 0.49 [b]	0.34 ± 0.2 [b]
ST	23.40 ± 10.21 [a]	2.01 ± 0.50 [a]	0.42 ± 0.2 [a]

SG1F: Diet incorporating freshly germinated soybeans (one day); **SG1S:** diet incorporating dried germinated soybeans (one day); **SC:** diet incorporating raw soybeans; **ST:** diet incorporating roasted soybeans; **WG:** weight gain; **SGR:** specific growth rate; **DGR:** daily growth rate. Values are expressed as mean ± standard deviation; (a, b, c, d) statistically different.

3.6. Feed Utilization Parameters

Figure 4 illustrates feed consumption across dietary treatments during the experimental period. At Day 14 (J14), feed intake was similar across all treatments. By Day 28 (J28), a slight increase in feed intake was observed for all diets. However, at Days 42 (J42) and 56 (J56), a decrease in feed consumption occurred, which was not correlated with prior intake levels. Notably, the analysis of variance revealed no significant differences between the treatments.

The feed conversion ratio (FCR) and feed efficiency rate (FER) are presented in Table 9. Analysis of variance revealed a significant effect ($p \leq 0.05$) of the dietary treatments on feed utilization parameters, which ranged from **3.87 ± 2.28** (SG1F) to **2.17 ± 1.80** (ST) for FCR, and from **0.30 ± 0.19%** (SG1S) to **0.41 ± 0.18%** (ST) for FER. According to Duncan's test, ST had a greater impact on both FCR and FER compared to the other diets (SC, SG1S, SG1F). Conversely, SG1F and SG1S exhibited less pronounced effects on feed utilization parameters (FCR and FER).

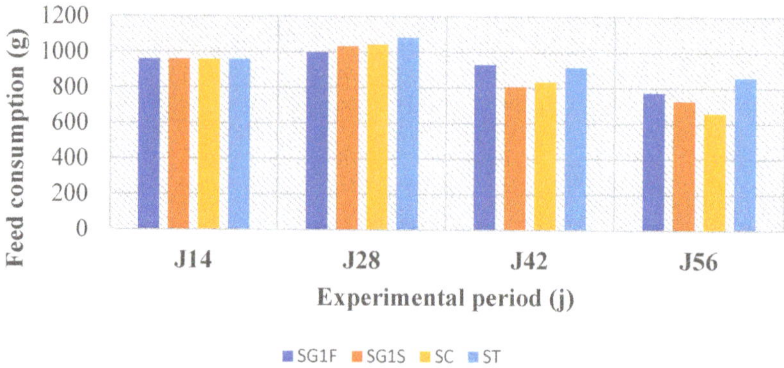

Figure 4. Evolution of feed consumption during the experimental period. SG1F: Diet incorporating fresh one-day germinated soybeans; SG1S: diet incorporating dried one-day germinated soybeans; SC: diet incorporating raw soybeans; ST: diet incorporating roasted soybeans; J14: first sampling; J28: second sampling; J42: third sampling; J56: fourth sampling.

Table 9. Mean values of feed utilization parameters.

Diets	Parameters	
	FCR	FER (%)
SG1F	3.87 ± 2.28 [b]	0.31 ± 0.11 [c]
SG1S	3.18 ± 2.90 [b]	0.30 ± 0.19 [c]
SC	2.75 ± 1.67 [a]	0.37 ± 0.17 [b]
ST	2.17 ± 1.80 [a]	0.41 ± 0.18 [a]

SG1F: Diet incorporating fresh one-day germinated soybeans; **SG1S**: diet incorporating dried one-day germinated soybeans; **SC**: diet incorporating raw soybeans; **ST**: diet incorporating roasted soybeans; **FCR**: feed conversion ratio; **FER**: feed efficiency rate. Values are expressed as means ± standard deviation; (a, b, c) statistically different.

The FCR values for ST and SC were statistically similar, as were those for SG1F and SG1S. Similarly, for FER, the effects of SG1F and SG1S were also statistically comparable.

3.7. Survival Rate

The survival rate of fish fed the various diets after the experimental period ranged from 80% to 100%, varying by treatment (Table 10). Specifically, the survival rate was 100% for SG1F, followed by 90.67% for SC, 89.33% for ST, and 88% for SG1S.

Table 10. Survival rate of fish fed the various diets after 56 days.

Diets	Survival Rate (%)
SG1F	100
SG1S	88
SC	90.67
ST	89.33

SG1F: Diet incorporating fresh one-day germinated soybeans; **SG1S**: diet incorporating dried one-day germinated soybeans; **SC**: diet incorporating raw soybeans; **ST**: diet incorporating roasted soybeans.

4. Discussion

The environmental conditions under which the fish were raised were consistent across all diets. In all experimental tanks, the values of the water's physicochemical parameters remained within the optimal range, except for dissolved oxygen levels [22–24]. The low dissolved oxygen levels observed can be attributed to the groundwater source used, combined with the absence of artificial aeration devices in the tanks. Nevertheless, these oxygen levels were not influenced by the type of diet. Thus, the fish were exposed to the same environmental conditions, and the differences in their growth performance cannot be attributed to the rearing environment.

Chemical composition analysis revealed that SG1S was richer in protein, followed by ST and SC. The lower protein levels in fresh samples may be explained by the high water content, which acts as a dilution factor. Dormancy breaking during germination affects the chemical composition of seeds by activating existing enzymes and synthesizing new ones, such as proteases for proteins, lipases for lipids, and amylases for starch [25,26]. These enzymatic activities likely increased the protein, lipid, and carbohydrate content in SG1S. Rapid drying halts enzymatic processes by swiftly removing water, resulting in a higher concentration of proteins in roasted and dried soybean samples.

In contrast to chemical composition, SG1F exhibited lower aflatoxin levels than SG1S, SC, and ST. This difference may be linked to storage or preservation conditions. According to Ahmed et al. [27], storing seeds with moisture content above 10% promotes mold proliferation.

Statistical analyses of feed utilization parameters revealed no significant differences among diets. Thus, at a 5% error margin, it can be concluded that the fish consumed the diets similarly without showing dietary preference.

Regarding feed utilization indices (FCR and FER), the results showed that ST exhibited the best performance (lowest FCR and highest FER), followed by SC, SG1S, and SG1F. These differences are due to the distinct physicochemical characteristics of the soybean treatments.

Growth performance results indicated that the reference diet ST had the best performance, followed by SC, SG1S, and SG1F. Variance analysis revealed a significant effect on weight gain at a 5% threshold ($p < 0.05$), with ST being significantly superior to other treatments (SC, SG1S, SG1F). Similar trends were observed for weight gain (WG), specific growth rate (SGR), and daily growth rate (DGR), which may be attributed to its higher organic content, including protein (38.42%), lipids (13.13%), and carbohydrates (37.08%).

The weight gain range observed in this study surpassed that reported by Boma et al. [10] at the same station using blood meal as a protein source over 42 days of feeding, but was lower than the results of Kirimi et al. [28] in Kenya, who used blood meal over a 100-day feeding period. The specific growth rate (SGR) obtained in this study was lower than that of Dibala et al. [24] in Burkina Faso, who used plant-based proteins over 56 days of rearing in plastic tanks, but higher than the results of Boma et al. [10]. The daily growth rate (DGR) was lower than those reported by Dibala et al. [24].

This study highlights the superiority of the roasted soybean (ST) diet compared to the SG1S diet, despite their theoretically comparable protein levels. Furthermore, our findings indicate that the aflatoxin content in SG1S meals was relatively high compared to ST. Numerous studies have demonstrated the detrimental effects of aflatoxins on fish growth [14,29]. This study suggests that optimal preservation of SG1S meals could minimize these factors, improving fish growth when fed this diet.

The survival rates observed during the experiment were high, ranging between 80% and 100%. This indicates that soybeans, in the forms used for this study's formulations, were non-toxic to fish. The high survival rate of fish fed raw soybean (90.67%), second only to SG1F (100%), demonstrates that incorporating raw soybeans at 30% is not toxic to Nile tilapia. However, our survival rate results exceeded those of Bamba et al. [30] in Côte d'Ivoire, who reported survival rates between 75% and 95%, but were slightly lower than those of Boma et al. [10], who reported rates between 96.67% and 100% for a study conducted at the same station in Agbodrafo.

5. Conclusions

The primary objective of this study was to evaluate the growth performance of Nile tilapia juveniles (*Oreochromis niloticus*) fed diets incorporating soybeans (roasted or germinated). The study concluded that the water's physicochemical parameters, such as temperature, pH, and salinity, remained within optimal ranges for Nile tilapia farming, resulting in high survival rates across the experimental diets. Among the diets tested, the best growth performance was observed with the reference diet (roasted soybeans), followed in descending order by raw soybeans, one-day dried germinated soybeans, and fresh one-day germinated soybeans. A similar trend was noted for feed utilization indices (FCR and FER). Daily growth rates showed a consistent upward trend for all diets from the start to the end of the experiment.

Author Contributions: Conceptualization, T.N. and S.B.; methodology, T.N., S.B., A.-M.K., E.T. (Ekanao Tedihou), N.L. and K.T.A.A.; software, T.N. and S.B.; validation, T.N., S.B., E.T. (Essodina Talaki), M.A.L. and A.-M.K.; writing—original draft preparation, T.N., S.B. and A.-M.K.; writing—review and editing, T.N., S.B. and A.-M.K.; visualization, E.T. (Essodina Talaki) and M.A.L.; supervision, M.A.L., E.T. (Essodina Talaki) and Y.L. All authors have read and agreed to the published version of the manuscript.

Funding: The authors are grateful to the Togolese Institute of Agronomic Research (ITRA) and Project to Support the Implementation of Higher Education Reform in Science and Engineering (PARESI, Phase 2) for financial support.

Institutional Review Board Statement: Not applicable.

Informed Consent Statement: Not applicable.

Data Availability Statement: Data are contained within the article.

Conflicts of Interest: The authors declare no conflicts of interest.

References

1. FAO. *Résumé de la Situation Mondiale des pêches et de l'aquaculture 2022: Vers une Transformation bleue*; FAO: Rome, Italy, 2022. [CrossRef]
2. Ministère de l'agriculture, de l'élevage et du développement rural. Plan D'action D'investissement de la Filière Poisson; Lomé, Togo Février. 2024; 33p. Available online: https://agriculture.gouv.tg/wp-content/uploads/2024/02/Plan_action_invest_filiere_Poisson_19_02_24.pdf (accessed on 2 March 2024).
3. Adjanke, A. Valorisation du tourteau de palmiste dans l'élevage du tilapia du Nil *Oreochromis niloticus*. Doctoral Thesis, Université de Lomé, Lomé, Togo, 2018.
4. N'feide, T.; Boma, S.; Kossoga, A.; Agbokousse, K.V.; Kangni, T.; Zoupoya, K.; Kpemoua, K.; Lombo, Y. Diagnostique des contraintes de production des fermes piscicoles du Togo. In Proceedings of the 4ème Colloque Scientifique International Organisé de l'Université de Kara, Kara, Togo, 18–22 October 2021.
5. Adjanke, A.; Tona, K.; Agbohessi, P.; Toko, I.; Gbeassor, M. Current situation of fish farming in Togo. *IJBCS* **2016**, *10*, 2015–2024. [CrossRef]
6. Batisse, K. Substitution de la farine de poisson par la farine d'asticots de mouches soldats noires dans l'alimentation du tilapia du Nil (*Oreochromis niloticus*) élevé en happas. Master's Thesis, Université de Lomé, Lomé, Togo, 2022.
7. Siddhuraju, P.; Becker, K. Comparative nutritional evolution of differentially processed mucuna seeds (*Mucuna pruriens* (L.) DC. Var. utilis (Wall ex Wight) (Baker ex Burck) on growth performance, feed utilization and body composition in Nile tilapia (*Oreochromis niloticus* L.). *Aquac. Res.* **2003**, *34*, 487–500. [CrossRef]
8. FAO. *Nourriture et Alimentation des Poissons*; FAO: Rome, Italy, 2015; Section 2, p. 42.
9. Boma, S.; N'feide, T.; Lombo, Y.; Batimsoga, B.; Kossoga, A.; Bataka, K.; Imorou, I.; Simion, O. Optimization of feed for Nile tilapia (*Oreochromis niloticus*) using an Excel programming model in small-scale feed-mixing operations in Togo. *Aquac. Int.* **2023**, *31*, 3545–3565. [CrossRef]
10. Boma, S.; N'feide, T.; Nakpa, T.; Batimsoga, B.; Bataka, K.; Lombo, Y.; Imorou, T.I. Substitution d'une farine de poisson locale par une farine de sang de bovin chauffée et performances d'alevins de tilapias du Nil au Togo. *Livest. Res. Rural. Dev.* **2022**. Available online: http://www.lrrd.org/lrrd34/3/3424bomas.html (accessed on 2 April 2022).

11. Francis, G.; Harinder, P.; Klaus, B. Antinutritional factors present in plant-derived alternate fish feed ingredients and their effects in fish. *Aquaculture* **2001**, *199*, 197–227. [CrossRef]
12. Kushwaha, M. Replacement of fish meal by soybean (*Glycine max*) in the formulation of fish feed ingredients essential for Immunostimulation and growth performance of carps. *Int. J. Fauna Biol. Stud.* **2013**, *1*, 35–38.
13. Ana, R.; Abdellatief, S.; Mahboub, H. Protection du tilapia du Nil, *Oreochromis niloticus* de la toxicité de l'aflatoxine B1 par une supplémentation alimentaire en huile essentielle de fenouil et Saccharomyces cerevisiae. *Egypte J. Aquat. Res.* **2017**, *43*, 235–240.
14. Daim, A.; Dawood, M.; Aleya, L. Effets du fucoïdane sur les indicateurs hématiques et les réponses antioxydantes des tilapias du Nil (*Oreochromis niloticus*) nourris avec des aliments contaminés par l'aflatoxine B1. *Environ. Sci. Pollut. Res.* **2020**, *27*, 12579–12586. [CrossRef]
15. FAO. *Évaluation des Pertes Après-Capture dans les Pêcheries Maritimes Artisanales du Togo*; FAO: Lomé, Togo, 2020; p. 14. [CrossRef]
16. CNUCED. SOJA: Un profil de produit de base par INFOCOMM, Fonds de la CNUCED pour l'information sur les marchés des produits de base agricoles. 2016. Available online: https://unctad.org/fr/system/files/official-document/INFOCOMM_cp10_SoyaBeans_fr.pdf (accessed on 15 May 2020).
17. Lukuyu, B.; Okike, I.; Duncan, A.; Beveridge, M.; Blümmel, M. *Use of Cassava in Livestock and Aquaculture Feeding Programs*; ILRI Discussion Paper 25; International Livestock Research Institute: Nairobi, Kenya, 2014.
18. IAFFD. IAFFD. 2020. Available online: https://app.iaffd.com/paff (accessed on 7 September 2021).
19. Adekambi, D.; Adechola, P.; Ifagbemi, B. Screening local feed ingredients of Benin, West Africa for fish feed formulation. *Aquac. Rep.* **2020**, *17*, 100386.
20. Ranirinaharilala, P.F. Amélioration de l'alimentation d'Oreochromis niloticus à travers la valorisation du soja; Mémoire de Fin d'études, Ecole supérieure polytechnique d'Antananarivo, Antananarivo, Madagascar. 2021. Available online: https://polytechnique.mg/these/RANIRINAHARILALA-Patricia-Fandresena.pdf (accessed on 6 June 2022).
21. Dimanche, P. Les exportateurs d'arachide de bouche des pays du Sud pénalisés par les normes sur l'aflatoxine édictées par l'Union Européenne. *OCL* **2001**, *8*, 237–238. [CrossRef]
22. Lazard, J. Synthèse Pisciculture de quelques espèces: La pisciculture des tilapias. *Cah Agric.* **2009**. Available online: https://revues.cirad.fr/index.php/cahiers-agricultures/article/view/30786 (accessed on 10 October 2021).
23. Mashaii, N.; Rajabipour, F.; Mohammadi, M.; Sarsangi, H.; Bitaraf, A.; Hossein-Zadeh, H.; Sharif-Rohani, M. Reproduction of Nile tilapia, *Oreochromis niloticus* in brackish water. *J. Appl. Aquac.* **2016**, *28*, 1–8. [CrossRef]
24. Dibala, C.; Yougbare, M.; Konate, K.; Coulibaly, N.; Dicko, M. Production du tilapia du Nil (*Oreochromis niloticus* linneaus, 1758) avec des aliments à base de protéines végétales. *J. Appl. Biosci.* **2018**, *128*, 12943–12952.
25. Aya, A.; Irie, V.; Patrice, L.; Irie, A. Bases génétiques et biochimiques de la capacité germinative des graines: Implications pour les systèmes semenciers et la production alimentaire. *Sci. Nat.* **2011**, *8*, 119–137.
26. CIVAM. Les Graines Germées, Agriculture Durable des Moyennes Montagnes. 2012. 4p. Available online: https://doczz.fr/doc/1074979/les-graines-germ%C3%A9es---agriculture-durable-de-moyenne-mont (accessed on 10 October 2021).
27. FAO. Seeds Toolkit-Module 6: Seed Storage. Rome, Italy. 2018. Available online: https://openknowledge.fao.org/server/api/core/bitstreams/cecaae8f-e7fd-4204-82cf-b79c2cdb39f8/content (accessed on 2 November 2021).
28. Kirimi, J.G.; Musalia, L.M.; Magana, A.; Munguti, J.M. Performance of Nile Tilapia (*Oreochromis niloticus*) Fed Diets Containing Blood Meal as a Replacement of Fish Meal. *J. Agric. Sci.* **2016**, *8*, 79–87. [CrossRef]
29. Tuan, N.; Grizzle, J.; Lovell, R.; Manning, B.; Rottinghaus, G. Croissance et lésions hépatiques de tilapia du Nil (*Oreochromis niloticus*) nourris avec des régimes contenant de l'aflatoxine B1. *Aquaculture* **2002**, *212*, 311–319.
30. Bamba, Y.; Ouattara, A.; Da Costa, K.S.; Gourène, G. Production de *Oreochromis niloticus* avec des aliments à base de sous-produits agricoles. *Sci. Nat.* **2008**, *5*, 89–99.

proceedings

MDPI

Proceeding Paper

Rice Value Chain Upgrading in Côte d'Ivoire: The Role of the Improved GEM System for Rice Parboiling †

Alban Landry Kanon [1,*], Rachidi Aboudou [2], Meougbé Ernest Depieu [1], Aminou Arouna [2] and Sali Atanga Ndindeng [2]

1 Centre National de Recherche Agronomique, Abidjan 01 BP 1740, Côte d'Ivoire; ernest.depieu@cnra.ci
2 Africa Rice Center (AfricaRice), Bouake 01 BP 2551, Côte d'Ivoire; r.aboudou@cgiar.org (R.A.); a.arouna@cgiar.org (A.A.); s.ndindeng@cgiar.org (S.A.N.)
* Correspondence: landry.kanon@cnra.ci
† Presented at the CORAF's 2023 Symposium on Processing and Transformation of Agricultural Products in West and Central Africa: Achievements and Opportunities for Private Sector Engagement, Lome, Togo, 21–23 November 2023.

Abstract: Rice is a staple food in Côte d'Ivoire, but parboiling remains a weak link in the local rice value chain. Parboiled rice has superior nutritional and physical properties compared with white rice. In 2017, AfricaRice introduced the "Grain quality enhancer, Energy efficient, and durable Material" (GEM) parboiling system in Côte d'Ivoire to address these challenges. This study evaluated the rice value chain, parboilers' perceptions of GEM technology, and its role in upgrading the rice value chain. Using two survey types (parboiler association level (focus group) and parboiler household level), 179 rice parboilers and 77 key players were randomly sampled and interviewed in Côte d'Ivoire. We used descriptive statistics and the parboiler budget for the profitability analysis. The results show that parboiling, primarily undertaken by women, is profitable in Côte d'Ivoire. Parboilers identified the key advantages of the GEM system, including improved rice quality, higher output rates, better nutritional value, and the ability to parboil larger volumes of rice. GEM technology enhanced rice quality, increased parboilers' incomes, and reduced energy costs by 22.3%, saving an average of 73,090 FCFA per ton of parboiled paddy. The study recommends scaling up GEM technology through local fabricator training and broader dissemination in rice-growing regions to further strengthen the rice value chain in Côte d'Ivoire.

Keywords: GEM; quality of rice; improved parboiling system; nutrition security; Côte d'Ivoire

1. Introduction

Food insecurity and childhood malnutrition remain persistent problems in sub-Saharan Africa, where rice is a staple food for more than half of the world's population [1]. The African continent consumes one-third of the world's rice imports annually. Rice consumption in West Africa surpasses that of any other region in sub-Saharan Africa (SSA), with regional demand increasing at a rate of nearly 6% per year. This growth is attributed to the expanding population, shifts in dietary preferences, and increasing urbanization [1,2]. Despite the growing demand, domestic rice production has not increased sufficiently. As a result, the shortfall is addressed by importing rice from Asian countries, which produce varieties that consumers find appealing [1,3]. To significantly reduce imports, the focus has been on improving the productivity and intrinsic quality of rice rather than the quality attributes acquired during processing (cleanliness and homogeneity) and marketing (moisture content linked to storage time). Small-scale rice processors in the region generally rely on traditional technologies for post-harvest activities. This has led to the production of low-quality rice, making it less competitive than imported rice. Inadequate post-harvest management primarily contributes to the quality of domestically produced rice [4]. Post-harvest activities are crucial for enhancing value, generating job opportunities, improving women's livelihoods, and minimizing food waste. The process of rice parboiling, which involves subjecting paddy (unprocessed rice) to hydrothermal treatment before husk removal and polishing, has been investigated as a method to enhance the physicochemical and nutritional properties of rice, including its digestibility [5].

In Côte d'Ivoire, the quality of local rice, in general, and parboiled local rice, in particular, does not always meet consumer expectations. This is why improving rice quality is at the heart of the National Rice Development Strategy, with greater attention paid to post-harvest processing strategies through the introduction of modern industrial rice mills and the definition of "made in Côte d'Ivoire" rice standards to increase the competitiveness of locally produced rice. This strategy is beginning to bear fruit, with local production growing from approximately 600,000 tons in 2012 to nearly 1.2 million tons of milled rice in recent years [6]. However, the use of improved technologies for processing paddy rice into quality parboiled rice has been less developed. Nevertheless, imported quality parboiled rice is starting to occupy a significant share of the market and is preferred by affluent urban consumers. In Côte d'Ivoire, parboiling remains the exclusive domain of women, and the process is still based on rudimentary, labor-intensive traditional techniques, classifying this type of rice as very low-quality, with the national slogan "le riz qui n'est pas doux là, malo woussou (parboiled rice in Malinké)". However, traditional parboiling processes are inefficient. Locally parboiled rice generally contains many impurities (stones, debris, and microbiological agents) and is sometimes produced under poor hygienic conditions. However, innovative technologies, such as AfricaRice's improved equipment called Grain quality enhancer, Energy efficient, and durable Material (GEM), help produce very high-quality parboiled rice. GEM helps present a quality consumable product to customers while creating added value to the final product. It contributes to the valorization of local rice and constitutes an undeniable niche for income generation and the empowerment of women in Africa. The introduction of GEM technology into the parboiled rice value chain in Côte d'Ivoire is already bearing fruit, with the repositioning of local parboiled rice in the local market and increased income for female processors. In Côte d'Ivoire, several groups have benefited since 2017 from the GEM-improved parboiling system in their production units to boost the production of high-quality local parboiled rice.

The technical performance of GEM systems has been tested in several studies [1,7].

However, no economic study has been conducted to evaluate the contribution of GEM to rice value chain upgrading in Cote d'Ivoire. This study aims to assess the technical and economic performance of the improved GEM parboiling system compared with that of the traditional parboiling system. Specifically, it (i) analyzes the rice situation, rice parboiling activities, and opportunities in the rice value chain in Côte d'Ivoire, and (ii) describes the socio-economic characteristics of parboilers in three rice parboiling groups benefiting from the GEM system in the departments of Bouaké, Daloa, and Odienné in Côte d'Ivoire, as well as the perceptions of parboilers about GEM technology. In addition, the quality of the parboiled rice produced and the economic profitability of the GEM were compared with those of the traditional parboiling system. Overall, this study contributes to the literature by bridging a crucial gap in the economic evaluation of GEM technology for rice parboiling, providing a comprehensive analysis of technical and economic aspects, and offering contextual insights specific to Côte d'Ivoire's rice value chain.

The remainder of this paper is organized as follows: First, we outline the GEM system and describe the implementation of an enhanced GEM parboiling system in Côte d'Ivoire in detail. Next, we present the methodology in the "Methodology" section. We then showcased and analyzed the findings in the "Results" and "Discussion" sections, respectively. Finally, we summarize the study and examine its policy implications in the "Conclusion and Recommendations" section.

2. Description and Dissemination of the Improved GEM Parboiling System Introduced in Cote d'Ivoire

The GEM parboiling system is an improved parboiling technology that combines the use of a uniform steam parboiler with an improved parboiling stove. Owing to the low capacity and quality of traditional parboiled rice, the GEM parboiling system was developed in Africa by AfricaRice and its partners [1]. The main components of the complex are parboilers (steaming tank and baskets), soaking vessels, stoves, labor-saving devices, a hot water siphoning system, drying surfaces, and a shade that accommodates the equipment. GEM technology is well described in [1]. The GEM system encompasses both the equipment and methodology. Implementation involves educating parboilers, primarily women, on utilizing the system to create high-quality parboiled rice. Training covers selecting appropriate rice varieties and paddies, focusing on slender grains and rough rice free from disease damage or threshing-induced de-husking. Instructions also include proper cleaning techniques through winnowing and washing to eliminate contaminants, as well as optimal soaking temperatures, steaming durations, drying procedures, and dehulling methods for the best outcomes. The system price varies based on the components and operational scale.

The GEM system has been disseminated in many African countries, including Côte d'Ivoire [1]. In the first stage of the GEM system dissemination, a dozen agricultural equipment manufacturers were trained. Women parboilers from the IP in Côte d'Ivoire were trained in the use and methods of rice parboiling using the GEM system. As of February 2024, the GEM system was introduced in 10 areas in Côte d'Ivoire: Daloa, Abidjan, Odienné, Man, Gagnoa, Korhogo, Boundiali, Abengourou, Bouaké marché de gros, and Dar-Salam (in Bouaké). Figure 1 highlights the 10 dissemination areas of the GEM parboiling system in Côte d'Ivoire.

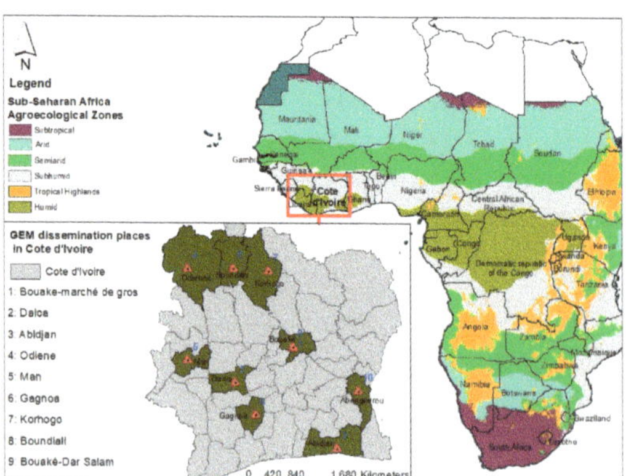

Figure 1. Map of Cote d'Ivoire highlighting the place of dissemination of the GEM system in Cote d'Ivoire by AfricaRice and CNRA.

3. Methodology

3.1. Study Area and Data Collection

This study was conducted in three districts in the Republic of Côte d'Ivoire. Three groups of rice parboilers benefited from the GEM-improved parboiling system in Bouaké, Daloa, and Odienné in Côte d'Ivoire (Figure 2). In these departments, the GEM system was first introduced through training and demonstration of female rice parboilers by AfricaRice and the Côte d'Ivoire National Center for Agronomic Research (CNRA).

Figure 2. Map of Cote d'Ivoire highlighting the study area.

Data were collected by enumerators who were selected based on their experience with data collection. The first stage of data collection involved visits to parboiling units and interviews with 77 key players (including 37 in Odienné, 20 in Daloa, and 20 in Bouaké) in the parboiled rice value chain as well as individual interviews with 179 female parboilers (including 88 parboilers in Odienné, 61 in Daloa, and 30 in Bouake), members of GEM beneficiary groups, on their sociodemographic characteristics, using a structured questionnaire. The second stage consisted of focus groups using an interview guide with members of these groups (20 parboilers per group) to gather information on the socioeconomic environment of parboiling and their perceptions of GEM equipment (advantages and disadvantages) and costs of using GEM and the traditional parboiling system (description of parboiling processes, parboiler perceptions of GEM, and determination of production costs with GEM and the traditional system). The third stage involved interviews with experts (from ADERIZ, AfricaRice, and ANADER) working in the study areas to obtain their opinions on rice production and marketing and rice parboiling practices (including four experts in Odienné, four experts in Daloa, and five experts in Bouake). These data were used to conduct a comparative study of the two parboiling systems. The primary data collected were related to the technical and economic performances of various technologies. More specifically, these include various technical operations, the costs of these operations, the quantities of paddy processed and their yields in husked parboiled rice, and the quantities of labor and other intermediate inputs used in the operations.

3.2. Data Collection on Parboiled Rice Quality

To assess the quality parameters of parboiled rice from each parboiling system, samples of husked parboiled rice were taken from markets (including the "grand marché de Odienné" in Odienné, "grand marché de Daloa" in Daloa, and "marché de gros" in Bouake) and GEM parboiling units in the study area. In each study area, five samples (0.5 kg) of each type of parboiling system were collected at different times from the main local markets. These samples were analyzed at the AfricaRice Grain Quality Laboratory at the M'bé Research Station (Bouaké). The main parameters assessed were as follows:

3.2.1. Whole Grain Rate

For each sample, 100 g of polished rice was weighed and classified into whole-grain and broken-grain fractions using a rotary cylinder grader (Satake, Hiroshima, Japan). Whole grains (intact grains and greater than 75% intact grains) were weighed and recorded as a percentage of whole grains.

3.2.2. Varietal Mixing Rate and Grain Color Alteration Rate

The varietal blend percentage, which is a measure of varietal purity, and color damage rate, which is a measure of heat or disease damage to grains, were measured using an S21 statistical rice analyzer (LKL Tecnologia Ltda, Santa Cruz do Rio Pardo, Brazil), as previously described [5].

3.2.3. Brightness of Grains

A colorimeter (CR-400, Minolta Co., Ltd., Tokyo, Japan) was used to measure the color values of the rice grains using the L, a, and b uniform color space procedures [5]. Approximately 5 g of rice grain was placed in the sampler, and the color was captured. The samples were analyzed in triplicate, and the mean L value was reported. The L value represents the brightness or luster of the grain, where 0 = dark and 100 = white.

3.2.4. Impurity Rate

The percentage of total impurities (organic and inorganic) was obtained from a 100 g sample of husked rice by separating the impurities from the rice grains. The samples were weighed and expressed as a percentage.

3.3. Data Analysis

Descriptive statistics of the socioeconomic characteristics of the women parboilers (age, ethnicity, level of education, etc.) were calculated. A comparative financial analysis of the GEM and traditional parboiling systems was conducted using focus groups and expert opinions. The costs of all production operations at the level of each parboiling system and the selling prices of the obtained products were evaluated. The results obtained for each parboiling system were used to calculate several technical and economic indicators, notably Gross Product (GP), Intermediate consumption (IC), depreciation of equipment used, value added (VA), and gross and net income (Table 1). For data concerning the whole grain rate, varietal purity rate, grain color alteration rate, grain brightness, and impurity rate, analysis of variance (ANOVA) using XLSTAT premium 2020 software was used to compare the quality characteristics of rice samples parboiled using GEM technology with those parboiled using traditional practice.

Table 1. Formulas for calculating some technical and economic indicators.

Some Technical and Economic Indicators	Formulas Used [8]
Gross Farm Product (GP) or Sales (SR)	=Quantity produced (by-products) × Unit price (unit produced)
Intermediate consumption (IC)	=∑Cost (Raw material, and other charges + Shelling cost)
Labor (L)	=∑ Cost (HJ/operation)
Depreciation (D)	=∑ Cost of depreciation/material for processing + Financial expenses
Total Cost (TC)	=IC + L + D
Cost of production per kg of unmilled parboiled paddy rice	=(TC − Hulling cost)/Quantity of parboiled paddy
Production cost per kg of milled parboiled rice	=CT/Quantity of milled parboiled rice
Gross Value Added (GVA)	=GB − IC
Net Value Added (NVA)	=GVA − D
Net income (NI)	=NVA − L
The cost–benefit ratio	=(NI/TC) ∗ 100

4. Results

We begin this section by presenting the socioeconomic characteristics of female parboilers. This was followed by an analysis of the rice situation and rice parboiling activities in Côte d'Ivoire. Next, we present the opportunities for rice parboiling and marketing in Côte d'Ivoire. Finally, we compare the profitability of the traditional parboiling and GEM systems.

4.1. Analysis of the Rice Situation in Cote d'Ivoire

Rice is a major source of livelihood for smallholder farmers, processors, and traders and is a major food for urban and rural people in Cote d'Ivoire, where the demand for this staple is growing. In Côte d'Ivoire, the annual per capita rice consumption also increased, reaching 104 kg from 2014 to 2018. Since 2019, the annual per capita rice consumption has been approximately 91 kg, as reported by the USDA data (Figure 3).

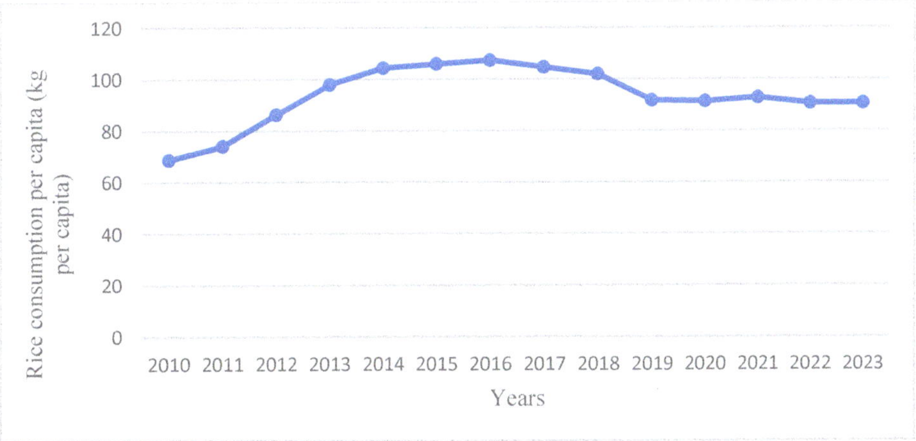

Figure 3. Rice consumption per capita in Cote d'Ivoire from 2010 to 2023. Source: USDA, 2024 (http://www.fas.usda.gov/psdonline/ (accessed on 24 January 2024)).

Rice consumption has grown in Côte d'Ivoire from 2010 to 2023, whereas production and importation have grown during the same period (Table 2). To meet the high demand for rice, imports were increased during this period. This situation provides an opportunity for rice farmers and post-harvest actors, such as millers and parboilers, to invest in rice production in Côte d'Ivoire.

Table 2. Rice area, production, importation, exportation, and domestic consumption in Cote d'Ivoire.

Attribute	2010	2011	2012	2013	2014	2015	2016	2017	2018	2019	2020	2021	2022	2023
Area Harvested (1000 HA)	550	380	664	792	846	900	900	862	875	750	635	710	720	730
Production (1000 HA)	784	567	1015	1257	1335	1399	1335	1377	1304	1225	962	1078	1108	1131
Rough Production (1000 HA)	1206	872	1562	1934	2054	2152	2054	2118	2006	1885	1480	1658	1705	1740
Imports (1000 HA)	850	1050	1150	800	1300	1250	1300	1370	1350	980	1350	1665	1456	1450
Exports (1000 HA)	24	27	30	40	50	70	70	80	100	50	25	30	30	50
Domestic Consumption (1000 HA)	1450	1600	1900	2200	2400	2500	2600	2600	2600	2400	2450	2550	2550	2550

Source: USDA, 2024.

While Côte d'Ivoire has been increasing its rice production, the demand has grown faster, leading to a reliance on imports to meet consumption needs, particularly during international disruptions (COVID-19 and the Russia–Ukraine war) (Figure 4). Figure 4 highlights the dynamic nature of rice production, imports, and consumption in Côte d'Ivoire, which is influenced by global events. By focusing on innovation, value addition, supply chain improvements, sustainability, and export potential, youth and enterprises can seize significant opportunities to contribute to and benefit from the growth of Côte d' Ivoire's rice sector.

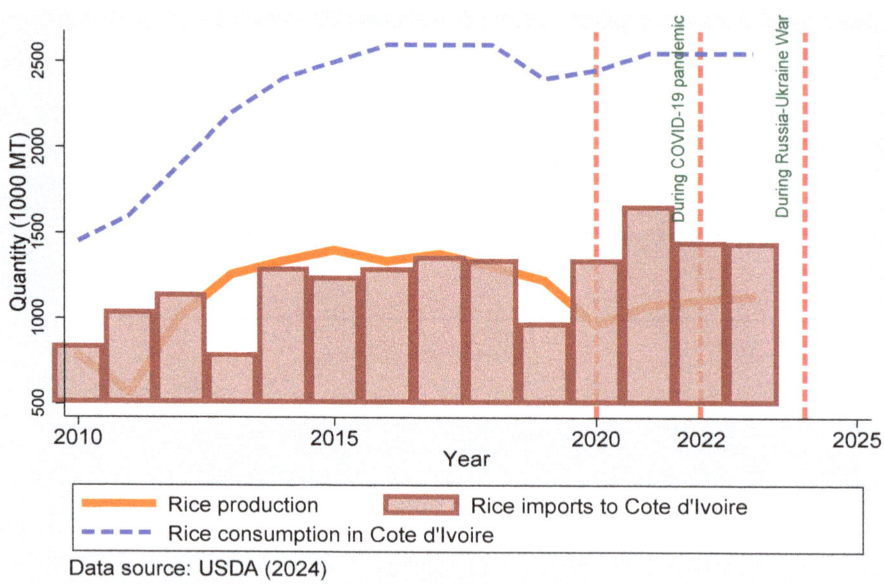

Figure 4. Rice production, importation, and consumption in Côte d'Ivoire from 2010 to 2023. Source: USDA, 2024.

Companies and enterprises can collaborate with local farmers to produce high-quality rice varieties suitable for export, thereby increasing foreign exchange earnings. Training programs in international trade can prepare young entrepreneurs to explore export opportunities, particularly in West African markets, where the demand for rice is high.

Encouraging young people to engage in agricultural innovation can increase the efficiency of rice production. Programs focused on agricultural technology, such as the use of drones, precision farming, and mobile applications for farm management, can attract young people.

4.2. Analysis of the Rice Parboiling Activities in Cote d'Ivoire

4.2.1. Rice Parboiling Activity and Equipment Used

Parboiling rice is an operation that consists of pre-cooking paddy with vapor. It is an operation that enhances the quality and yield of rice husks. Women in Cote d'Ivoire participated in this activity. According to the parboilers, the yield of pre-cooked rice after husking was shown to be higher than that of the non-pre-cooked paddy. In addition, the nutritional value (rich in vitamin A and protein), organoleptic quality, and appearance of pre-cooked rice are much better. Parboilers in Côte d'Ivoire have a good understanding of the advantages of parboiling rice, including higher head rice yield, milling recovery, and price. They also pointed out that cooked parboiled rice was less sticky and tasted better and had a higher number of vitamins than non-parboiled rice. They also pointed out that parboiled rice is good for diabetics, but did not know why this was the case. Parboilers have shown a lot of interest in producing and marketing high-quality and high-quantity parboiled rice, but they lack the equipment and methods to achieve this goal. The women indicated that the equipment and procedures currently used do not allow them to produce high-quality and high-quantity parboiled rice in a manner comparable to that of imported rice. This makes their products less competitive with imported rice, and as such, they are forced to sell at a cheaper price. The most commonly used equipment for rice parboiling in Côte d'Ivoire is traditional equipment and improved GEM equipment.

Traditional parboiling equipment consists of barrels in which paddy is soaked, pre-heated, and steamed (Figure 5). The stainless-steel GEM technology consists of a 400 kg capacity paddy soaking tank, two rice husk pans with accessories, a lidded pot with a strainer inside, and a 500–1000 kg hoist hung on a support with a pulley wheel to lift loads of 100 kg (Figure 5).

(a) Barrels for soaking (b) Firewood as an energy source (c) Barrels for steaming (d) Unprotected drying of parboiled paddy

Figure 5. Traditional parboiling system equipment.

The GEM equipment is shown in Figure 6. The GEM system encompasses both the equipment and methodology. During its implementation in Côte d'Ivoire, female parboilers received comprehensive training in various aspects of the system. This included instructions on producing high-quality parboiled rice, selecting appropriate rice varieties and paddies for parboiling, and identifying rough rice free from disease damage or de-husking during threshing. Additionally, they learned proper cleaning techniques, such as winnowing and washing, to eliminate impurities. The training also included optimal soaking temperatures (85 °C for most varieties and aged rough rice), steaming duration (20–25 min), and drying and dehulling procedures to achieve the best outcomes.

(a) Soaking tank (b) Rice husk fireplace (C) Boiler and hoist (d) Protected drying of parboiled paddy

Figure 6. Improved GEM system.

4.2.2. Opportunities for Youth and Enterprises for Rice Parboiling and Marketing in Cote d'Ivoire

Several opportunities exist in Côte d'Ivoire for women, youth, and enterprises to take advantage of better livelihoods. The key areas revealed in this study are discussed below.

Demand for local rice exists in Côte d'Ivoire: The annual per capita rice consumption in Côte d'Ivoire is approximately 91 kg (Figure 3). Although slow, the demand for parboiled rice is growing. Improving the quality and packaging it attractively in different sizes could further increase the demand for commodities in the domestic markets of Côte d'Ivoire. However, upgrading quality to appreciable levels will not occur unless conscious efforts are made to increase investment in the local rice value chain. Increasing investment in post-harvest infrastructure and improving research and development on rice varieties locally preferred by consumers is critical to sustaining the local rice industry in Côte d'Ivoire.

Presence of the Côte d'Ivoire school feeding program, especially in remote places: The school feeding program currently running in most schools in Côte d'Ivoire is a good avenue that can be used to create a market for parboiled rice. A government policy that makes it mandatory for suppliers under the school feeding program to supply the schools with mostly parboiled rice could go a long way in creating a market and demand for local rice. Linking processors to other government institutions (such as hospitals, senior high schools, and prison services) under contract supply agreements/arrangements could promote the consumption of local parboiled rice, improve processor incomes, and trigger rice production in Côte d'Ivoire.

Presence of research institutions: The National Centre for Agronomic Research (CNRA), Agency for the Development of the RIZ Sector (ADERIZ), International Research Institute (AfricaRice, IITA, etc.), Research Laboratory from the University of Felix Houphouet-Boigny, University of Alassane Ouattara, and other universities and institutions can continue to develop and scale up improved rice varieties, such as WITA, JT11, NERICA, and ORYLUX, which are preferred by consumers because of their aroma.

Well-organized women's groups or associations: The presence of existing female groups engaged in rice processing and marketing in rice hubs in Côte d'Ivoire provides an opportunity for capacity building and training in rice parboiling to be integrated into the value chain. Strengthening these groups to undertake community advocacy programs in Côte d'Ivoire could improve the local rice industry. Women and youth can take advantage of the paddy available in the communities and process it for a higher income.

Packaging and labeling of parboiled rice to make it more attractive: Poor packaging of local rice makes it less competitive in the market. Therefore, there is a need to package smaller sizes (1, 2, 5, 10, 25, 50, and 100 kg), as is performed for imported rice from Asia, Thailand, etc., so that many people can afford it conveniently. The use of quality packaging materials with good labels in Côte d'Ivoire will also help consumers make correct decisions.

Micro-project subsidies: The presence of government input subsidies and micro-enterprise finance programs in Côte d'Ivoire provides an opportunity for women and youth to go into rice cultivation for increased income and better livelihoods. The positive effects of input subsidy programs on production and gender outcomes have been documented. Improving the efficiency of production, processing, and transportation is an effective way to improve the system.

4.2.3. Paddy Supply Mode and Marketing of Local Parboiled Rice in Cote d'Ivoire

Parboilers buy paddy rice from paddy collectors and producers. Some parboilers are also paddy producers. The system of pre-financing paddy production through contracts is rare in the Odienné departments compared to Daloa and Bouaké. Mixing varieties is a common practice among suppliers. However, in Bouake and Daloa, parboilers are increasingly buying paddy by variety, and the varieties that are frequently parboiled are being improved. The main varieties were WITA 9, Bouaké AM, C15, C26, CY2 or JT2, CG 11 or JT 11, and ORYLUX6. The purchase price per kilogram of paddy rice is between 125 and 150 FCFA during the harvest period, and 160 to 200 or even 275 FCFA during shortage periods.

Traditionally, parboiled rice is less popular than imported rice. In local markets, traditionally parboiled and milled rice is sold at between 325 and 350 FCFA per kilogram, with an average price of 340 FCFA/kg. GEM parboiled rice is sold at an average of 400 FCFA per kilogram in markets in Odienné and Daloa and between 450 and 500 FCFA in Bouake, where women have greater mastery of the GEM system, as they were the first to acquire parboiling units equipped with GEM. The average market price of GEM parboiled rice was 425 FCFA/kg. For imported parboiled rice, the prices vary between 350 and 500 FCFA per kilogram. Parboiled and milled rice is sold in hulling units and local markets. In Odienné, customers are drawn from the local population and Guinea. In Daloa and Bouake, in addition to the local population, whose main customers are Malinké, customers come from Abidjan.

4.3. Some Socio-Economic Characteristics of Women Parboilers in the Study Areas

Table 3 describes the characteristics of the surveyed parboiler households in Côte d'Ivoire. Specifically, descriptive statistics showed that the women in the study areas were predominantly of Malinké ethnicity, mostly illiterate (76%), and predominantly married (79%), with an average of seven dependents per household. Approximately 83% of the women were aged $\leq$ 50 years. This highlights that parboiling is an activity practiced by relatively young women who generally have less than 20 years of practical experience in parboiling (94%). Parboiling was the main activity performed by the women (Table 3). Before the introduction of GEM in Côte d'Ivoire, the farmers had no contact with extension agents and research structures and had received no training in improved parboiling techniques. However, they had stronger relationships with millers, paddy collectors, and rice producers. Parboiled rice is intended for domestic consumption and marketing. This situation is common among professional parboilers in West Africa. However, in Odienné, parboiling is a common household practice, and parboiled rice is primarily oriented towards self-consumption as a matter of dietary habits.

Table 3. Socio-economic characteristics of rice parboilers.

Variables	Odienné (N = 88)	Daloa (N = 61)	Bouake (N = 30)	Pooled (N = 179)
Marital Status				
Married (%)	75	100	53	79.89
Divorced (%)	14.80	0	0	7.26
Widow (%)	9	0	33	10.06
Single (%)	1.20	0	14	2.79
Age (year)				
Under 30 (%)	22.70	3.30	13.30	14.52
Between 30 and 50 (%)	63.60	93.40	30	68.16
Over 50 (%)	13.70	3.30	56.70	17.32
Level of education of parboilers				
None (%)	72.62	91.51	56.70	76.30
Primary school (%)	7.14	6.78	16.70	8.67
Junior high school (6e-3e) (%)	15.48	0	16.60	10.40
Senior high school (2d-Tle) (%)	4.70	1.70	6.70	4.05
University (%)	0	0	3.30	0.58
Parboiling experience				
Less than 10 years (%)	71	44	60	59.89
Between 10 and 20 years (%)	22	56	26.70	34.46
Over 20 years (%)	7	0	13.30	5.65
Main activities				
Rice parboiling (%)	88	85	80	85.55
Agriculture (%)	3	0	0	1.67
Household size	5.4	10.6	6.23	7.41

4.4. Perception of the GEM System by Rice Parboilers in Cote d'Ivoire

The results of the focus group with parboilers, leaders of the association, and rice parboilers in each area of Côte d'Ivoire showed that women parboilers expressed great interest in using GEM technology and had very positive perceptions of the improved GEM parboiling system.

Women parboilers who used the GEM system appreciated it for several reasons. From a technical perspective, GEM is rust-resistant (stainless steel). Drying with the GEM is quicker than drying with conventional equipment for multi-batch proofing of large quantities of samples. Working with the GEM system requires less physical effort than working with the traditional system. There is no exposure to excessive heat or smoke, similar to traditional systems that use firewood. The GEM technique produces clean rice products. The breakage rate was low, and the appearance of the parboiled and milled rice was cleaner and more attractive. Rice parboiled with the GEM attracts new types of customers (middle-class, non-Malinké civil servants, and administrative executives). Consequently, it sells faster than rice parboiled using the traditional system. The GEM system repositioned local parboiled rice in the market compared to imported parboiled rice. Furthermore, GEM parboiling units have encouraged women parboilers to join formal cooperative and association actions to benefit from better quality of service and credit.

4.5. Profitability of Parboiling Activity with GEM and Traditional Equipment

Differences between rice parboiled using GEM technology and the traditional system were observed for the varietal mixture rate (5.5% vs. 20.8%, respectively), grain color alteration (27.9% vs. 59.1%, respectively), grain brightness (54.1 vs. 51.2, respectively), and impurity rate (0.2% vs. 0.8%, respectively). No differences were observed in the whole grain content (Table 4). The low level of varietal blending in the GEM parboiled rice samples is because the GEM technology training manual recommends using paddy with a high degree of varietal purity for parboiling. This is not the case in the traditional system, where the paddy is raw and unprocessed. Grain color damage was also lower for GEM parboiled rice samples than for those prepared using the traditional system. This is due to the low level of heat-damaged grains, which is related to the uniform steam distribution in the GEM system and the users' respect for the optimal cooking time, which was set to 20 min for the GEM system. In contrast, in a traditional system, the steam distribution is non-uniform, and some parts of the system receive more heat than others, resulting in uneven steaming and burnt grains.

Table 4. Quality parameters of parboiled rice using GEM technology and the traditional system.

Variables	Whole Grain Content (%)	Varietal Mix Rate (%)	Rate of Grain Color Alteration (%)	L (Brightness of Grains)	Rate of Impurities (%)
Traditional system	84.1 a *	20.8 a	59.1 a	51.2 b	0.800 a
GEM technology	85.8 a	5.5 b	27.9 b	54.1 a	0.200 b
Pr > F(Model)	0.686	0.006	0.001	0.002	0.002
Significance at 5% level	No	Yes	Yes	Yes	Yes

* Values followed by different letters in the same column are significantly different at the 5% level. Identical letters indicate no significant difference.

Moreover, the steaming time is not fixed and is generally greater than 60 min for the same quantity of sample as in the GEM system. This longer steaming time also explains why the samples parboiled in the traditional system were darker (L = 51.2) than those in the GEM system (L = 54.1).

The reduction in total impurities in GEM parboiled rice samples compared to samples parboiled according to traditional practice is due to the cleaning (winnowing and washing) of the paddy before soaking and the use of improved drying surfaces consisting of tarpaulin placed on a raised cement surface enclosed in a fence.

A comparative economic analysis of the GEM (capacity 400 kg paddy) and traditional systems was conducted for a rice parboiling session, starting from the acquisition of a 400 kg paddy to obtain the finished product (milled parboiled rice). The depreciation of parboiling equipment (GEM and traditional) was calculated based on two parboiling sessions per week (Table 5). The results show that the production cost of 1 kg of non-machined parboiled paddy is lower than that of the traditional system (189 FCFA/kg paddy) and that of the GEM system (198 FCFA/kg paddy). However, the production cost for 1 kg of milled parboiled rice was higher in the traditional system (325 FCFA/kg of milled parboiled rice) than in the GEM system (307 FCFA/kg of milled parboiled rice). This gives parboilers using the GEM system an average net additional gain of 73.09 FCFA/kg of parboiled paddy, or 73,090 FCFA per ton of parboiled paddy, compared with the traditional system (Table 6). The cost–benefit ratio results also show that for every 100 FCFA invested in GEM parboiling, the profit is 38 FCFA, compared to 4 FCFA for the traditional parboiling system (Table 6). This is a largely beneficial system for parboilers, with a 22.3% reduction in energy costs (from 175 to 39 FCFA per 100 kg of parboiled paddy) per cycle.

Table 5. Depreciation of equipment used in traditional and GEM systems.

Items	Technical Variables		Traditional System	Improved GEM System
	Total Cost (FCFA)	**Useful Time**	**Depreciation Cost in FCFA/Session**	**Depreciation Cost in FCFA/Session**
Large-scale equipment				
Traditional steaming equipment (six barrels)	75,000	2 years	391	
GEM soaking tank	1,300,000	20 years		677
GEM steaming kettle	1,000,000	20 years		521
GEM pulley and rail system	700,000	10 years		729
Two GEM rice husk stoves with accessories	500,000	5 years		1042
Small tools				
Three 60 L basins per soaking tank	24,000	6 months	500	500
Three baskets per soaking tank	9000	2 years	47	47
Three rakes per grouping	6000	2 months	375	375
One drying tarpaulin 7 m × 7 m	280,000	3 years	97	97
Total Depreciation			1410	3988

Table 6. Comparative operating account of traditional and GEM systems.

Items	Unit	Traditional System	Improved GEM System
Intermediate consumption (IC)		79,700	79,856
Raw materials (paddy)		70,000	70,000
Quantity of paddy	kg	400	400
Cost per kg paddy	FCFA/kg	175	175
Energy		700	156
Wood	FCFA/session	700	
Rice husks	FCFA/session		0
Electricity (fan)	FCFA/session		156
Husking (QP × MY × HC)		6300	7000
Milling yield (MY) (%)		63	70
Quantity of paddy (QP)	kg	400	400
Hulling cost per kg milled parboiled rice (HC)	FCFA/kg	25	25
Other costs		2700	2700
Transport (factory-point of sale)	FCFA	500	500
Equipment maintenance	FCFA	200	200
Paddy transport	FCFA	1000	1000
Transport of parboiled paddy to the factory	FCFA	1000	1000
Labor for parboiling operations (L)		900	2250
Winnowing	FCFA/session	0	600
Washing	FCFA/session	0	750
Soaking	FCFA/session	900	900
Depreciation of equipment (D)		1410	3988
Total expenses (TC)		82,010	86,094
Cost of production per kg of unmilled parboiled paddy rice	FCFA/kg	189	198
Production cost per kg of milled parboiled rice	FCFA/kg	325	307
Products			
Quantity of husked parboiled rice	kg	252	280
Price per kg of husked parboiled rice	FCFA/kg	340	425
Gross Product (GP)		85,680	119,000
Gross added value GVA = GP − IC		5980	39,144
Net value added NVA = GVA − D		4570	35,156
Net processing income NPI = NVA − L		3670	32,906

Table 6. *Cont.*

Items	Unit	Traditional System	Improved GEM System
Net additional gain with the GEM system	FCFA/session for 400 kg of parboiled paddy		29,236
	FCFA/kg of parboiled paddy		73.09
Cost/benefit ratio (NPI/TC) × 100 (%)		4.47	38.22

5. Discussion

AfricaRice introduced the GEM system to improve the physicochemical and nutritional value of paddy rice produced in Côte d'Ivoire. This study aimed to assess the technical and economic performance of the improved GEM parboiling system compared with that of the traditional parboiling system. The results showed that rice parboiling in Côte d'Ivoire is an activity practiced exclusively by women, as in other countries in West Africa, and has been passed down through generations. This result is in line with other studies showing that rice parboiling activity is mainly exerted by women on rice parboiling in Africa [1,7,9]. Women parboilers expressed a high interest in using GEM technology and had positive perceptions of the improved GEM parboiling equipment in Côte d'Ivoire. Parboilers perceived that the good quality of rice obtained, better nutritional value, and parboiling of a large quantity of paddy rice were the major advantages of the GEM system. The GEM system mainly aims to improve physicochemical and nutritional quality. Training and recent studies have highlighted the nutritional aspects of an improved GEM system [5,9–11]. The findings revealed that parboiling activity is profitable for different equipment; however, the most profitable parboiling system is the GEM. This study shows that GEM technology improves the quality of parboiled rice and increases the profit of parboilers. The GEM system reduced energy costs by 22.3%, representing an average saving of 73,090 FCFA per ton of parboiled paddy rice. The cost–benefit ratio shows that parboilers using GEM earn ten times more than those using the traditional system. These results are consistent with those of [1], who highlighted the impact of GEM on the income of rice parboilers in Benin. GEM technology enables parboilers to achieve a production level of 25 tons of high-quality parboiled rice monthly. This increased capacity can help meet the growing market demand and reduce the reliance on imports in Côte d'Ivoire. Furthermore, the GEM system has been found to have significant socio-economic benefits. These results are consistent with those reported previously [1,11]. The improved efficiency and profitability of GEM technology mean that women who traditionally engage in parboiling can achieve higher incomes, thereby improving their livelihoods and economic status. Women's empowerment aligns with broader gender equality and economic development objectives.

The ability of technology to produce rice of consistent and superior quality also opens new market opportunities, both domestically and internationally. High-quality rice with better nutritional value can attract premium prices, thereby boosting the overall competitiveness of the rice industry.

The environmental benefits of GEM technology should not be overlooked. The reduction in energy consumption by 22.3% not only translates to cost savings but also contributes to the reduction in greenhouse gas emissions. This makes GEM technology a more sustainable option, aligning with global efforts to combat climate change and promote sustainable agricultural practices in the future.

The introduction of the GEM system in Côte d'Ivoire represents a significant advancement in rice parboiling technology. The technical and economic benefits, combined with the socio-economic and environmental advantages, make GEM technology a compelling choice for rice parboilers. Continued support for the adoption and implementation of GEM technology, along with targeted training and capacity-building programs, can further enhance its impact, leading to a prosperous and sustainable rice sector in Côte d'Ivoire.

The results from understanding the perception of rice parboilers on GEM systems suggest that the GEM parboiling system should be promoted among parboilers in Côte d'Ivoire, as parboilers perceived the good quality of rice obtained, output rate, better nutritional value, and parboiling of a large quantity of paddy rice as the major advantages of the improved GEM system [1].

The presence of established female groups involved in rice processing and marketing within rice hubs in Côte d'Ivoire [9] presents an opportunity to integrate capacity building and training in rice parboiling into the value chain of rice. Additionally, grain color damage was lower in GEM parboiled rice samples than in those prepared using the traditional system. This improvement is attributed to the reduced incidence of heat-damaged grains, which results from the GEM system's uniform steam distribution [8] and adherence to an optimal cooking time of 20 min.

In general, significant improvements in rice quality using GEM technology present compelling opportunities to enhance the productivity, profitability, and sustainability of the rice industry. Embracing this technology could lead to better market positioning, economic gains, and a stronger agricultural sector in Côte d'Ivoire. Young entrepreneurs can start businesses focused on using GEM technology for rice processing, capitalizing on the improved quality and market demand for high-quality rice. Youth can innovate by creating value-added rice products, such as rice snacks, flour, and fortified rice, catering to niche markets and health-conscious consumers. The adoption of GEM technology can create new job opportunities in the rice processing sector, from machine operations to quality control and logistics. By engaging with GEM technology, youth can contribute to rural development, reduce urban migration, and support local economies. Enterprises can engage in contract farming agreements with local farmers and provide them with GEM technology and technical support to ensure a consistent supply of high-quality rice products. Enterprises can engage in contract farming agreements with local farmers, providing them with GEM technology and technical support to ensure a consistent supply of high-quality rice products.

Policymakers should intensify the transfer of GEM system activities in all parboiled rice production areas of Côte d'Ivoire through actions such as training local fabricators. This could be an opportunity for young entrepreneurs to invest in parboiling activities and make them a business venture.

6. Conclusions and Recommendations

This study assessed the technical and economic performance of an improved GEM parboiling system compared with that of a traditional parboiling system. The results show that female rice parboilers have a good perception of the GEM parboiling equipment, and that parboiling is a profitable business. The improved GEM system contributed to an increase in the profit of the parboilers. Indeed, parboilers using GEM technology earned an average of 73,090 FCFA (USD 146.18) more per ton of parboiled paddy rice than those using the traditional system did. GEM parboiling technology can achieve a production level of 25 tons of high-quality husked rice (lighter, more uniform color, absence of damaged grains and impurities, and low levels of broken fractions) monthly. The GEM system can contribute to upgrading the rice value chain in Côte d'Ivoire. The production of high-quality rice with consistent and superior characteristics (e.g., lighter color, more uniform color, and fewer impurities) can open new domestic and international market opportunities. Enterprises using GEM technology can achieve higher profitability, with parboilers earning an average of 73,090 FCFA (USD 146.18) more per ton of parboiled paddy rice than those using traditional methods. Young entrepreneurs can establish new businesses focused on rice processing using GEM technology. These startups can leverage the improved efficiency and profitability of GEM technology to capture market share. Youth can innovate by creating and marketing value-added rice products, such as rice flour, rice snacks, and ready-to-eat rice dishes, which cater to niche and premium markets. The adoption of GEM technology will create new jobs in rice processing plants, offering employment opportunities for the youth in rural and urban areas. The adoption of GEM technology in rice processing presents numerous opportunities for youth and enterprises in Côte d'Ivoire. By embracing these opportunities, they can contribute to the growth of the agricultural sector, enhance food security, and drive economic development.

To meet the growing demand for quality parboiled rice, we need to (i) raise producers' awareness of quality paddy production, (ii) create parboiling centers equipped with the GEM system and high-performance hullers, and (iii) strengthen the technical capacities of parboilers through training in good parboiling practices, labeling, marketing, financial and accounting management, and cooperative life. In other words, given the potential of this new technology and the enthusiasm of women for its use, the government should step up its efforts to transfer GEM technology to all rice-producing areas across Côte d'Ivoire.

Studies also need to be conducted to ascertain consumer perceptions and preferences (organoleptic assessment) of GEM versus traditional parboiled rice and their willingness to pay for rice parboiling with GEM technology.

Author Contributions: Conceptualization, A.L.K., R.A. and M.E.D.; methodology, A.L.K., R.A., M.E.D. and A.A.; formal analysis, A.L.K. and R.A.; data curation and analysis, A.L.K., R.A., M.E.D., A.A. and S.A.N.; writing—original draft preparation, A.L.K., R.A. and M.E.D.; writing—review and editing, A.L.K., R.A., A.A. and S.A.N. All authors have read and agreed to the published version of the manuscript.

Funding: This research was funded by FIRCA (Fonds Interprofessionnel pour la Recherche et le Conseil Agricoles), AfricaRice (Africa Rice Centre), the "Transforming Agri-Food System for West and Central Africa (TAFS-WCA)" initiative.

Institutional Review Board Statement: Not applicable.

Informed Consent Statement: Consent of parboilers were obtained before proceeding with the data collection.

Data Availability Statement: The raw data supporting the conclusions of this study will be made available by the authors upon request.

Acknowledgments: The authors would like to thank the FIRCA (Fonds Interprofessionnel pour la Recherche et le Conseil Agricoles), AfricaRice (Africa Rice Centre), the "Transforming Agri-Food System for West and Central Africa (TAFS-WCA)" initiative, and ANADER (Agence Nationale d'Appui au Développement Rural) for their invaluable contributions to this project. The financial support for this endeavor was provided by the FCIAD (Fonds Compétitif pour l'Innovation Agricole Durable) through the C2D (Contrat de Désendettement et de Développement) programme.

Conflicts of Interest: The authors declare that the research was conducted in the absence of any commercial or financial relationships that could be construed as potential conflicts of interest.

Abbreviations

AfricaRice	Africa Rice Centre
ADERIZ	Agence pour le Développement de la filière Riz
ANADER	Agence Nationale d'Appui au Développement Rural
CNRA	Centre National de Recherche Agronomique
C2D	Contrat de Désendettement et de Développement
FCIAD	Fonds Compétitif pour l'Innovation Agricole Durable
FIRCA	Fonds Interprofessionnel pour la Recherche et le Conseil Agricoles
GEM	Grain quality enhancer, Energy efficient and durable Material
TAFS-WCA	Transforming Agri-Food System for West and Central
USDA	United States Department of Agriculture

References

1. Arouna, A.; Aboudou, R.; Ndindeng, S.A. The adoption and impacts of improved parboiling technology for rice value chain upgrading on the livelihood of women rice parboilers in Benin. *Front. Sustain. Food Syst.* **2023**, *7*, 1066418. [CrossRef]
2. Arouna, A.; Fatognon, I.A.; Saito, K.; Futakuchi, K. Moving toward rice self-sufficiency in sub-Saharan Africa by 2030: Lessons learned from 10 years of the Coalition for African Rice Development. *World Dev. Perspect.* **2021**, *21*, 100291. [CrossRef] [PubMed]
3. Demont, M.; Rutsaert, P.; Ndour, M.; Verbeke, W. Reversing urban Bias in African Rice markets: Evidence from Senegal. *World Dev.* **2013**, *45*, 63–74. [CrossRef]
4. Zohoun, E.V.; Ndindeng, S.A.; Soumanou, M.M.; Tang, E.N.; Bigoga, J.; Manful, J.; Sanyang, S.; Akissoe, N.H.; Futakuchi, K. Appropriate parboiling steaming time at atmospheric pressure and variety to produce rice with weak digestive properties. *Food Sci. Nutr.* **2018**, *6*, 757–764. [CrossRef] [PubMed]
5. Ndindeng, S.A.; Futakuchi, K.; Ndjiondjop, M.N. Screening of Rice germplasm and processing methods to produce low glycemic Rice. *Rice Sci.* **2022**, *29*, 101–104. [CrossRef]
6. Fall, A.A. Chaine de valeur riz en Afrique de l'Ouest: Performance, enjeux et défis en Côte d'Ivoire, Guinée, Libéria, Mali, Sierra Léone et Sénégal. *J. Appl. Biosci.* **2018**, *130*, 13175–13186. [CrossRef]
7. Dandedjrohoun, L.; Diagne, A.; Biaou, G.; N'cho, S.; Midingoyi, S. Determinants of diffusion and adoption of improved technology for rice parboiling in Benin. *Rev. Agric. Environ. Stud.* **2012**, *93*, 171–191.
8. Sodjinou, E. *Guide Pratique D'analyse Financière d'une Entreprise Agricole: Théorie et Application à la Pisciculture*; Institut National des Recherches Agricoles du Bénin (INRAB) Eds.: Ministère de l'Agriculture, de l'Elevage et de la Pêche:: Porto-Novo, Benin, 2016; 68p.
9. Depieu, M.E.; Kanon, A.L.; Adolphe, M.G.; Koffi, C.; Ndingeng, S.A. L'inventivité des femmes transformatrices du riz: Les unités d'étuvage de Bouaké, Daloa et Odienné (Côte d'Ivoire). *Afr. Contemp.* **2023**, *275*, 125–136. [CrossRef]
10. Etoa, J.M.A.; Ndindeng, S.A.; Owusu, E.S.; Woin, N.; Bindzi, B.; Demont, M. Consumer valuation of an improved rice parboiling technology: Experimental evidence from Cameroon. *Afr. J. Agric. Resour. Econ.* **2016**, *11*, 8–21. [CrossRef]
11. Zossou, E.; Fiamohe, R.; Vodouhe, S.D.; Demont, M. Experimental auctions with exogenous and endogenous information treatment: Willingness to pay for improved parboiled rice in Benin. *J. Agric. Econ.* **2022**, *73*, 806–825. [CrossRef]

proceedings

MDPI

Proceeding Paper

Post-Harvest Losses Along the Main Value-Added Chains and Strategies for Reduction in the Soybean Sector in Benin †

Daniel Missimahou Noukpozounkou [1,*], Roméo Sossou [1], Hervé Comlan Sossou [1], Juvénal Privaël Koffi [2], Abel Hotegni [3], Valère Dansou [3], Alfred Akpado Oluwatogni Ayedoun [1], Symphorien Dossouhoui [1] and Soul-Kifouly Midingoyi [1]

1 Agricultural Policy Analysis Program (PAPA), Agonkanmey Agricultural Research Center, National Agricultural Research Institute of Benin (INRAB), Cotonou 01 BP 884, Benin; sossour25@gmail.com (R.S.); chsossou@gouv.bj (H.C.S.); alfredayedoun@gmail.com (A.A.O.A.); dossouhouisymphorien@gmail.com (S.D.); zoulkm2000@yahoo.fr (S.-K.M.)

2 Department of Economics, Socio-Anthropology, and Communication for Rural Development (EESAC), Faculty of Agronomic Sciences, University of Abomey-Calavi, Cotonou 01 BP 526, Benin; koffi.pjuvenal@gmail.com

3 Agricultural and Food Technologies Program (PTAA), Agonkanmey Agricultural Research Center, National Agricultural Research Institute of Benin (INRAB), Cotonou 01 BP 884, Benin; abel2511@live.fr (A.H.); vadansou@gmail.com (V.D.)

* Correspondence: danimaster81@yahoo.fr

† Presented at the CORAF's 2023 Symposium on Processing and Transformation of Agricultural Products in West and Central Africa: Achievements and Opportunities for Private Sector Engagement, Lome, Togo, 21–23 November 2023.

Abstract: The aim of this study was to provide information on post-harvest losses (PHLs) along the main value-added chains in the soybean industry, and on farmers' strategies for reducing these losses in Benin. Data were collected using a subjective method (questionnaires) from 152 direct actors in the soybean value chain and an objective method (technological monitoring using technological monitoring sheets) from 27 processing units using a two-stage sampling technique in eighteen of the country's municipalities. The data were analyzed using descriptive statistics. The results revealed that post-harvest losses in soybean value chains ranged from 0.03% to 7.98%. The highest loss percentages were 7.98% (production), 10.0% (processing), and 4.27% (marketing), and were obtained during the subjective measurements. The link most affected by PHL was the production link. These losses were mainly observed during threshing, winnowing, sorting, and harvesting operations. The main causes of these losses were poor mechanization and a lack of skills for post-harvest operations. The main strategies developed by players to cope with this situation remained traditional (timely harvesting, use of appropriate equipment, and good storage practices). However, additional efforts in terms of infrastructure, adapted subsidies/credits, and recruitment of specialists are needed to mechanize the operations most sensitive to losses and to reinforce the capacities of agricultural actors in order to considerably reduce post-harvest losses.

Keywords: post-harvest losses; technological monitoring; soybean value chains; strategies; Benin

1. Introduction

Population growth, which is more prevalent in developing countries, is driving up food demand. Paradoxically, more than half the food produced is wasted or lost in post-harvest operations [1]. One of the crucial strategies to enhance food security and to promote

https://doi.org/10.3390/proceedings2025118010

sustainable development mainly in rural areas is to mitigate such wastes and losses. Indeed, reducing post-harvest losses (PHLs) in this way would make a lasting contribution to fighting hunger, raising incomes, and improving food security in developing countries [2]. However, the FAO and the World Bank estimated that between 10% and 20% of the volume of grain produced in sub-Saharan Africa is lost every year, for an amount of USD 4 billion corresponding to the minimum annual needs of around 48 million people [3]. In Benin, preliminary surveys suggest a considerable soybean PHL due to inadequate storage and processing infrastructure, though precise figures remain undocumented. This underscores the need for urgent and well-targeted strategies that can boost post-harvest efficiencies.

In Benin, soybean is a highly important commercial crop. It accounted for around 62% of the national legume production [4,5]. Production rose from 57,000 tons in 2009 to 253,954 tons in 2020, for a total soybean export value of USD 66.6 million in 2020 [6]. This production makes Benin the second largest soybean producer in West Africa after Nigeria [7]. Soybeans rank fourth among Benin's export crops, after cotton, pineapple, and cashew nuts [8]. Soybean has emerged as a pivotal sector in enhancing rural livelihoods and stimulating economic activities across Benin. Moreover, this legume plays an important role in reducing malnutrition, as it is a source of vegetable protein for human and animal consumption. In rural areas, given its richness in magnesium, vitamins, and mineral salts, soy is a substitute for meat, fish, and eggs [9]. Due to its nutritional diversity and economic importance, soybean is a core component of the nutritional and socio-economic framework of rural communities.

Given soybean's place in Benin's socio-economic sphere, finding a solution to reduce or even prevent PHL is imperative. Beyond farmers, PHL in soybeans jeopardizes the livelihoods of traders, processors, and rural households who are dependent on affordable protein, amplifying economic vulnerabilities across Benin's agrarian economy. However, the available literature on this crop focuses more on soil fertility improving properties [10–12], improved varieties developed [13,14], variety adoption factors [15,16], and its contribution to improving farmers' livelihoods [17,18]. This specific focus reveals a significant research gap concerning the critical post-harvest stage of the soybean value chain. Thus, there is a lack of knowledge regarding the aspects related to post-harvest losses of soybeans in Benin. Soybeans have been the subject of very limited research on post-harvest losses despite the critical importance of generating information to identify sensitive stages requiring intervention. Such information is also useful in political decision making and designing effective strategies to reduce both quantitative and qualitative losses. Addressing this gap is then imperative for the formulation of evidence-based policies that can transform the efficiency of soybean processing and distribution networks in Benin. The present article fills this gap in the literature on PHL in the soybean sector. Its aim was to provide insights into the percentages and critical loss stages of post-harvest losses (PHLs) along the main soybean value chains, as well as farmers' strategies for reducing these losses in Benin. This study aligns with emerging frameworks on sustainable food systems, viewing PHL reduction as a nexus of economic, nutritional, and social outcomes in rural developing contexts.

2. Materials and Methods

The general methodology for this study was inspired by the FAO [2] recommendations, emphasizing the combination of subjective and objective approaches to measure post-harvest losses (PHLs). A two-stage sampling method was used: purposive sampling (to select municipalities) and random sampling (to select producers, processors, and traders). The choice of municipalities was based on their soybean production levels within the Agricultural Development Poles (PDAs).

2.1. Study Area

The study was conducted in the main soybean-producing municipalities in Benin, distributed across six of the country's seven Agricultural Development Poles—PDAs (Figure 1). Eighteen municipalities were selected based on their significant roles in various soybean value chains (Table 1). Within each municipality, villages and/or urban neighborhoods were randomly selected for collecting subjective data, while objective measurements were purposively conducted.

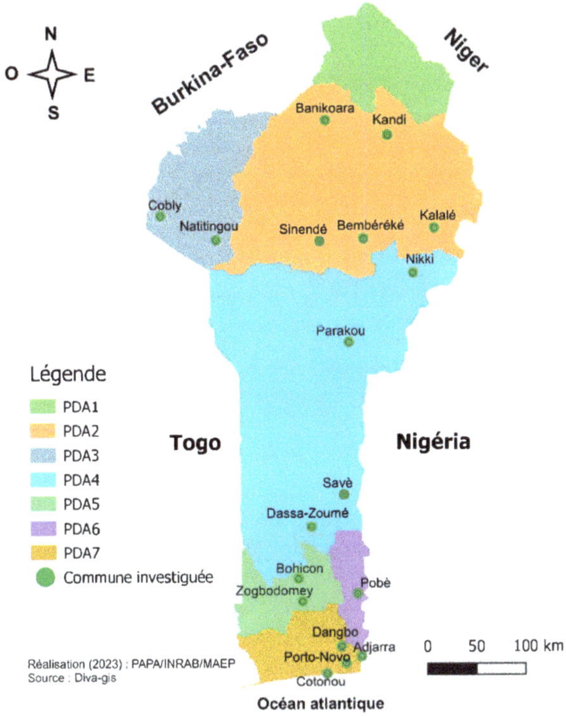

Figure 1. Map of area covered by the study.

Table 1. Sampling procedure and sample size.

PDA	Municipalities	Number of Respondents		
		Producers	Processors	Traders
PDA2 « Alibori-Sud, Borgou-Nord et 2KP »	Bembéréké, Kalalé, Banikoara, Sinendé, and Kandi	38	5	5
PDA3 «Atacora-Ouest»	Cobly and Natitingou	40	5	4
PDA4 «Borgou-Sud, Donga et Collines»	Nikki, Parakou, Savè, and Dassa-Zoumé	10	3	2
PDA5 « Zou et Couffo»	Bohicon and Zogbodomey		9	
PDA6 «Plateau»	Pobè		5	5
PDA7 «Ouémé, Atlantique et Mono»	Cotonou, Porto-Novo, Dangbo, and Adjarra		9	12
Total	18	88	36	28

2.2. Sampling

Simple random sampling was used to select a sample size of 152 respondents for this study, calculated using Yamane's formula [19] (Equation (1)). A total of 88 producers, 36 processors, and 28 traders were surveyed for subjective assessments. This sampling method was chosen due to the accessibility, collaboration, and spatial distribution of the actors across the PDAs. The sampling process involved two steps:

1. Based on the list of soybean-producing municipalities obtained from the Department of Agricultural Statistics, those with high production levels were selected (as shown in Table 1). The research unit consists of producers, processors, and traders from the selected municipalities.
2. The 152 respondents were distributed as outlined in Table 1.

Among the 152 respondents, five soybean processors per value chain unit were selected for technological monitoring to collect objective data, excluding the soybean grain value chain since the practices are nearly identical [20]. The municipalities of Dassa-Zoumè, Zogbodomey, Bohicon, Porto-Novo, and Adjarra were chosen for objective measurements within the processing segment.

$$n_o = \frac{N}{1 + N(e^2)} \tag{1}$$

where

n_o: Minimum sample size for each actor category;
N: Population size for each actor category;
e^2: Margin of error, fixed at 5%.

2.3. Selection of the Value Chains

According to the FAO [2], the evaluation of post-harvest losses (PHLs) along value chains must be preceded by the characterization and comprehensive analysis of the chains, which allows the identification of the processes most likely to result in post-harvest losses. The inventory of all the soybean value chains (SVCs) led to the identification of nine SVCs: (i) soybean grain SVC, (ii) soybean cheese SVC, (iii) soybean flour SVC, (iv) soybean milk SVC, (v) soybean mustard (afitin) SVC, (vi) soybean oil SVC, (vii) soybean cake SVC, (viii) soybean biscuit SVC, (ix) soybean meal SVC [21,22]. For the selection of SVCs in this study, two criteria were used: the priorities defined in the National Soybean Development Program (PNDF) and the significance of the SVCs within the soybean sector in terms of their contribution to local and even national wealth creation. Consequently, four SVCs were selected for this study: soybean grain SVC, soybean cheese SVC, soybean flour SVC, and soybean milk SVC.

2.4. Data Collection and Analysis Methods and Tools

Four data collection techniques were employed. First, a literature review grid was used to conduct a document review. Second, an exploratory survey was carried out using a semi-structured interview guide with around twenty respondents randomly selected by SVC outside the main sample. The key points of the interviews and the information gathered focused on the main value chains and the stakeholders involved in each sector, the stages and operations prone to post-harvest losses (PHLs) at each SVC link, storage and preservation practices, the causes and/or factors contributing to post-harvest losses, the estimated values of these losses, and strategies for reducing them. The results from this phase were used to finalize the questionnaires and objective measurement sheets for the actual data collection process. Thirdly, a quantitative survey was conducted using a structured questionnaire to collect subjective data from soybean producers, processors, and traders. Finally, a technological monitoring sheet was used during participatory observations (technological monitoring) within processing units to collect objective measurements. During these technological monitoring sessions conducted in March 2022, quantitative losses were recorded at each processing operation. This combination of subjective and objective approaches is one of the FAO's recommendations for measuring post-harvest losses (PHLs). For each SVC, three repetitions were carried out during the technological monitoring in different municipalities. The measurement equipment, consisting of a 50 kg capacity scale (Zhongshan Camry Electronic Co., Ltd., Guangdong, China) with 1 g precision (Figure 2), was used to measure the initial and final quantities of the product at each operation. At the end of each transformation process step, the quantities of products left behind or discarded by the processor were collected and weighed using the 1 g precision scale (Yongkang Nengzhi Industry & Trade Co, Yongkang, China). These collected or discarded quantities then represent the objective losses recorded at that operation.

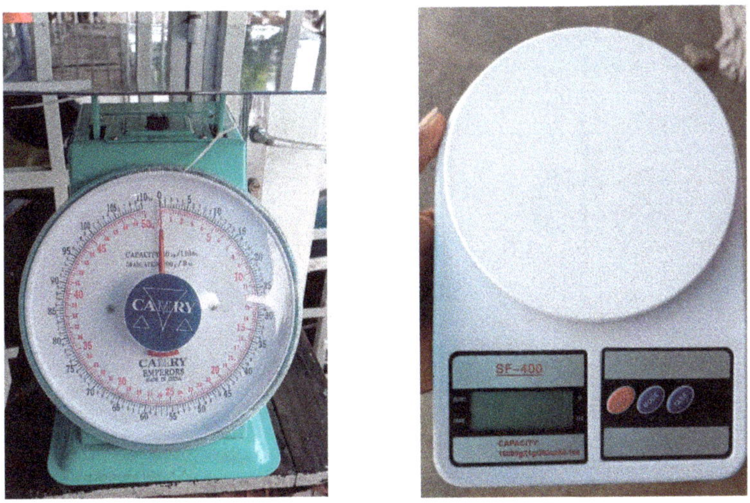

Figure 2. Photos of a 50 kg range scale (**left**) and a 1 g precision scale (**right**).

The collected data were subjected to statistical analysis using IBM SPSS Statistics v.25.0 software. This analysis involved calculating frequencies, means, and standard deviations.

2.5. Determination of Post-Harvest Loss (PHL) Indicators

The loss indicators were calculated following the FAO's recommendations for the different stages and selected value chains [23]. Thus, both quantitative (or weight-based) losses and relative losses (loss rate) were determined. Quantitative losses, expressed in kilograms, were collected from agricultural operations (subjective measurements) and obtained through objective (or physical) measurements.

Relative losses (loss rate) are expressed as percentages and measure the intensity of losses occurring during various harvesting and post-harvest operations. They are obtained by dividing the quantity lost at each operation by the quantity handled. For instance, the drying loss rate is calculated by dividing the quantity of grains lost during drying by the quantity of grains dried (Table 2). The use of the quantities handled (in this case, the quantities subjected to drying) as the denominator ensures that the loss rates are between 0 and 100 percent (Formula (2)).

Table 2. T Formulas for determining PHLs by level.

Levels	Formulas		Variables
Operation	$TPO = \frac{QPO}{QMO} \times 100$	(2)	TPO: Rate of loss during operation QPO: Quantity lost during operation QMO: Quantity handled during operation
Value chain	$TPCVA = \sum_{i=1}^{n} TPO$	(3)	TPCVA: Loss rate per value chain n: Number of value chain transactions TPO: Loss rate per transaction

Source: [22].

The determination of the most sensitive stages to losses was performed. The value chains most sensitive to losses were evaluated using Formula (3). For each value chain, losses were determined by summing the losses recorded at each individual operation.

3. Results

The socio-economic profiles of the respondents, the key values and loss hotspots across the value chain stages of the soybean sector, and the strategies employed by stakeholders to mitigate post-harvest losses constitute the main findings of this study.

3.1. Socio-Economic and Demographic Characteristics of Respondents

The socio-economic and demographic characteristics are presented by the value chain stage or stakeholder category (producers, processors, and traders).

3.1.1. Soybean Producers

Soybean production was predominantly a male-dominated activity (87.1%). Most soybean producers were married (92.1%) and lacked any form of literacy (76.1%) (Table 3). The respondents had, on average, seven years of experience in soybean production and were generally not affiliated with any soybean producer organization (4.6%). Additionally, soybean production contributed slightly less than half (46.4%) to the agricultural income of producers. This contribution was slightly higher for producers in PDA3 (55.7%). Across all the surveyed PDA zones, about four out of ten household members were engaged in agricultural activities.

Table 3. Socio-economic and demographic characteristics of soybean producers.

Variables		PDA2	PDA3	PDA4	Total
Gender	Female	15.79	12.50	0	12.50
	Male	84.21	87.50	100.00	87.50
Age (mean)		43.66 ± 12.61	36.48 ± 9.21	49.4 ± 14.62	41.05 ± 12.18
Marital status (%)	86.84	95.00	100.00	92.05	
	2.63	5.00	0	3.41	
	10.53	0.00	0	4.55	
Average contribution of soybean to farm income (%)		36.20	55.70	48.00	46.40
Literacy level (%)	None	71.05	85.00	60.00	76.14
	Ability to read	0	2.50	0.00	1.14
	Ability to write	0	2.50	10.00	2.27
	Ability to read and write	28.95	10.00	30.00	20.45

Table 3. *Cont.*

Variables	PDA2	PDA3	PDA4	Total
Average number of dependents	12.82 ± 9.26	5.95 ± 3.27	14.8 ± 6.37	9.92 ± 7.69
Average number of farm workers in the household	4.32 ± 3.04	2.78 ± 1.98	4.3 ± 2.58	3.61 ± 2.64
Years of experience in soybean production	9.24 ± 8.64	4.8 ± 2.91	9.5 ± 3.86	7.25 ± 6.50
Membership of cooperative or producer group (%)	10.53	0.00	0.00	4.55

3.1.2. Soybean Processors

Soybean processing was primarily undertaken by women (94.4%) (Table 4). Both male and female processors were adults, with an average age of 42 years, and approximately two-thirds were illiterate. The respondents had an average of 11 years of experience in soybean processing, which contributed about 68% to their agricultural income. The main final products obtained from processing were soybean cheese (69.4%) and soybean porridge (16.7%).

Table 4. Socio-economic and demographic characteristics of soybean processors.

Variables		PDA2	PDA3	PDA4	PDA5	PDA6	PDA7	Total
Gender	Female	100	100	66.67	88.89	100	100	94.44
	Male	0.00	0.00	33.33	11.11	0.00	0.00	5.56
Age (mean)		41 ± 12.18	32.8 ± 4.96	53 ± 9.64	43.67 ± 8.37	39.8 ± 7.43	46.60 ± 11.33	42.08 ± 10.21
Marital status (%)	Married							
	Single							
	Widowed							
Average contribution of soybean processing to agricultural income (%)		60.0	68.0	36.66	72.22	74	74.44	67.77
Literacy level (%)	None	19.23	15.38	0.00	23.08	15.38	26.92	72.22
	Ability to read	0.00	0.00	33.33	33.33	0.00	33.33	8.33
	Ability to read and write	0.00	14.29	28.57	28.57	14.29	14.29	19.44
Years of experience in soybean processing		8.80 ± 4.54	5.88 ± 3.80	11.67 ± 4.16	9.22 ± 2.11	9.80 ± 5.45	16.0 ± 15.64	10.67 ± 8.70
Membership of processors organizations (%)		0.00	0.00	100.00	22.22	0.00	0.00	13.89
Main final product sold (%)	Soy cheese	80.00	100	33.33	77.78	80.00	44.44	69.44
	Soy milk	0.00	0.00	33.33	11.11	0.00	0.00	5.56
	Soy porridge	0.00	0.00	0.00	0.00	20.00	55.56	16.67
	Soy flour	20.00	0.00	33.33	11.11	0.00	0.00	8.33

3.1.3. Soybean Traders

The majority of the traders were illiterate (82.1%), and three-quarters (75.0%) were women (Table 5). Their average experience in soybean trading was 14 years across all the PDA zones, although it was relatively lower in PDA3 (5.5 years). On average, soybean trading contributed 50% to the agricultural income of the respondents, and membership in trader organizations was notably low (3.6%).

Table 5. Socio-economic and demographic characteristics of soybean traders.

	Variables	PDA2	PDA3	PDA4	PDA6	PDA7	Total
Gender	Female	60.00	75.00	50.00	100.00	75.00	75.00
	Male	40.00	25	50.00	0.00	25.00	25.00
Age (mean)		41.82 ± 2.19	27.50 ± 2.08	54.50 ± 7.77	37.80 ± 2.17	46.08 ± 13.98	41.82 ± 12.19

Table 5. *Cont.*

Variables		PDA2	PDA3	PDA4	PDA6	PDA7	Total
Average contribution of soybean to agricultural income (%)		53.42	47.50	50.00	64.00	54.10	53.20
Literacy level (%)	None	80.00	100.00	100.00	60.00	83.33	82.14
	Ability to read	20.00	0.00	0.00	0.00	0.00	3.57
	Ability to read and write	0.00	0.00	0.00	40.00	16.67	14.29
Years of experience in soybean marketing		13.50 ± 9.21	5.5 ± 2.65	15 ± 14.14	11.4 ± 4.83	17.00 ± 8.40	13.57 ± 9.11
Membership of cooperatives or organizations (%)		0.00	0.00	50.00	0.00	0.00	3.57

3.2. *Values and Critical Stages of PHL*

At the three stages of the soybean value chain, post-harvest losses ranged from 0.03% to 7.98%. The production and marketing stages recorded the highest (7.98%) and lowest (0.03%) loss rates, respectively.

3.2.1. Soybean Production Stage

The losses at the production stage of the soybean value chain are estimated at 7.9% (Figure 3). The highest loss rates occurred during threshing, winnowing, and harvesting (2.0%, 1.7%, and 1.4%, respectively). The lowest loss rates were recorded during sorting, drying, transportation from field to home, storage/packaging, and transportation to market/distribution (0.8%, 0.8%, 0.7%, 0.4%, and 0.1%, respectively).

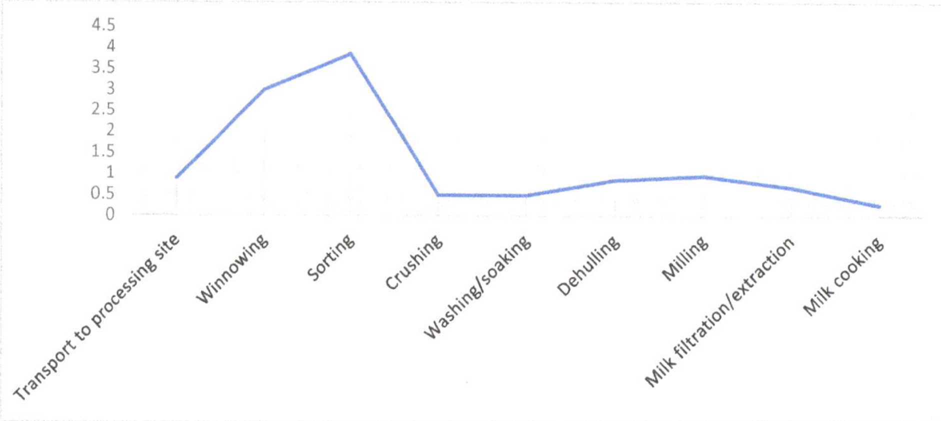

Figure 3. PHL in the soybean production operations.

3.2.2. Soybean Processing Stage

- Soybean cheese value chain

The subjective and objective post-harvest loss (PHL) rates for the soybean cheese value chain are illustrated, with significant losses observed during winnowing (3% for subjective losses and 0.5% for objective losses) and sorting (3.9% for subjective losses and 1.2% for objective losses) (Figure 4).

- Soya milk value chain

The subjective and objective post-harvest loss (PHL) rates for the soya milk value chain show that for the subjective measures, losses were observed during grain transport, sorting/winnowing, soaking, filtration/milk extraction, and sterilization, with rates of 2%, 2.8%, 1%, 2.2%, and 4%, respectively. For the objective measures, losses were primarily seen during sorting/winnowing and sterilization, with rates of 1.4%, 1.1%, and 2.4%, respectively (Figure 5).

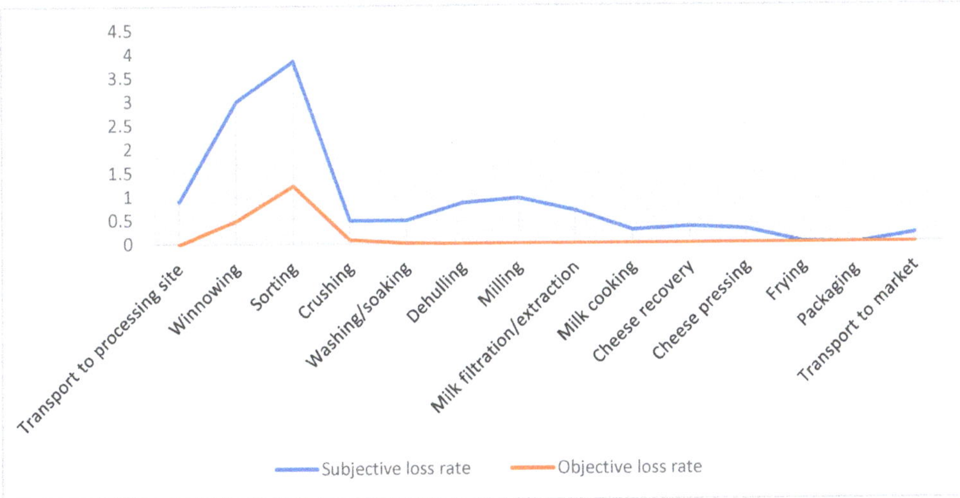

Figure 4. PHL values in the soybean cheese value chain processing operations.

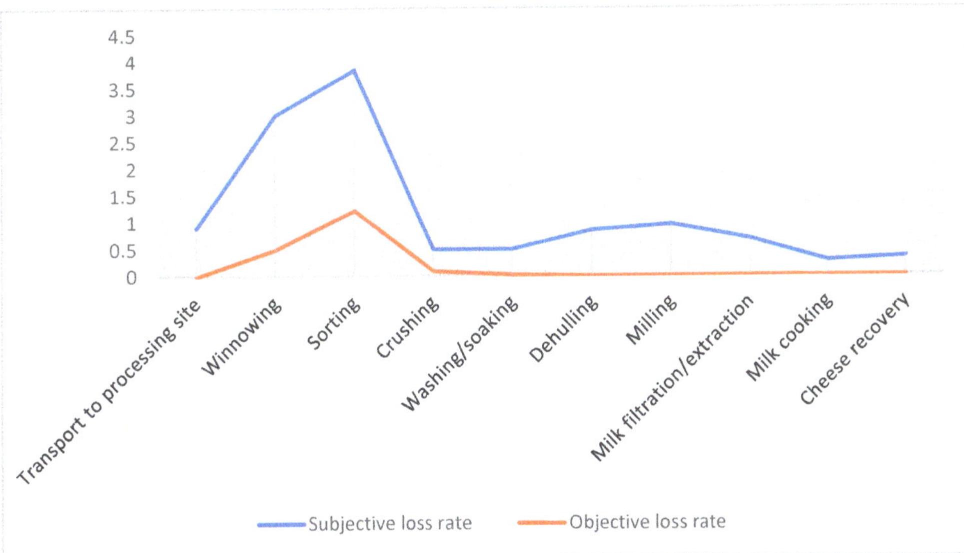

Figure 5. PHL values in the soya milk value chain processing operations.

- Soybean flour value chain

The subjective and objective post-harvest loss (PHL) rates for the soybean flour value chain reveal that subjectively, the stages of sorting/winnowing, roasting, dehulling, and milling were the most prone to losses, with rates of 1.7%, 1%, 1%, and 1.8%, respectively. Objectively, the highest losses occurred during winnowing/sorting (0.7%) and milling (0.4%) (Figure 6).

Figure 6. PHL values in the soya flour value chain processing operations.

3.2.3. Soybean Trading Stage

Soybean traders were grouped into retailers, semi-wholesalers, and wholesalers. Figure 7 illustrates the PHL rates during the marketing stage. The semi-wholesalers recorded the highest average loss rate (0.9%) compared to the retailers (0.2%) and wholesalers (0.3%). The semi-wholesalers experienced the highest losses during sorting (6.5%), whereas the wholesalers and retailers recorded lower losses, not exceeding 1%. The wholesalers experienced their highest losses during procurement (0.9%), while the retailers were the least loss-prone, with 0.5% recorded during sorting.

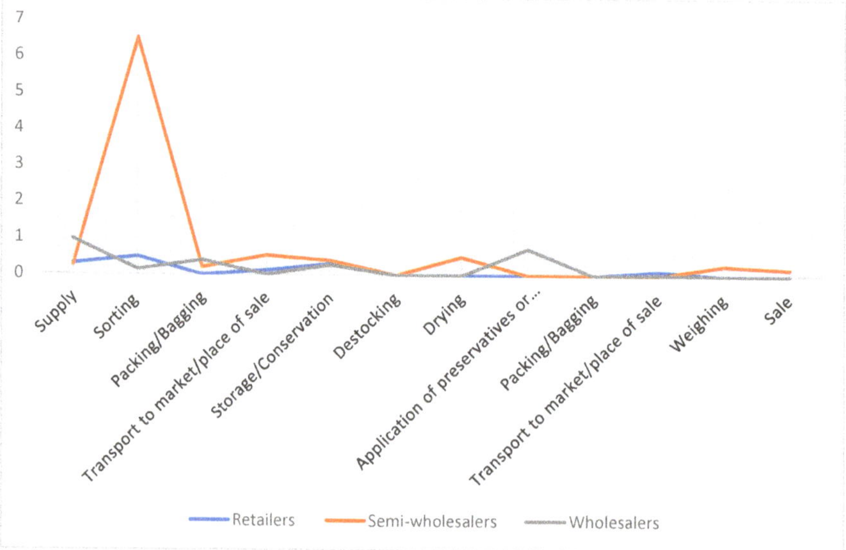

Figure 7. PHL values in the soybean value chain in trading operations.

3.3. *Strategies for Reducing PHL at Critical Stages*

3.3.1. Soybean Production Stage

The majority of the soybean producers (86.4%) identified timely harvesting as the best strategy for reducing PHL (Figure 8). Other strategies included using proper bags (preferably jute) and ensuring that storage areas were free from direct contact with walls, roofs, or other structures to prevent temperature fluctuations. Proper winnowing, timely threshing, and using appropriate tarpaulins were also cited as effective measures.

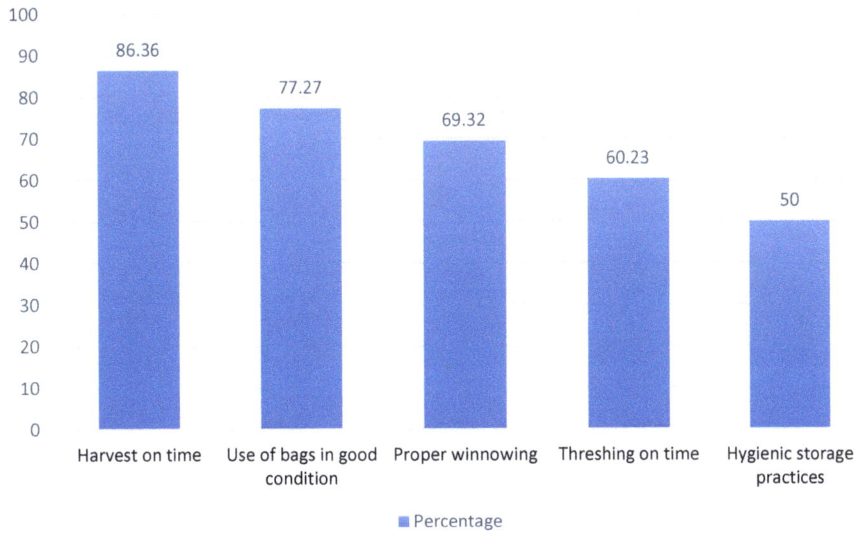

Figure 8. Main strategies for reducing losses in the soybean production stage.

3.3.2. Soybean Processing Stage

Using appropriate equipment (66.7%), mechanizing operations, and completing tasks quickly (45.5%) were the most used strategies by the processors (Figure 9).

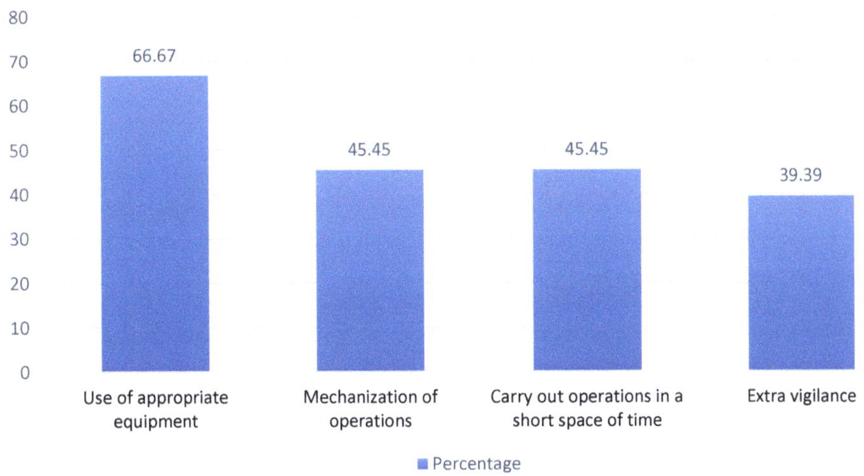

Figure 9. Main strategies for reducing losses in the soybean processing stage.

3.3.3. Soybean Trading Stage

To reduce PHL in this sector (Figure 10), the traders made more use of good storage practices (39.3%). This included ventilation if the stocks were to be used as seed. To reduce losses even further, good collection was necessary (28.6%), with the aim of sorting good grain from bad. It was essential to use bags that were in good condition (25.0%) in order to reduce transport losses. On-time delivery (21.4%) reduced losses when the customer was satisfied on time. It was also very important, according to some retailers, to protect the bags with tarpaulins during transport (21.4%) to prevent them from deteriorating during the journey.

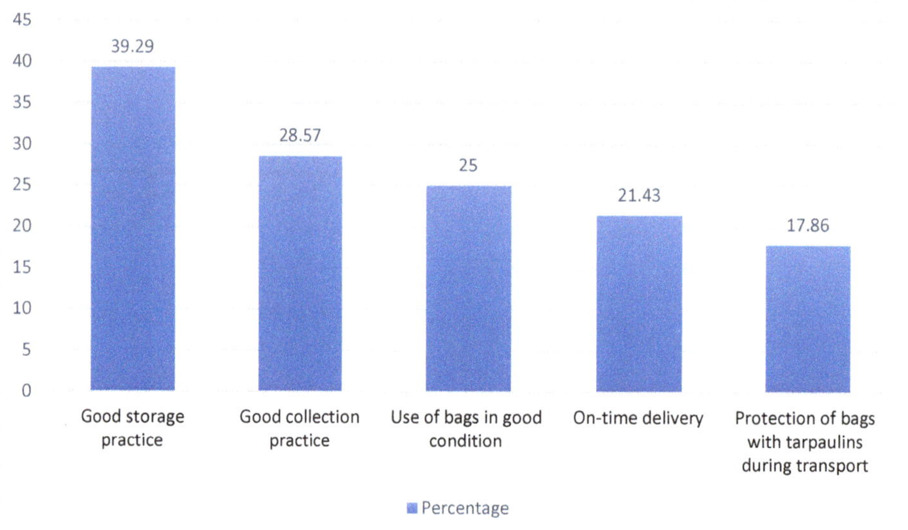

Figure 10. Main strategies for reducing losses in the soybean trading stage.

4. Discussion

Reducing post-harvest losses (PHLs) is an effective method for ensuring food security. It also enables farmers in disadvantaged regions to fully exploit the potential of their production.

4.1. Values and Critical Points for PHL

The study revealed that the operations most responsible for post-harvest losses in soybean production, regardless of the chain segment (production, processing, and marketing), include threshing, winnowing, sorting, and harvesting. These results align with Atiim [24], who found that more than half of the soybean farmers assessed in Ghana identified the aforementioned steps in soybeans as sensitive.

Similarly, research by Paulsen et al. [25] on the causes of post-harvest losses in developing countries identified threshing and winnowing as the most vulnerable stages of PHL. This issue is due to the lack of mechanization in post-harvest operations. According to these authors, the mechanization of operations with skilled labor can reduce soybean post-harvest losses by 75% to 100%. In Benin, producers still rely on traditional equipment and materials for these operations, and these activities remain largely manual.

It is important to note that winnowing and sorting operations are critical points for losses at all levels of the soybean production chain. This highlights the issue of insufficiently qualified personnel dedicated to these tasks, which also contributes to higher loss rates, as noted by Paulsen et al. [25]. The work of Arends-Kuenning et al. [26] corroborates this observation, showing that a lack of skill in using harvesters was the main cause of post-harvest losses in Brazil.

However, contrary to the findings of Caixeta-Filho and Péra [27], the present study did not identify transportation as a critical point for post-harvest losses. This may be attributed to policy efforts aimed at improving the condition of roads connecting various regions in the country or the proximity of markets.

4.2. Strategies for Reducing Critical Points for PHL

This study highlights several significant challenges in minimizing post-harvest losses in the soybean production chain. First, although timely harvesting is recognized as an effective strategy, the use of inappropriate tools during harvesting is a major cause of losses. This indicates the need to create and promote appropriate harvesting tools that minimize product damage. Additionally, processors wish to improve the quality of equipment and mechanize various steps in the processing process. However, these improvements require significant investment, which is beyond the capacity of most actors in the chain. As demonstrated by Paulsen et al. [25], most non-mechanized threshing/cleaning systems in developing countries like Benin lack adequate means for preserving harvested grains. These results underscore the need to develop financing and support mechanisms that make these investments more accessible.

It is evident that the current approaches implemented by traders are not sufficiently effective in reducing post-harvest losses. Several factors may contribute to this situation. On the one hand, the harvesting methods themselves may not be effective enough to reduce post-harvest losses. This suggests that a combination of methods is needed to be effective. For example, research by Coradi et al. [27] evaluated different sustainable strategies for managing soybean grain mass based on moisture content, optimizing drying and storage operations to improve grain flow and quality in large-scale storage units. They found that combined drying and dry aeration storage systems were the best option for achieving optimal soybean quality. Their experiments demonstrated that low-temperature drying and storage produced the best quality results for the grains. On the other hand, traders may lack the necessary knowledge or training, which could lead to inadequate implementation. Furthermore, external factors such as environmental or economic conditions may make it difficult to effectively apply these techniques. Therefore, it is crucial to continue research to identify the precise causes of this inadequacy and develop strategies for improvement.

5. Conclusions

Post-harvest losses are a major cause of economic and financial losses in the soybean sector, which is the fourth largest export crop in Benin. The aim of the study was to identify the most vulnerable stages and assess strategies for reducing post-harvest losses (PHLs) in the Benin soybean value chain in Benin. The findings revealed that the stages most critical stages along the value chains are in proportions ranging from 0.03% to 7.98%. Specifically, production recorded the highest losses at 7.98%, while marketing recorded the lowest at 0.03%. Threshing, winnowing, sorting and harvesting operations are the most sensitive to PHL. These losses are mainly due to inadequate mechanization and the lack of qualified personnel for post-harvest activities. The main coping strategies developed by stakeholders remain traditional.

Policymakers can set up an integrated post-harvest conservation and storage system that spans from the farmer's field to the final consumer. This system could include appropriate and accessible harvesting and storage equipment or infrastructure, paired with a well-trained workforce skilled to conduct post-harvest operations efficiently. It would be desirable for future research to look at the other links not covered here such as transportation, packaging, and quality losses, and to include medium and large soybean-processing companies in Benin.

Author Contributions: Conceptualization, D.M.N., R.S. and H.C.S.; methodology, R.S., D.M.N., V.D. and S.D.; formal analysis, A.A.O.A., D.M.N., R.S. and S.-K.M.; investigation, J.P.K., S.D. and A.H.; data curation, S.D., A.A.O.A. and S.-K.M.; writing—original draft preparation, D.M.N., R.S. and J.P.K.; writing—review and editing, R.S., S.-K.M., H.C.S. and A.H. All authors have read and agreed to the published version of the manuscript.

Funding: This research was funded by Programme d'Appui au Développement Durable du Secteur Agricole" (PADDSA), grant number 003/FED/DPC/C-PADDSA/2020 and The APC was funded by CORAF/WECARD.

Institutional Review Board Statement: Not applicable.

Informed Consent Statement: Informed consent was obtained from all the subjects involved in the study.

Data Availability Statement: Data Raw data supporting the conclusions of this article will be made available by the authors upon request.

Acknowledgments: The authorities of the Ministry of Agriculture, Livestock and Fisheries (MAEP), the Institut National des Recherches Agricoles du Bénin (INRAB), and the Centre de Recherches Agricoles d'Agonkanmey (CRA Agonkanmey) for their administrative and scientific support. To the soybean producers, processors, and traders surveyed, as well as their umbrella organizations, for their facilitation and participation in the survey.

Conflicts of Interest: The authors declare no conflicts of interest.

Abbreviations

The following abbreviations are used in this manuscript:

PHL	post-harvest loss
SVC	soybean value chain
PDAs	Agricultural Development Poles

References

1. Kumar, D.; Kalita, P. Reducing postharvest losses during storage of grain crops to strengthen food security in developing countries. *Foods* **2017**, *6*, 8. [CrossRef]
2. FAO. *Directives sur la Mesure des Pertes Post-Production- Recommandations sur la Conception d'un Système Statistique de Calcul des Pertes à la Récolte et Après Récolte de Grains Vivriers (Céréales et Légumes Secs)*; FAO: Rome, Italy, 2018; p. 140. Available online: https://openknowledge.fao.org/server/api/core/bitstreams/dbad87cb-58f7-40ad-a5d2-9c78e7b17d26/content (accessed on 5 February 2021).
3. FAO. *La Situation Mondiale de L'alimentation et de L'agriculture, Aller Plus Loin dans la Réduction des Pertes et Gaspillages Alimentaires*; FAO: Rome, Italy, 2019. Available online: https://openknowledge.fao.org/server/api/core/bitstreams/090695a9-56e0-489a-a29b-8f3ba7c88bfa/content (accessed on 5 February 2021).
4. Direction de la Statistique Agricole. *Les Chiffres Definitifs de la Campagne Agricole 2023-2024*; DSA: Cotonou, Bénin, 2024; p. 20. Available online: https://instad.bj/images/docs/insae-statistiques/agricoles/production-agricole/Evolution_de_la_production_agricole_2023/R%C3%A9sultat_d%C3%A9finitif%20de%20la%20campagne_VERSION_FINALE.pdf (accessed on 19 May 2025).
5. Baco, M.N. La place actuelle du maïs dans les exploitations agricoles au Bénin. *Bull. Rech. Agron. (BRAB)* **2019**, *11*, 19–30.
6. International Atomic Energy Agency (IAEA). Le Bénin Améliore la Production et L'exportation du Soja à L'aide de Bio Engrais et de la Technologie Isotopique. 2020. Available online: https://www.iaea.org/fr/newscenter/news/le-benin-augmente-sa-production-et-ses-exportations-de-soja-au-moyen-dengrais-biologiques-et-de-la-technologie-isotopique (accessed on 12 March 2023).
7. Loko, Y.L.E.; Zandjanakou-Tachin, M.; Montcho, D.; Toffa, J.; Agolo, A.; Okpeicha, R.; Orobiyi, A.; Gavoedo, D.; Dansi, A. On-Farm Management of Soybean (Glycine max) Varietal Diversity in Southern and Central Regions of the Republic of Benin. *Agric. Res.* **2022**, *11*, 359–372. [CrossRef]
8. *Les Chiffres de la Campagne Agricole 2019–2020 et les Perspectives de la Campagne Agricole 2020–2021*; Direction de la Statistique Agricole (DSA): Cotonou, Benin, 2021.

9. Fresan, U.; Mejia, M.A.; Craig, W.J.; Jaceldo-Siegl, K.; Sabate', J. Meat analogs from different protein sources: A comparison of their sustainability and nutritional content. *Sustainability* **2019**, *11*, 3231. [CrossRef]
10. Chabi, F.O.; Dagbenonbakin, G.D.; Agbangba, C.E.; Oussou, B.; Amadji, G.L.; Ahoton, E.L.; Saïdou, A. Soil fertility level and cropping practices determining soybean yield in northern east and center of Benin. *Int. J. Plant Soil Sci.* **2019**, *30*, 1–10. [CrossRef]
11. Houngnandan, H.B.; Adandonon, A.; Akplo, T.M.; Zoundji, C.C.; Kouelo, A.F.; Zeze, A.; Akinocho, J. Effect of rhizobial inoculation combined with phosphorus fertilizer on nitrogen accumulation, growth and yield of soybean in Benin. *J. Soil Sci. Environ. Manag.* **2020**, *11*, 153–163.
12. Houngnandan, H.B.; Adandonon, A.; Adoho, T.S.; Bossou, L.D.; Fagnibo, A.H.; Gangnon, O.S.; Houngnandan, P. Diversity of Arbuscular Mycorrhizal Fungi Species Associated with Soybean (*Glycine max* L. Merill) in Benin. *Am. J. Plant Sci.* **2022**, *13*, 686–701. [CrossRef]
13. Obafémi, O.M.; Ibidon, A.F.; Afouda, Y.J. Impact Environnemental de l'adoption des variétés ameliorées de soja dans le département du Borgou au Nord du Bénin. *Eur. Sci. J.* **2019**, *15*, 11–20. [CrossRef]
14. Akpo, F.I.; Dohou, M.D.; Houessingbe, Z.; Yabi, J.A. Analyse comparative des systèmes de production de soja basés sur l'utilisation de l'inoculum dans un contexte de gestion durable des terres au Centre du Bénin. *Int. J. Innov. Appl. Stud.* **2021**, *32*, 230–239.
15. Oloumilade, M.O.; Yabi, J.A. Facteurs explicatifs de l'adoption des varietés améliorées de soja dans le département du Borgou au Nord du Bénin. *Les Cahiers Du Cread* **2019**, *35*, 51–76.
16. Ollabode, N.; Tovihoudji, P.; Labiyi, A.; Aihounton, G.; Adimi, O.; Yabi, J. Déterminants du rendement de soja dans la commune de N'Dali au Nord-Bénin. *Ann. UP Série Sci. Nat. Agron. Hors Série* **2017**, 35–42.
17. Chogou, S.K.; Okry, F.; Santos, F.; Hounhouigan, D.J. Efficacité technique des producteurs de soja du Bénin. *Ann. Sci. Agron.* **2018**, *22*, 93–110.
18. Tossou, B.W.; Chogou, S.K.; Sossou, C.H. Analyse de la compétitivité de la production de soja au Bénin. *Sci. Afr.* **2022**, *19*, e01491. [CrossRef]
19. Yamane, T. *Statistics, An Introductory Analysis*, 2nd ed.; Harper and Row: New York, NY, USA, 1967. Available online: https://discovered.ed.ac.uk/permalink/44UOE_INST/110jsec/alma99381573502466 (accessed on 14 February 2021).
20. Programme Alimentaire Mondial (PAM). *Développement de la Chaîne de Valeur, Genre et Autonomisation des Femmes au Ghana, VAM Etude Genre et Marchés, PAM 2016–2017*; PAM: Rome, Italy, 2017; p. 72. Available online: https://tapipedia.org/sites/default/files/developpement_de_la_chaine_de_valeur_genre_et_autonomisation_des_femmes_au_ghana.pdf (accessed on 23 March 2021).
21. CRADIB-ONG. *Rapport D'activités*; CRADIB-ONG: Bénin, Germany, 2017; p. 14.
22. Sossou, C.H.; Ayedoun AO, A.; Dossouhoui, S.; Noukpozounkou, M.D.; Sossou, R.; Dansou, V.; Koffi, J.P.; Houssou, A.P. Evaluation des pertes post-récolte dans la filière maïs: Cas du maillon production au Bénin. *Int. J. Biol. Chem. Sci.* **2023**, *17*, 1392–1403. [CrossRef]
23. Atiim, P. Assessment of Postharvest Losses in Soybeans Production in the Builsa District in the Upper East Region and Savelugu District in the Northern Region of Ghana. Master's Thesis, Kwame Nkrumah University of Science and Technology College of Agriculture and Natural Resources Faculty of Agriculture Department of Horticulture, Kumasi, Ghana, November 2011.
24. Paulsen, M.R.; Kalita, P.K.; Rausch, K.D. Postharvest Losses due to Harvesting Operations in Developing Countries: A Review. In Proceedings of the ASABE Annual International Meeting, New Orleans, LA, USA, 26–29 July 2015; p. 152176663.
25. Arends-Kuenning, M.; Garcias, M.; Kamei, A.; Shikida, P.; Romani, G.E. Factors Associated with Harvest and Postharvest Loss among Soybean Farmers in Western Paraná State, Brazil. *Food Policy* **2021**, *112*, 102363. [CrossRef]
26. Caixeta-Filho, J.V.; Péra, T.G. Post-harvest losses during the transportation of grains from farms to aggregation points. *Int. J. Logist. Econ. Glob.* **2018**, *7*, 209–247. [CrossRef]
27. Coradi, P.C.; de Oliveira, M.B.; de Oliveira Carneiro, L.; de Souza, G.A.C.; Elias, M.C.; Brackmann, A.; Teodoro, P.E. Technological and sustainable strategies for reducing losses and maintaining the quality of soybean grains in real production scale storage units. *J. Stored Prod. Res.* **2020**, *87*, 101624. [CrossRef]

proceedings

MDPI

Proceeding Paper

Characterizing Polycyclic Aromatic Hydrocarbons in Smoked Chicken from Ilorin and Implications for Human Health †

Idorenyin Nwaehujor [1], Fatimah Badmos [1], Mercy Ige [1], Tawakalitu Ahmed [1], Damilola Ariyo [1], Saburi Atanda [1,*], Adijat Okunlade [1], Adetutu Adediji [1] and Lateef Sanni [1,2]

1 Nigerian Stored Products Research Institute, Ilorin 240242, Nigeria; nwaehujoriu@nspri.gov.ng (I.N.); badmosfo@nspri.gov.ng (F.B.); igeim@nspri.gov.ng (M.I.); ahmedt@nspri.gov.ng (T.A.); ariyodo@nspri.gov.ng (D.A.); okunladeaf@nspri.gov.ng (A.O.); adedijiay@nspri.gov.ng (A.A.)

2 Department of Food Science and Technology, College of Food Science and Human Ecology, Federal University of Agriculture, Abeokuta 110124, Nigeria

* Correspondence: the19icon@gmail.com; Tel.: +234-8023635432

† Presented at the CORAF's 2023 Symposium on Processing and Transformation of Agricultural Products in West and Central Africa: Achievements and Opportunities for Private Sector Engagement, Lome, Togo, 21–23 November 2023.

Abstract: Smoked chicken products from the Nigerian Stored Products Research Institute and three locations in Ilorin, Nigeria, were analyzed for polycyclic aromatic hydrocarbons (PAHs) using gas chromatography. Total PAH levels ranged from 490.893 to 509.064 μg/kg across samples, with benzo(α)pyrene levels (0.947–1.072 μg/kg) within the safe limit of 2 μg/kg. However, PAH4 levels (477.771–491.757 μg/kg) exceeded the European Union's safe limit of 30 μg/kg. Mean Estimated Daily Intake, Carcinogenic Risk, and Toxicity Equivalent Factor were 0.299 μg/kg/day, 45.341, and 1.443, respectively. Regular monitoring and enforcement of quality control standards are crucial for consumer safety while promoting best practices across the industry.

Keywords: polycyclic hydrocarbons; food safety; smoked chicken; health risk; carcinogenic risk

1. Introduction

Smoking meat remains essential for food preservation in regions with limited refrigeration, particularly in developing countries [1]. However, this process introduces polycyclic aromatic hydrocarbons (PAHs)—compounds formed during incomplete combustion that pose significant health risks [2].

Polycyclic aromatic hydrocarbons comprise over 200 compounds characterized by fused aromatic ring structures. The US Environmental Protection Agency has classified 16 PAHs as priority pollutants based on toxicity profiles [3]. These PAHs are further categorized based on their structural characteristics and health impacts. Light PAHs (2–3 rings) typically exhibit acute toxicity but lower carcinogenicity, while heavy PAHs (4–6 rings) demonstrate greater carcinogenic and mutagenic potential. Benzo(α)pyrene (BaP), a 5-ring structure, serves as a marker compound and is classified as a Group 1 human carcinogen by the International Agency for Research on Cancer [4]. The PAH4 group consists of four high-concern compounds—benzo[a]pyrene, benzo[a]anthracene, chrysene, and benzo[b]fluoranthene—and their combined presence is assessed using Toxicity Equivalent Factors to determine cumulative carcinogenic risk [4].

In Nigeria, smoked chicken is popular among street vendors, yet data on PAH levels, estimated daily intake (EDI), and associated health risks remain scarce [5]. This study

https://doi.org/10.3390/proceedings2025118011

analyzes BaP, total PAHs, and PAH4 in smoked chicken samples from Ilorin, Kwara State, with results compared against safety limits established by the Joint FAO/WHO Expert Committee on Food Additives [6]. The findings will help evaluate potential health implications from consuming PAH-exposed smoked chicken in this region.

2. Methodology

2.1. Study Area and Sampling Points

The study was conducted in Ilorin, Kwara State, Nigeria (8.5373° N, 4.5444° E), in June 2023. The map of Ilorin metropolis and the sample collection areas are shown in Figures 1 and 2, respectively. The areas include NSPRI (Sample A), Maraba (Sample B), Al Hikmah University Area (Sample C), and Tanke Junction (Sample D). The maps were created using Google Earth Pro, with all geographic coordinates verified against satellite imagery.

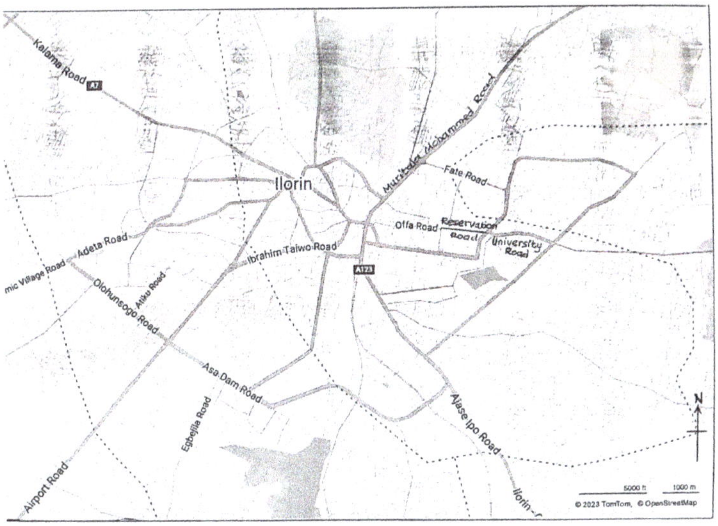

Figure 1. Map of Ilorin metropolis providing an overview of the study areas.

Figure 2. Map illustrating sample collection areas across Ilorin, with locations pinned and numbered A–D.

2.1.1. NSPRI Processed Chicken Sample

Two broiler chickens were slaughtered, cleaned, and seasoned (with a 1:2:2 mixture of salt, ginger, and seasoning powder, respectively). The seasoned samples were smoked using kiln technology at the Nigerian Stored Products Research Institute (NSPRI). Hot, moist air at temperatures ranging from 90 to 110 °C was applied for six hours, followed by an additional six hours of drying. NSPRI smoked chicken (Sample A) serves as the control (Figure 3).

Figure 3. Smoked and dried chicken prepared using NSPRI kiln technology, indicating the uniform texture and golden-brown appearance achieved through controlled smoking and drying.

2.1.2. Street-Sampled Chicken

Smoked chicken samples were collected from three locations in Ilorin: Maraba (Sample B), Figure 4; Al Hikmah University Area (Sample C), Figure 5; and Tanke Junction (Sample D), Figure 6. These samples were separately homogenized and kept encased in foil at 4 °C for further use.

Figure 4. Traditional open smoking of chicken in Maraba, highlighting the use of charcoal and wood for flavour development in street-smoked chicken.

Figure 5. Traditional direct-heat smoking of chicken utilizing charcoal and wood combustion at Al-Hikmah University Area.

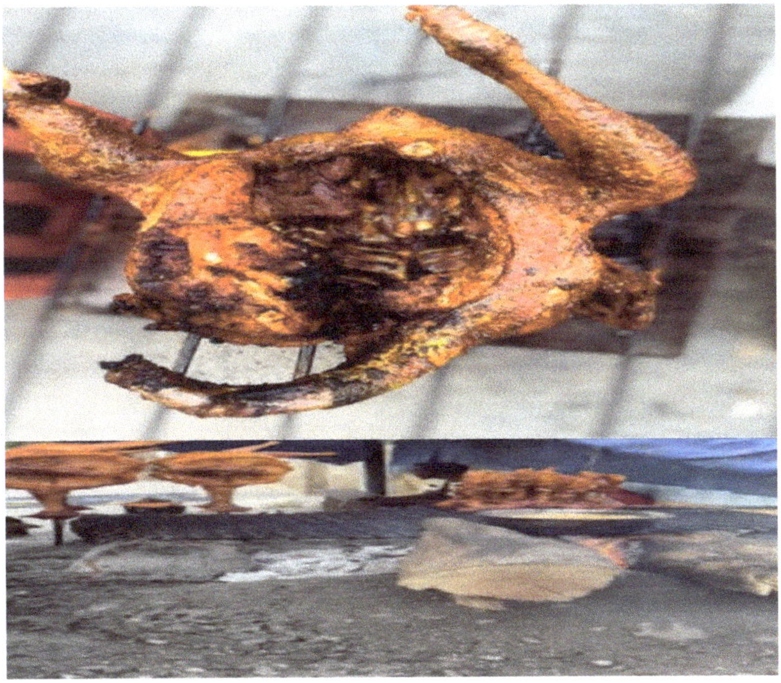

Figure 6. Street-smoked chicken at GRA Tanke Junction, highlighting open-flame smoking with wood and charcoal to process and flavour chicken.

2.2. Sample Preparation, Extraction, and Detection of PAHs

2.2.1. Extraction and Cleanup

Following Smith et al. [7], 2 g of homogenized smoked chicken samples were mixed with analytical-grade (Merck, Boston, MA, USA) anhydrous sodium sulphate (Na_2SO_4)

to remove moisture. Analytical-grade (Merck, USA) dichloromethane (CH_2Cl_2) was used for extraction. The extract was cleaned using a Na_2SO_4-packed column, concentrated via rotary evaporation, and re-dissolved in 1 mL of CH_2Cl_2.

2.2.2. Gas Chromatography Analysis

A Hewlett-Packard 5890 Series II gas chromatograph, equipped with a flame ionization detector (FID), was utilized to analyze the polycyclic aromatic hydrocarbons (PAHs) in the samples. PAH compounds were identified by comparing their retention times to those of a standard PAH mixture. This mixture included a range of PAHs: Naphthalene, Acenaphthylene, Acenaphthene, Fluorene, Phenanthrene, Anthracene, Fluoranthene, Pyrene, Benzo(α)anthracene, Chrysene, Benzo(β)fluoranthene, Benzo(κ)fluoranthene, Benzo(α)pyrene, Indeno(1,2,3-cd)pyrene, Dibenzo(a,h)anthracene, and Benzo(g,h,i)perylene. Then, the concentrations of the PAHs were quantified using external calibration curves. This method ensured the accurate identification and quantification of the PAHs present in the samples.

2.2.3. PAH Classification

In our study, PAHs were categorized as carcinogenic or non-carcinogenic based on International Agency for Research on Cancer (IARC) and US Environmental Protection Agency (EPA) evaluations. The non-carcinogenic group includes Fluorene, Naphthalene, Pyrene, Acenaphthylene, Anthracene, and Phenanthrene, which are not classified as significant human carcinogens by IARC. The carcinogenic group consists of Indeno(1,2,3-cd)pyrene, Benzo(g,h,i)perylene, Benzo(a)pyrene, Chrysene, Benzo(b)fluoranthene, Benzo(a)anthracene, Dibenzo(a,h)anthracene, and Benzo(k)fluoranthene. These compounds are classified by IARC as either Group 1 (known human carcinogens) or Group 2B (possible human carcinogens). This classification enables more accurate health risk assessment and supports targeted risk management strategies for smoked chicken products.

2.2.4. Method Validation

Analytical validation of the method was conducted through calibration curves, limits of detection (LOD), limits of quantification (LOQ), and recovery studies to ensure reliability and accuracy. Calibration curves were established using three standard solutions with concentrations ranging from 0.2 to 10 mg/L, providing the basis for quantifying the PAH congeners. The LOD and LOQ for each PAH class were determined to evaluate the method's sensitivity and ensure accurate detection and quantification at low concentration levels. Recovery studies were performed by spiking the samples with a known concentration of PAHs (0.05 mg/L) and analyzing them in triplicate.

2.2.5. Health Risk Assessment

The health risks from PAH exposure in smoked chicken were evaluated using standardized methods from multiple studies [8–11]. The assessment incorporates four key parameters:

The Estimated Daily Intake (EDI), which measures how much PAH-contaminated food a person consumes daily, accounting for their body weight, is calculated using the following expression [8]:

$$EDI = C \times \frac{IR}{BW} \quad (1)$$

where C = PAH concentration, *IR* = ingestion rate, and *BW* = body weight (70 kg).

The Toxic Equivalent Concentration (TEQ) evaluates overall toxicity by considering both the amount of each PAH present and its relative toxicity compared to a reference compound and is evaluated using the following formula [9]:

$$TEQ = \sum \text{Ci} \times \text{TEFi} \quad (2)$$

where Ci = PAH concentration, and TEFi = toxicity factor.

The sum of four key carcinogenic PAHs, specifically looking at the total amount of four major cancer-causing PAHs (BaA, Chr, BbF, and BaP) present in food, is calculated using the following equation [10]:

$$\text{PAH4} = \text{B}(\alpha)\text{A} + \text{Chr} + \text{B(b)FL} + \text{B}(\alpha)\text{P} \quad (3)$$

The Cumulative Cancer Risk (*CR*), which estimates lifetime cancer risk by combining daily exposure levels with established cancer slope factors, is evaluated by the following equation [11]:

$$CR = \text{Slope factor} \times \text{EDI} \times \text{Exposure duration} \quad (4)$$

The final assessment uses carcinogenic toxicity equivalents (TEQs) to estimate the overall cancer risk by summing the potencies of individual PAHs [12].

3. Statistical Analysis

Data on LOD, LOQ and contamination indices were analyzed for significance using ANOVA (SPSS v2.0, IBM, Chicago, IL, USA).

4. Results of Validation Method, Total Polyaromatic Hydrocarbons, Benzo (α) Pyrene, and the Four Polyaromatic Hydrocarbons

4.1. Validation and Quality Control of PAH Analysis

Table 1 shows the limits of detection (LOD) and limits of quantification (LOQ) for the 16 PAH classes. Table 2 shows that the ANOVA analysis revealed a *p*-value well below the 0.05 significance level, indicating significant differences in mean PAH concentrations. This finding validates our method's ability to distinguish between various PAHs.

Table 1. Limits of detection (LOD) and limits of quantification (LOQ) for the 16 PAHs (means and standard deviation of triplicates).

S/N	PAH	LOD (mg/L)	LOQ (mg/L)
1.	Naphthalene	0.0000023 ± 0.0000001	0.0000015 ± 0.0000001
2.	Acenaphthylene	0.0000034 ± 0.0000002	0.0000025 ± 0.0000001
3.	Acenaphthene	0.0000030 ± 0.0000001	0.0000021 ± 0.0000001
4.	Fluorene	0.0000026 ± 0.0000001	0.0000017 ± 0.0000001
5.	Phenanthrene	0.0000041 ± 0.0000002	0.0000032 ± 0.0000002
6.	Anthracene	0.0000084 ± 0.0000004	0.0000052 ± 0.0000003
7.	Fluoranthene	0.0000091 ± 0.0000005	0.0000063 ± 0.0000003
8.	Pyrene	0.0000075 ± 0.0000004	0.0000049 ± 0.0000002
9.	Benzo(α)anthracene	0.0000069 ± 0.0000003	0.0000044 ± 0.0000002
10.	Chrysene	0.0000042 ± 0.0000002	0.0000036 ± 0.0000002
11.	Benzo(β)fluoranthene	0.0000066 ± 0.0000003	0.0000039 ± 0.0000002
12.	Benzo(κ)fluoranthene	0.0000033 ± 0.0000002	0.0000022 ± 0.0000001
13.	Benzo(α)pyrene	0.0000052 ± 0.0000003	0.0000039 ± 0.0000002
14.	Indeno(1,2,3)pery-lene	0.0000036 ± 0.0000002	0.0000028 ± 0.0000001
15.	Dibenzo(α,h)anthracene	0.0000014 ± 0.0000001	0.0000011 ± 0.0000001
16.	Benzo(g,h,i)perylene	0.0000014 ± 0.0000001	0.0000012 ± 0.0000001

Table 2. ANOVA results for comparing LOD and LOQ (mean concentrations) of 16 PAHs, showing F-statistic, degrees of freedom, and *p*-value.

Source of Variation	Degrees of Freedom (df)	Sum of Squares (SS)	Mean Square (MS)	F-Statistic	*p*-Value
Between Groups	15	1.23×10^{-10}	8.20×10^{-12}	82.0	<0.0001
Within Groups	32	3.20×10^{-12}	1.00×10^{-13}	-	-
Total	47	1.26×10^{-10}	-	-	-

The recovery rates for PAH analyses, ranging from 95.3% to 104.8%, demonstrate high accuracy and reliability of the method. These values, close to the ideal recovery of 100%, indicate that the method consistently performs well across different PAHs, ensuring precise and dependable results. Chromatograms from the recovery studies conducted on the four sample sets are shown in Figures 7–10.

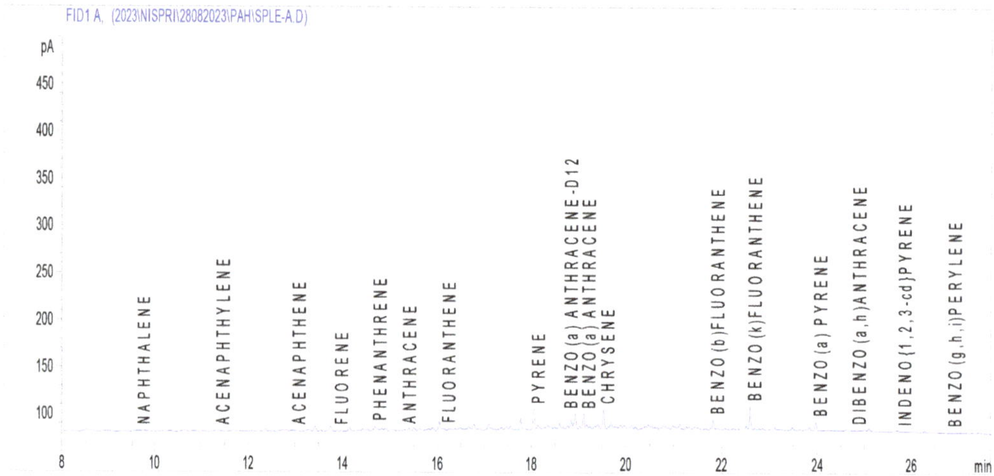

Figure 7. Recovery profile for PAHs (A), indicating the percentage recovery across different concentration levels.

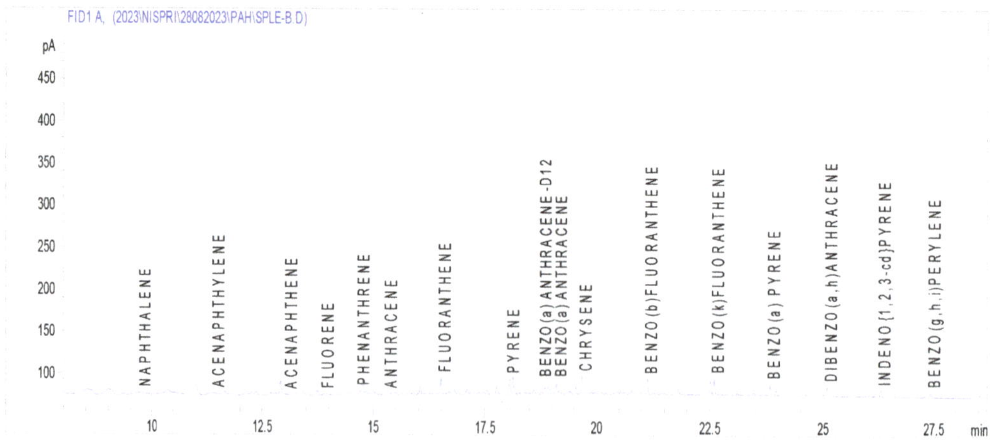

Figure 8. Recovery profile for PAHs (B), indicating the percentage recovery across different concentration levels.

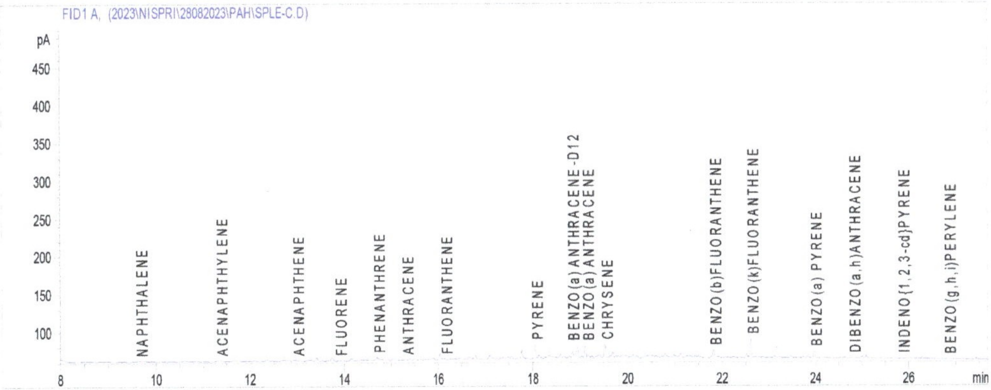

Figure 9. Recovery profile for PAHs (C), indicating the percentage recovery across different concentration levels.

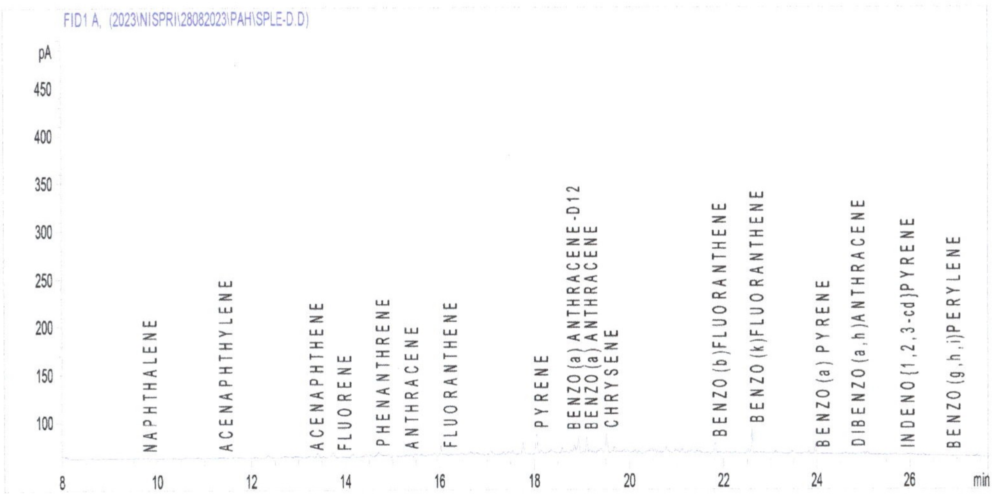

Figure 10. Recovery profile for PAHs (D), indicating the percentage recovery across different concentration levels.

4.2. Total Polyaromatic Hydrocarbons, Benzo (α) pyrene, and the Four Polyaromatic Hydrocarbons

Figure 11 illustrates the total concentrations of polycyclic aromatic hydrocarbons (PAHs), Benzo(α)pyrene (BaP) and PAH4 (BaA, Chr, BbF, and BaP) in smoked chicken samples, with the data summarized as follows.

Total PAHs: The total PAH content varied significantly across samples, with values ranging from 490.88 μg/kg to 509.06 μg/kg ($p < 0.05$). Sample A had the highest concentration, while Sample D had the lowest.

Benzo(α)pyrene (BaP): Levels of BaP, a key carcinogenic congener, ranged from 0.94 μg/kg to 1.07 μg/kg. Sample A exhibited the highest BaP concentration, and Sample B had the lowest.

PAH4: The combined concentration of the four major carcinogenic PAHs (BaA, Chr, BbF, and BaP) was highest in Sample A (6.97 μg/kg) and lowest in Sample B (5.34 μg/kg). PAH4 levels generally exceeded those of BaP, except in Samples B and D, where no significant difference was observed ($p > 0.05$).

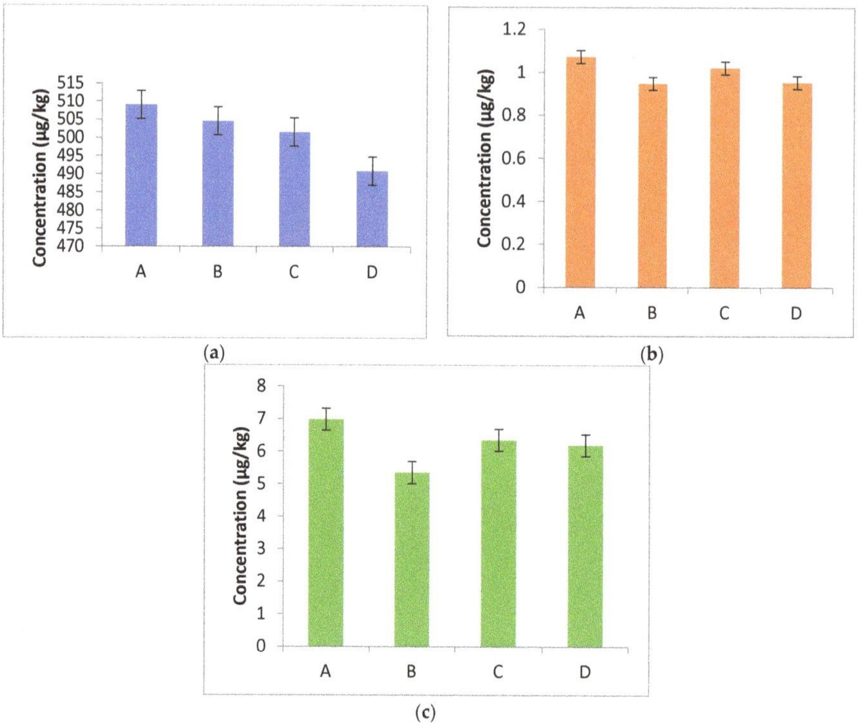

Figure 11. (**a**) Levels of total PAHs, (**b**) Benzo(α)pyrene, and (**c**) PAH4.

4.3. Assessment of Carcinogenic Risk Indices

Figure 12 presents the estimated carcinogenic risk indices of PAHs in smoked chicken samples, which exhibited significant variation ($p < 0.05$). The indices assessed include the Estimated Daily Intake (EDI), Toxicity Equivalent Factor (TEQ), and Carcinogenic Risk (CR).

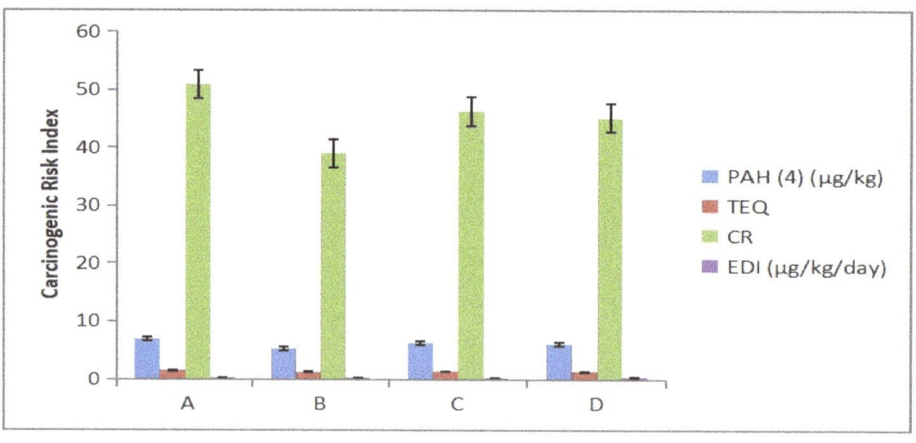

Figure 12. Estimated carcinogenic risk indices of PAHS in smoked chicken samples. **EDI:** Estimated Daily Intake; **TEQ:** Toxicity Equivalent Factor; **CR:** Carcinogenic Risk.

Toxicity Equivalent Factor (TEQ): The TEQ values, which measure the overall toxicity of PAHs relative to benzo(α)pyrene, ranged from 1.33 to 1.57 across the samples. Sample C had the highest TEQ, indicating a greater relative toxicity compared to the other samples.

Carcinogenic Risk (CR): The CR levels, which estimate the lifetime cancer risk, varied between 39.01 and 50.91. Sample B had the lowest CR, suggesting a reduced cancer risk, while Sample A had the highest value. However, Samples C and D did not show a significant difference in CR ($p > 0.05$), indicating similar cancer risk profiles for these samples.

Estimated Daily Intake (EDI): The EDI, which estimates the daily consumption of PAHs, was generally low in all samples, including the control. Sample D had the highest EDI at 0.36, while Sample B had the lowest at 0.23.

5. Discussion and Health Risk Implications

Total PAH concentrations in smoked chicken from Ilorin ranged between 490.88 and 509.06 μg/kg, aligning with levels found in other smoked meats [13], although differences in smoking methods make direct comparisons challenging [14,15]. Benzo(α)pyrene (BaP) levels (0.94 to 1.07 μg/kg) were found below the EU limit, suggesting safety, but local practices may affect PAH formation [16]. Existing research primarily focuses on other meats, highlighting the need for more studies on PAH contamination in street-vended smoked chicken [17].

6. Conclusions

PAHs were found in all smoked chicken samples, with BaP levels below the EU limit but total PAH and PAH4 concentrations exceeding safe levels, posing health risks. TEQ and CR values indicated increased carcinogenicity risk, highlighting the need for improved processing, vendor training, and quality control.

7. Limitations and Future Directions

This study's limitations include a focus on Ilorin and a lack of detailed consumer data on smoked chicken consumption. Future research should expand the sampling, source additional data on consumption in children and adults, explore PAH formation, and calculate the EDI and TEQ for each individual PAH compound and for each specific subgroup of PAHs, as well as develop standardized protocols and local safety guidelines to improve public health and smoking practices.

Author Contributions: S.A.: conceptualization, original and final draft preparation, research design, execution of research design, and manuscript preparation; I.N.: supervision and manuscript preparation and review and editing; D.A.: conceptualization and research design, data interpretation, sourcing of materials, supervision, and manuscript preparation. F.B.: data interpretation, supervision, and manuscript preparation. T.A.: data interpretation, sourcing of materials, supervision, and manuscript preparation. M.I.: supervision and manuscript preparation. A.O.: supervision and manuscript preparation. A.A.: data interpretation, supervision, initial draft preparation, and manuscript preparation. L.S.: supervision and initial draft preparation. All authors have read and agreed to the published version of the manuscript.

Funding: No funding, grants, or support were received from any agency for this study.

Institutional Review Board Statement: Not applicable.

Informed Consent Statement: Not applicable as the studies did not involve humans hence no ethical approval was required.

Data Availability Statement: The authors declare that the data supporting this study are available within the manuscript. Should any data files be needed, they will be made available by the corresponding author upon reasonable request.

Acknowledgments: The management of Nigerian Stored Products Research Institute (NSPRI) is appreciated for their support; Akeem Bello of Multi Environment Laboratory, Lagos, Nigeria, is worthy of recognition for his technical guidance in the analysis of PAHs. We thank Olufemi Olaleye-Otunla, Scientific Illustrator at the A.G. Leventis Museum of Natural History, OAU, Ile-Ife Nigeria for his effort in creating the study maps.

Conflicts of Interest: The authors have no relevant financial or non-financial interests to disclose.

References

1. Tongo, I.; Ezemonye, L.; Akpeh, K. Distribution, characterization, and human health risk assessment of polycyclic aromatic hydrocarbons (PAHs) in Ovia River, Southern Nigeria. *Environ. Monit. Assess.* **2017**, *189*, 247. [CrossRef] [PubMed]
2. Rigling, M.; Höckmeier, L.; Leible, M.; Lüthy, M.; Gasser, C.A.; Morasch, B.; Schloter, M.; Spörri, C.; Graf, F. Characterization of the aroma profile of food smoke at controllable pyrolysis temperatures. *Separations* **2023**, *10*, 176. [CrossRef]
3. Chen, Y.W.; Cheng, Y.H.; Hsu, C.Y. Characterization of the sources and health risks of polycyclic aromatic hydrocarbons in PM2.5 and their relationship with black carbon: A case study in northern Taiwan. *Environ. Pollut.* **2023**, *336*, 122427. [CrossRef] [PubMed]
4. International Agency for Research on Cancer. *Some Polycyclic Aromatic Hydrocarbons, Coal Tar Pitches, and Some Related Exposures*; IARC Monographs on the Evaluation of Carcinogenic Risks to Humans; International Agency for Research on Cancer: Lyon, France, 2022; Volume 133, Available online: https://publications.iarc.fr/Book-And-Report-Series/Iarc-Monographs-On-The-Identification-Of-Carcinogenic-Hazards-To-Humans (accessed on 8 November 2024).
5. Onyeaka, H.; Ekwebelem, O.C.; Eze, U.A.; Onwuka, Q.I.; Aleke, J.; Nwaiwu, O.; Chionuma, J.O. Improving Food Safety Culture in Nigeria: A Review of Practical Issues. *Foods* **2021**, *10*, 1878. [CrossRef] [PubMed]
6. Joint FAO/WHO Expert Committee on Food Additives (JECFA). *Evaluation of Certain Food Additives and Contaminants*; WHO Technical Report Series No. 1021; World Health Organization: Geneva, Switzerland, 2021. Available online: https://www.who.int/publications/i/item/9789240054585 (accessed on 14 May 2025).
7. Wang, X.; Li, Y.; Chen, Z.; Liu, J.; Zhang, Y. Advancements in Analytical Techniques for Polycyclic Aromatic Hydrocarbons in Environmental Matrices. *Anal. Chem.* **2022**, *94*, 5678–5689. [CrossRef]
8. Otoo, E.A.; Ocloo, F.C.K.; Appiah, V.; Nuviadenu, C.; Asamoah, A. Reduction of polycyclic aromatic hydrocarbons concentrations in smoked guinea fowl (*Numida meleagris*) meat using gamma irradiation. *CyTA J. Food.* **2022**, *20*, 343–354. [CrossRef]
9. Coroian, C.O.; Coroian, A.; Becze, A.; Miresan, V.; Pop, F.A.; Cozma, A.; Vodnar, D.C. Polycyclic aromatic hydrocarbons (PAHs) occurrence in traditionally smoked chicken, turkey, and duck meat. *Agriculture* **2023**, *13*, 57. [CrossRef]
10. Şahin, Ş.; Ulusoy, H.; Alemdar, S.; Erdogan, S.; Agaoglu, S. The presence of polycyclic aromatic hydrocarbons (PAHs) in grilled beef, chicken, and fish by considering dietary exposure and risk assessment. *Food Sci. Anim. Resour.* **2020**, *40*, 675. [CrossRef] [PubMed]
11. Zhen, H.; Zhang, F.; Cheng, H.; Hu, F.; Jia, Y.; Hou, Y.; Shang, M.; Yu, H.; Jiang, M. Association of Polycyclic Aromatic Hydrocarbons Exposure with Child Neurodevelopment and Adult Emotional Disorders: A Meta-Analysis Study. *Ecotoxicol. Environ. Saf.* **2023**, *255*, 114770. [CrossRef] [PubMed]
12. Wang, L.; Zhang, S.; Wang, L.; Li, Y.; Zhang, H. Concentration and risk evaluation of polycyclic aromatic hydrocarbons in urban soil in the typical semi-arid city of Xi'an in Northwest China. *Int. J. Environ. Res. Public Health* **2018**, *15*, 607. [CrossRef] [PubMed]
13. Warind, J.; Ahmad, N.; Noor, N. Profiling of polycyclic aromatic hydrocarbons in traditional smoked foods: A comprehensive review of sampling, analysis, levels, and health risk assessment. *Food Control* **2021**, *127*, 108123.
14. Mohammed, A.B.; Rahman, M.S.; Sarker, M.R. Polycyclic aromatic hydrocarbons (PAHs) in traditionally processed fish and meat products from Bangladesh: A comprehensive food safety assessment. *Food Control* **2021**, *124*, 107918.
15. Seyda, A.; Yilmaz, I.; Meral, R. Investigation of polycyclic aromatic hydrocarbons content in commercial smoked meat and fish products. *Food Sci. Technol. Int.* **2020**, *26*, 751–762.
16. European Commission. Commission Regulation (EU) 2021/1317 of 9 August 2021 amending Regulation (EC) No 1881/2006 as regards maximum levels of lead in certain foodstuffs. *Off. J. Eur. Union.* **2021**, *L285*, 1–4.
17. Onopiuk, A.; Kołodziejczak, K.; Szpicer, A.; Wojtasik-Kalinowska, I. Polycyclic aromatic hydrocarbons in grilled meat products: Critical review. *Int. J. Food Sci. Technol.* **2021**, *56*, 2215–2227.

proceedings

MDPI

Proceeding Paper

Comparative Study of the Impact of Stearin-Modified Starches from Maize, Ginger, and Sweet Potato on the Physicochemical Properties of Low-Fat Mayonnaise †

Emmanuel Panyoo Akdowa [1,*], Astaharam Didi [1] and Tchiégang Clergé [2]

1 Department of Food Science and Nutrition, ENSAI, University of Ngaoundere, Ngaoundere BP 454, Cameroon; astaharamdidi@univ-ndere.cm
2 Department of Food Engineering and Quality Control, UIT, University of Ngaoundere, Ngaoundere BP 454, Cameroon; tclerge@yahoo.fr
* Correspondence: emmanuelpanyoo@gmail.com or panyooemmanuel@univ-ndere.cm
† Presented at the CORAF's 2023 Symposium on Processing and Transformation of Agricultural Products in West and Central Africa: Achievements and Opportunities for Private Sector Engagement, Lome, Togo, 21–23 November 2023.

Abstract: The utilization of modified starches derived from ginger, sweet potatoes, and maize has been employed as a strategy to reduce the oil content in mayonnaise formulations. Amylose–lipid complexes and the native starches were synthesized, characterized, and subsequently incorporated into various mayonnaise formulations, replacing 50 and 80 percent of the oil. The sensory analysis showed that the mayonnaises produced with 50% and 80% modified corn starch were particularly well received. However, when compared to conventional mayonnaises, the viscosity of the mayonnaises prepared with 50% and 80% modified maize starch was found to be remarkably low. This outcome demonstrates that when preparing low-fat mayonnaise, it is not possible to substitute tuber starches for fat.

Keywords: mayonnaise; fat substitute; viscosity measurements; ginger; maize; sweet potato

1. Introduction

Mayonnaise is an emulsion based on egg yolk, with vegetable oil and vinegar as unavoidable ingredients. It is considered a high-calorie food, as it can contain over 70% oil [1]. It is widely used as a sauce and condiment in a variety of dishes and in most fast-food outlets [2]. Consumption of high-fat foods like mayonnaise may have a direct impact on human health through the onset of certain diseases such as cardiovascular disease, hypertension, obesity, and diabetes [3].

In view of the high fat content of mayonnaise and the health problems associated with its consumption, low-calorie alternatives have been developed due to the positive association between a high-calorie diet and lifestyle-related diseases [4]. The design of hypocaloric foods requires the use of fat replacers to mimic some of the functional properties of fat; otherwise, low-calorie foods will have inferior sensory properties [5]. Starch is a natural biopolymer which can be used as a fat substitute [6]. Cereals and tubers are sources of starch acquired through their different cultivation, structure, and functional properties [7]. It has been suggested that cereal starch is a good fat mimetic [8]. Although native starch can sometimes be used as a fat substitute, it is usually modified by physical or chemical methods for this purpose.

Indeed, native starch presents certain limitations for food use, such as low heat resistance, a strong tendency to retrograde, and a tendency to form low-bodied rubbery gels [9]. Chemical and physical modification of starch can improve properties such as thickening, binding, mouthfeel, gelation, dispersion, or cloud formation [10]. Stearic acid has been found to modify the pasting properties of starch at a level of around 1.5% (w/w starch) and can increase pasting viscosity twice during pasting for more than 1 h [11]. D'Silva et al. [12] also found that modified cereal with added stearic acid pastes exhibited a non-gelling behavior compared to the gelling behavior of unmodified starches. Corn and teff starches modified with stearic acid could substitute oil in mayonnaise at a level of 80% [8]. However, this has been proven exclusively with cereal starches, and among these cereals, teff starches are not available locally in Cameroon. This raises the question of whether it would be possible to use another available source of starch? Hence, the aim of the present study is to compare the use of different starch sources for the development of a low-calorie mayonnaise by substituting fat with modified starches from local tubers and cereals.

2. Material and Methods

2.1. Starch Extraction

Maize, potato, and ginger were cleaned, cut into small pieces, and soaked in distilled water containing sodium metabisulfite (0.2% v/v) for 24 h while refrigerated. To extract the starch, the raw material was crushed for five minutes in an industrial mixer using a 0.2% (v/v) sodium metabisulfite solution. After homogenization, the material was sieved through 200 mesh (0.074 mm). Following a 24 h decantation period, the sample was resuspended in 0.2% sodium metabisulfite solution (v/v), and the supernatant was extracted following 12 min of centrifugation at 1105× g. The residue's surface was cleared of the mucilage that had developed there. After that, the starch residue was dried in oven at 40 °C [13]. The yield of starch was calculated according to the following formula:

$$Yields = \frac{Me(g) * 100}{Mc\ (g)} \tag{1}$$

where *Me* is the mass of dry starch after extraction and *Mc* is the mass of the raw material.

2.2. Determination of Amylose Content

An iodine binding method was used to assess the amylose content of starches [14]. A total of 5 g of starch was weighed into a 25 mL volumetric flask to the nearest 0.1 mg in order to determine the amount of amylose. Afterward, 1 mL of ethanol was used to spread the starch; then, 2.7 mL of 0.1 M NaOH was added and swirled to further enhance the dispersion. The contents were brought to a boil at 75 °C for fifteen minutes, during which the starch fully gelatinized. After cooling, distilled water was added to the flask until it reached the 25 mL threshold using a vortex mixer. Separate test tubes were used to collect duplicate 2.5 mL samples, which were neutralized with 2 mL of 0.15 M citric acid, followed by 1 mL of fresh iodine solution (0–2 g 12 + 2 g KI + 250 mL distilled water). After adding 14.5 mL of water, the sample was refrigerated for 20 min. After the tubes were combined using a vortex mixer, the duplicate subsamples' optical densities were measured at 620 nm using a spectrophotometer.

2.3. Production of Starch–Lipid Complexes

The Figure 1 shows the method of starch–lipid complex production according to D'Silva et al. (2011) [12].

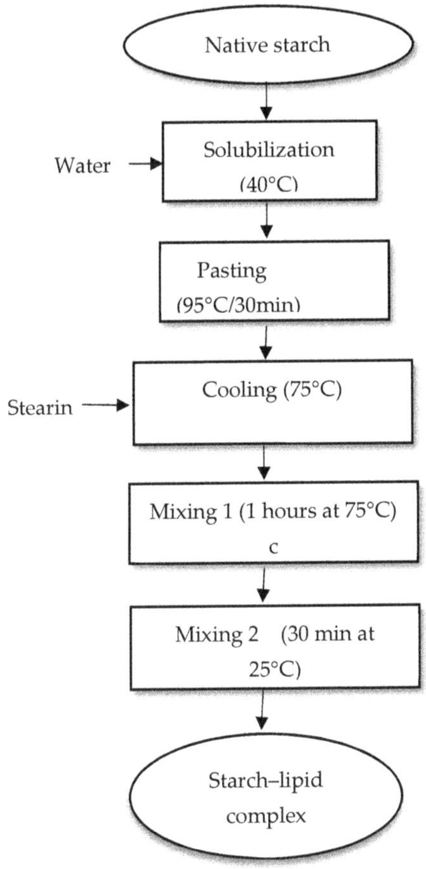

Figure 1. Method of starch–lipid complex production according to D'Silva et al. (2011) [12].

2.4. Complexation Indexes (CIs) of Amylose–Lipid Complexes

The complexation index (CI) was measured using an iodine solution to assess the degree of starch in the lipid complex formation [15]. A total of 2 g of potassium iodine and 1.3 g of I_2 were dissolved in 50 milliliters of distilled water and left to dissolve overnight to create the iodine solution utilized for the analysis. The capacity was then increased to 100 mL by adding distilled water. In a test tube, a 5 g sample of the amylose–lipid complex was combined with 25 mL of distilled water. The test tube was centrifuged at $1500\times g$ for 20 min after being vortexed for two minutes. In a 15 mL test tube, distilled water (7.5 mL) was combined with the supernatant (250 μL) and 1 mL of iodine solution. At 690 nm, the absorbance was measured after the tube was vortexed. The following formula was used to determine the CI:

$$\mathrm{CI}(\%) = \frac{\text{control absorbance} - \text{sample absorbance}}{\text{control absorbance}}$$

2.5. Texture

Texture is the set of rheological and structural properties perceptible by mechanical, tactile, visual, and auditory receptors. It represents a property of the food that is perceived at a sensory level and which evolves from the moment the food is placed in the mouth until the formation of the food bolus conducive to swallowing. The parameter measured in this work is the hardness of native starch. Hardness (H) is the force opposed by the product to the advancement of the probe. It can also be defined as the force in Newtons required to obtain a given deformation or penetration. In this work, a texturometer equipped with various probes was used. The following conditions were chosen for this test: a cylindrical probe with a diameter of 1mm, a detection force on the surface of the product of 5 g, and a probe displacement speed of 1mm/deformation of 40%.

2.6. Syneresis

Syneresis is the expulsion or rejection of water after starch cooking. The syneresis protocol was described by Gonzalez et al. [16]. A total of 10 g of native starch was mixed in 100 mL of distilled water and then heated to 90 °C for 10 min. The gel obtained was left to stand for 24 h at 4 °C and the water released was measured using a test tube or pipette.

2.7. Paste Viscosity (RVA)

Using a Rapid Visco Analyzer 4 RVA (Newport Scientific Pty Ltd., Warriewood, Australia), the paste viscosity of the starch suspension (14% moisture on a wet basis) (3 g starch in 25 mL of water) was measured. The time–temperature profile involved blending using the spoons, rotating for the first ten seconds at 960 rpm, and finishing at 160 rpm. After being heated steadily at a rate of 6 °C per minute between 50 and 90 °C, the samples were cooled to 50 °C. The maximum peak viscosity, paste temperature, rupture viscosity, final viscosity, and regress or retrograde viscosity were the values obtained from the bonding curve.

2.8. Viscosity

A BROOKFIELD DV-III rheoviscosimeter coupled to a thermostatic water bath was used to carry out the rheological analyses. A duvet geometry was observed in which a volume of 7 mL was used. The following quantities were evaluated to deduce the subsequent behavior of the pastes: the viscosity and flow curves. These viscosities were evaluated at a different shear rate of 0.4–186 s^{-1}.

2.9. Sensory Analysis

The sensory analysis is defined as incorporating the five senses (sight, hearing, smell, taste, and touch); human beings are the measuring instrument for sensory analysis methods to characterize and evaluate products. In this section, consumer preference is the focus of the study. It thus involves hedonic tests intended for naive consumers made up of 15 to 20 individuals who are in the town of Ngaoundéré. The approach used for this study is the hedonic ranking test.

The test took place in the sensory analysis room, somewhat isolated from the outside environment (sounds, people, and smells). After the instructions were explained, the subject was presented with eleven samples at the same time, which they had to classify (after tasting) in order of their preference. They had to analyze the samples in the order given and as many times as they deemed necessary to establish a ranking. Each sample was ordered in the same way, in the same quantity, and at the same temperature. They were all represented by a code. The panelists were not trained in sensory analysis. The panelists then entered the product numbers in the code section of each sheet provided.

2.10. Statistical Analysis

For the determination of the level of significance and comparison of means, statistical analysis was applied to the data that were obtained from each parameter. Experimental data were analyzed using two-way ANOVA. The significant difference between means was revealed by the LSD method. Principal component analysis (PCA) were used to plot the sensory analysis data. The data were analyzed using Minitab 16 (manufactured by UNIFI company, headquartered in Greesboro, N.C) software to validate the outcomes of the study.

3. Result and Discussion

3.1. Sweet Potato, Maize, and Ginger Starch Production Yields

The starches were produced using successive unit operations to obtain the yields shown in Figure 2. Maize, sweet potato, and ginger starch yields vary widely. Ginger had a higher starch yield than maize and sweet potatoes. The results for maize and sweet potato starches are significantly higher than the yields of Chinese sweet potatoes varieties reported by Wang et al. [17]. The starch yield of ginger is similar to the starch yield reported by Madeneni et al. [18]. This variation may be due to genetic and environmental factors [19].

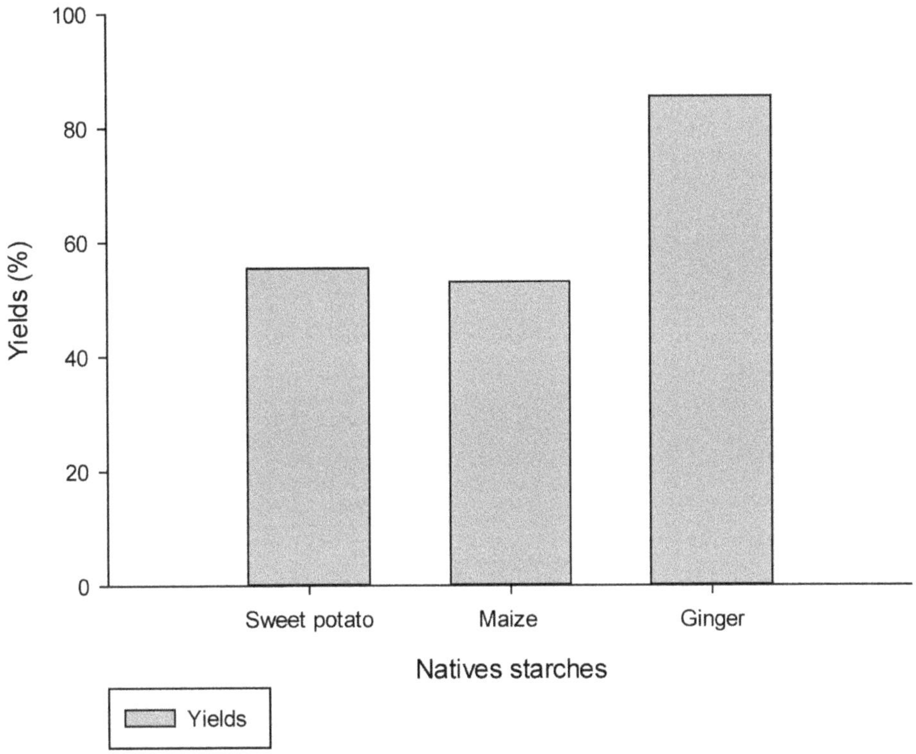

Figure 2. Yields of native starches of sweet potato, maize, and ginger.

3.2. Amylose Content of Native Starches

The amount of amylose in starches affects their functional characteristics; when amylose content rises, starch's transparency and swelling decreases [20]. Table 1 shows the amylose content of cereals (maize) and tubers (sweet potato and ginger); the values range from 21.35% to 29.45%. Indeed, the highest value was observed for maize with 29.45%, followed by ginger with a value of 28.91%, and finally, potatoes had the lowest value of 21.35%. Emmambux and Taylor [7] have shown that the amylose content of cereals and tubers can range from 18 to 30%. This variation may be due to the starch source or molecular variation in the starch granules.

Table 1. Amylose content, volume of syneresis, and gel hardness of native starches (maize, sweet potato, and ginger).

Starch	Amylose Content	Volume of Syneresis (mL)	Gel Hardness (N)
Maize	29.45 ± 0.38 [b]	17.15 ± 0.21 [b]	0.10 ± 0.03 [a]
Sweet potato	21.35 ± 1.91 [a]	1.3 ± 0.14 [a]	0.07 ± 0.03 [a]
Ginger	28.91 ± 0.38 [b]	36.25 ± 0.35 [c]	1.27 ± 0.54 [b]

Means with different superscripts differ significantly per sample ($p < 0.05$).

3.3. Rapid Viscosity Analysis (RVA)

Figure 3 shows the pasting properties of maize, ginger, and potato starches during RVA. Pasting properties are the most important properties when considering starches for use as gelling and thickening agents. Starches with relatively high peak viscosities, like sweet potato starch, could be used as thickening or gelling agents. However, low peak viscosities, like those seen in ginger and maize starches, would be suitable for the manufacture of weaning foods, where food ingredients with a low paste viscosity are required. It can be seen that the gelatinization temperature of sweet potato starch was 75 °C, ginger was 72 °C, and maize was 90 °C. The peak viscosity of potato starch is higher than that of ginger and maize starch, and consequently, potato starch retrogrades more than ginger and maize starch. This could be explained by the fact that potato starch has a low amylose content. Studies have shown that the peak viscosity of starch is influenced by its amylose content [21]. Since linear amylose chains can reassociate at a faster rate than highly branched amylopectin molecules, a high amylose content is linked to a higher rate of retrogradation in starches. Several authors have shown that starches with longer amylopectin chains (potatoes and peas) retrograde faster than those with short amylopectin chains (cereals) [22].

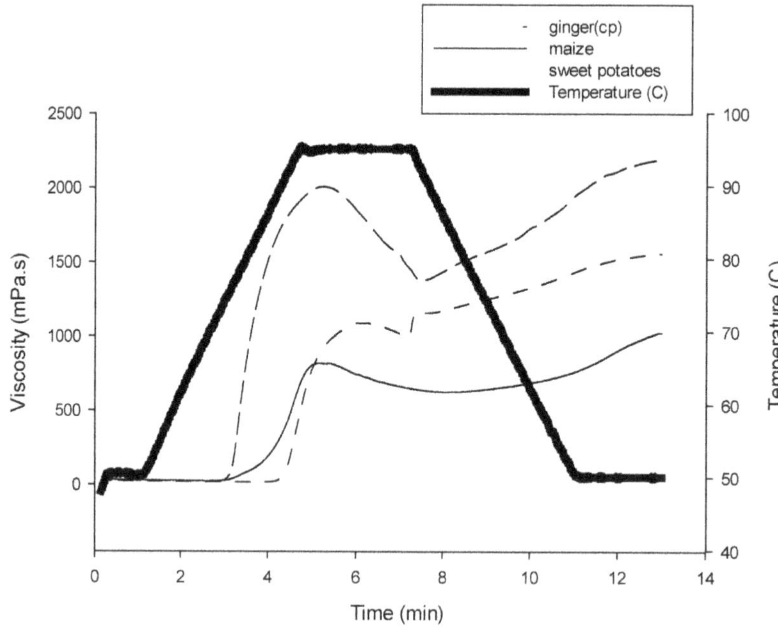

Figure 3. Rapid viscosity analysis (RVA) of ginger, maize, and sweet potatoes starches.

3.4. Syneresis

Table 1 shows the different volumes obtained after syneresis. Indeed, during amylose reorganization, the water that had been absorbed during cooking is released [23]; the syneresis of the various samples varies between 36.25 mL and 1.3 mL. However, the lowest value was observed in sweet potato with a volume of 1.3 mL, which was lower than for maize (17.15 mL) and ginger (36.25 mL). These differences in values could be explained by the fact that sweet potato has a low amylose content, as found in Table 1, and therefore expels less water into the medium after prolonged storage compared to ginger and maize which, being rich in amylose, expel very high quantities of water; this is because the amylose molecule presents a rigid and compact structure which limits the gel's capacity to retain water, resulting in significant syneresis [24].

3.5. Gel Hardness

Table 1 shows the different gel hardnesses of maize, ginger, and sweet potato starches. The hardnesses of the different samples vary between 1.27 and 0.07. The lowest value was observed for sweet potato, while the highest value was for ginger. These results may be due to the fact that sweet potatoes have a low amylose content, favoring the formation of a brittle, less compact gels, whereas ginger and maize have high amylose contents, favoring the formation of a rigid, more compact gel.

3.6. Complexation Indexes of the Amylose–Lipid Complexes

Table 2 shows the complexation indexes of starches according to the different starch sources. The complexation index represents the ability of starches to form a complex with the lipid. The complexation index values varied from 43% for sweet potato starches to 69% for ginger starches. This variation may be due to the amylose content. Indeed, the higher a starch source's amylose percentage, the more it will be able to form complexes with lipids [25].

Table 2. Texture and complexation index (CI) of the amylose–lipid complex (ALC).

Sample	Hardness (N)	IC (%)
ALC ginger	0.097 ± 0.001^{b}	69.13 ± 1.41^{c}
ALC sweet potato	0.081 ± 0.002^{a}	43.98 ± 1.55^{a}
ALC maize	0.101 ± 0.002^{a}	61.92 ± 0.28^{b}

Means with different superscripts differ significantly per sample ($p < 0.05$).

3.7. Viscosity of the Modified Starches with Lipids and Without Lipids

Figure 4 shows the viscosity results for starches with and without complexation. The viscosity of complexed ginger starch is high compared with the viscosity of uncomplexed ginger starch, and the same applies to corn starch. On the other hand, complexed sweet potato starch has a low viscosity compared with the viscosity of non-complexed potato starch. This could be explained by the fact that the more complex a starch is, the higher its viscosity will consequently be [12]. Specifically, potato starch with an amylose–lipid complex has a high viscosity compared to ginger and corn starches with amylose–lipid complexes but a low viscosity compared to the same potato starch without an amylose–lipid complex. As potatoes have a low amylose content, this is in agreement with the work of Chung and Liu [26], who stated that the formation of an amylose–lipid complex requires the presence of high amylose content in a cereal or tuber in order to promote interaction between the lipid ligand and the amylose helices.

removes about 40% to 50% of moisture from cassava mash [35]. This significant moisture reduction might have been responsible for the TCC concentration in dewatered mashes when compared to the equivalent weight of fresh storage root or grated mash (where the TCCs are diluted by a higher volume of moisture). Varieties IBAI070539 and IBAI011368 had the minimum and maximum percentage of TCC deviation from peeling to grating, and from grating to fermentation and dewatering stages, suggesting that IBAI070539 retained most of its carotenoids during these processing stages, while carotenoids in TMSI011368 were less stable and more readily released through tissue disruption and processing treatments such as grating, fermentation, and dewatering.

Garification can reduce the moisture content of *gari* to as low as 7% [36]. Thus, the heat generated during garification simultaneously dehydrated the bio-fortified *gari* granules, concentrated their inherent TCC, and broke the membranes of large proteins, which might still be enclosing some carotenoids, thereby increasing their availability. Earlier, Gomes et al. [37] reiterated that while domestic food preparation like garification increases the availability of PVACs, it also simultaneously causes their loss through oxidation and/or isomerization. Hence, the increase in TCC of *gari* from IBAI070593 after garification when compared to an equivalent mass of the dewatered sample suggests that variety IBAI070593 has good carotenoid stability potential after exposure to heat during garification. On the contrary, the decrease in TCC of *gari* from IBAI070539 and TMSI011368 following garification implies a poor carotenoid stability capacity of these varieties after exposure to heat. This finding corroborates the submission of Maziya-Dixon et al. [30] that the variability in *gari* TCC (immediately after garification) is a function of its genotypic retention potential. Meanwhile, the disparity in the effect of garification on the TCC of *gari* from IBAI070539 in the two seasons affirms the earlier claim of Lee and Coates [38] that the stability of carotenoids in food is variable even within the same variety. The significant seasonal variation recorded for the TCC of the cassava varieties during processing corroborates the analogous finding of Bechoff et al. [31] and might be due to the difference in initial TCC of the YRCVs. However, the positive relationship between the TCC of fresh storage root and *gari* suggests that the amount of TCC in bio-fortified *gari* is dependent on the TCC in the fresh storage root it was obtained from. Hence, the highest TCC of *gari* from IBAI070593 in this study can be linked to the highest TCC of its fresh storage root, which was also enhanced by the stability of its TCC after exposure to heat. This finding negates the previous submission of La Frano et al. [39] who opined that the final concentration of carotenoid retained in *gari* after garification is not a function of the TCC of the fresh storage root it was obtained from. This might be due to the discrepancies in varieties used in the two studies.

5. Conclusions

From an overview of this study, it can be concluded that the total carotenoid content in the evaluated yellow root cassava varieties fluctuated during *gari* processing. The bio-fortified cassava varieties responded differently to each *gari* processing step, but the total carotenoid content in bio-fortified *gari* is dependent on the total carotenoid content in fresh storage roots and the year of production. Variety TMSI070593 with a high concentration of total carotenoid content in its fresh storage roots and high TCC retention during processing will enhance the total carotenoid contents of *gari* and ensure the delivery of higher pro-vitamin A carotenoids to end-users. The cultivation and processing of variety TMSI070593 are therefore recommended due to its highest *gari* total carotenoid content.

Author Contributions: Conceptualization, I.O.U. and B.O. Methodology, I.O.U., B.O. and P.I.; Software, I.O.U. and B.O.; Validation, I.O.U., B.O. and P.I.; Formal Analysis, I.O.U. and B.O.; Investigation, I.O.U. and B.O. Resources, B.O., P.I. and I.O.U.; Data Curation, I.O.U. and B.O.; Writing—Original

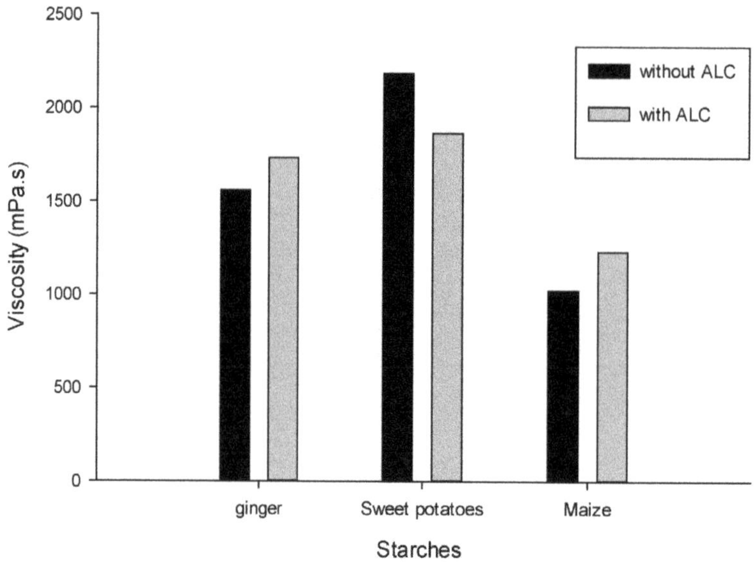

Figure 4. Viscosity of modified and native starches. Legend: ALC: amylose–lipid complex. Legend: ALC: amylose–lipid complex.

3.8. *Texture of the Amylose–Lipid Complex (ALC)*

Table 2 shows the different textures of the corn, ginger, and sweet potato amylose–lipid complexes. The textures of the different samples varied between 0.101 and 0.081. However, the lowest value was observed in the sweet potato amylose–lipid complex (0.081), while the highest value was for maize (0.101). As potatoes have a low amylose content, the amylose–lipid complex had a brittle, less compact texture; meanwhile, maize and ginger starches have a high amylose content, favoring a rigid, more compact amylose–lipid complex texture in maize starch and a less rigid, less compact texture in ginger starch.

3.9. *Influence of Modified Starch Substitution on Mayonnaise Viscosity*

Figure 5 shows the viscosity of the various mayonnaises formulated as a function of shear stress. Viscosity describes the physical property of a liquid's resistance to flow. The figure shows that apparent viscosity decreases with increasing shear rate for all samples. Please note that MS was our reference sample. It can be seen that the highest viscosity is observed in ginger with 50% oil substitution and the lowest viscosity is observed in the sample with water. This may be explained by the fact that, at the time of formulation, sample G50 contained 50% oil, having been reduced by the addition of modified ginger starch (amylose–lipid complex), which made the viscosity more resistant than that of the other samples, and also the formation of the complex was more accentuated on ginger starch due to its high amylose content. This sample also showed phase separation immediately after preparation, with an oily phase at the top, an aqueous phase at the bottom, and a small phase in the middle showing emulsion instability. This shows the importance of adding modified starches during mayonnaise preparation.

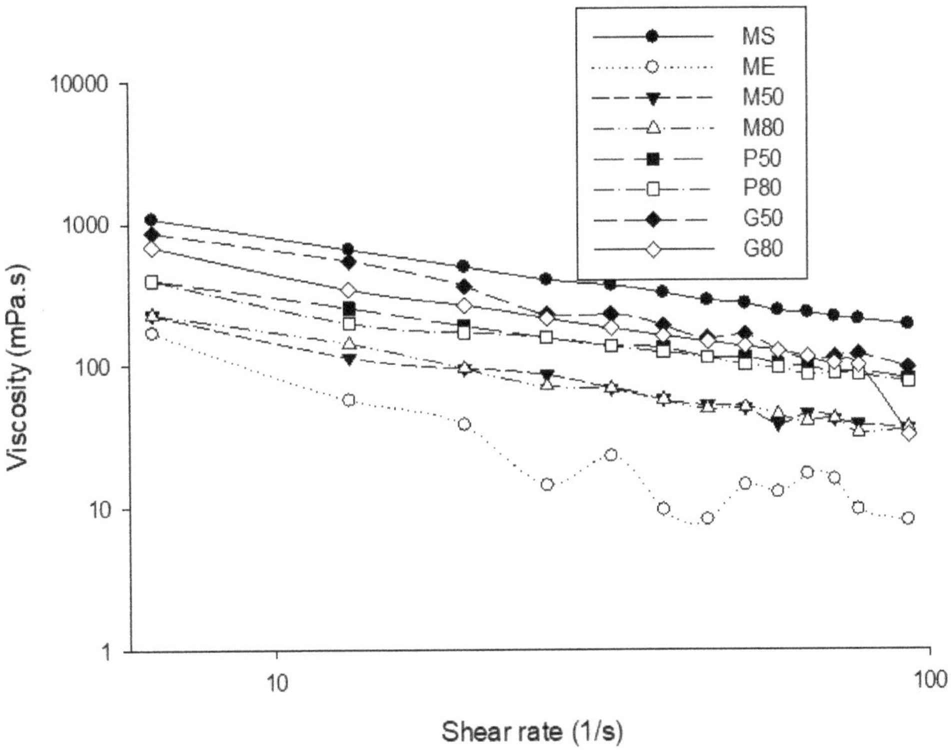

Figure 5. Influence of modified starch substitution on mayonnaise viscosity. **Legend: MS**: full-fat mayonnaise; **ME**: mayonnaise with 50% oil replacement + water; **M50**: mayonnaise with 50% oil replacement + modified maize starch; **M80**: mayonnaise with 80% oil replacement + modified maize starch; **P50**: mayonnaise with 50% oil replacement + modified sweet potato starch; **P80**: mayonnaise with 80% oil replacement + modified sweet potato starch; **G50**: mayonnaise with 50% oil replacement + modified ginger starch; **G80**: mayonnaise with 80% oil replacement + modified ginger starch.

3.10. Influence of Substitution on Sensory Analysis of Mayonnaises

Figure 6 shows the effect of substitutions on the sensory analysis of mayonnaises. In order to select the most widely accepted formulation, a hedonic test based on seven criteria, namely homogeneity, color, viscosity, creaminess, aroma, texture, and general acceptability, was carried out. According to Figure 6, the formulations were organized into two main components (F1 and F2) which explain 79.85% of the variability between different foods, i.e., 52.46% for F1 and 27.39% for F2. Indeed, consumer acceptability of a food product is extremely closely linked to its organoleptic properties. Using the hedonic scores for mayonnaises, as well as principal component analysis, helped us to visualize correlations and understand consumer preferences. The following analysis reflects panelists' behavior during the sensory evaluation. It shows that the ME sample was the least appreciated by the panel. On the other hand, the M50 and M80 samples represent those validated according to the descriptors. Sample M50 was accepted for its smooth texture, homogeneity, and viscosity, while sample M80 was accepted for its aroma and general acceptability.

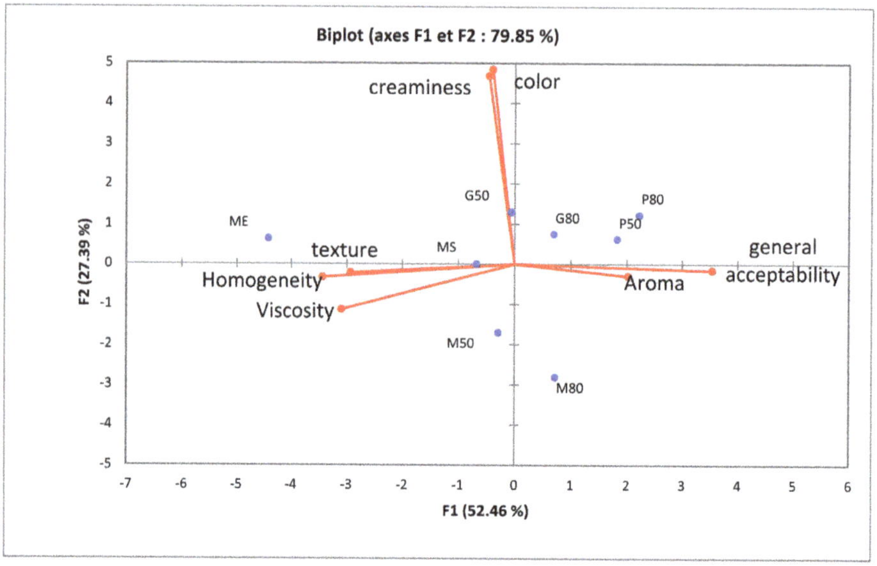

Figure 6. Influence of substitutions on sensory analysis of mayonnaises. Legend: MS: full-fat mayonnaise; ME: mayonnaise with 50% oil replacement + water; M50: mayonnaise with 50% oil replacement + modified maize starch; M80: mayonnaise with 80% oil replacement + modified maize starch; P50: mayonnaise with 50% oil replacement + modified sweet potato starch; P80: mayonnaise with 80% oil replacement + modified sweet potato starch; G50: mayonnaise with 50% oil replacement + modified ginger starch; G80: mayonnaise with 80% oil replacement + modified ginger starch.

4. Conclusions

The aim of the present study was to contribute to the valorization of local tubers and cereals through the formulation of mayonnaises made with modified starches. Low-calorie mayonnaise can be made from modified and unmodified sweet potato, ginger, and corn starches, but the properties depend on the percentage of oil reduction. Ginger and corn starches modified with stearin can produce low-calorie mayonnaise with 50% oil replacement, as physicochemical and rheological analyses revealed that corn cereals and ginger tubers had high amylose contents of 29.45% and 28.91%, respectively, consequently enabling better complexation of amylose with stearin. According to the sensory analysis, panelists preferred the mayonnaise with corn starch modified to replace 50% and 80% of the oil content, respectively. However, their viscosities were not as low as that of standard mayonnaise. Low-calorie mayonnaise made with no modified starch was more unstable and did not resemble a mayonnaise product compared with low-calorie mayonnaise made with modified starch.

Author Contributions: Conceptualization, E.P.A. and A.D.; writing—original draft preparation, E.P.A. and A.D.; supervision, T.C.; project administration, T.C. All authors have read and agreed to the published version of the manuscript.

Funding: No funding was provided for this research.

Institutional Review Board Statement: Not applicable.

Informed Consent Statement: Informed consent was obtained from all subjects involved in the study.

Data Availability Statement: Not applicable.

Acknowledgments: We acknowledge the support of the university of Ngaoundere in providing the materials used for experiments.

Conflicts of Interest: The authors declare no conflicts of interest.

References

1. Souza, C.J.F.D.; Rojas, E.E.G. Emulsion of systems containing egg yolk, polysaccharides and vegetable oil. *Cienc. Agrotec* **2012**, *36*, 543–550. [CrossRef]
2. Mirzanajafi-Zanjani, M.; Yousefi, M.; Ehsani, A. Challenges and approaches for production of a healthy and functional mayonnaise sauce. *Food Sci. Nutr.* **2019**, *7*, 2471–2484. [CrossRef] [PubMed]
3. Panchal, S.K.; Poudyal, H.; Iyer, A.; Nazer, R.; Alam, A.; Diwan, V.; Kauter, K.; Sernia, C.; Campbell, F.; Ward, L.; et al. High-carbohydrate, High-fat Diet–induced Metabolic Syndrome and Cardiovascular Remodeling in Rats. *J. Cardiovasc. Pharmacol.* **2011**, *57*, 611–624. [CrossRef] [PubMed]
4. Mohammed, N.K.; Ragavan, H.; Ahmad, N.H.; Hussin, A.S.M. Egg-free low-fat mayonnaise from virgin coconut oil. *Foods Raw Mater.* **2022**, *10*, 76–85. [CrossRef]
5. Schädle, C.N.; Sanahuja, S.; Bader-Mittermaier, S. Influence of Fat Replacers on the Rheological, Tribological, and Aroma Release Properties of Reduced-Fat Emulsions. *Foods* **2022**, *11*, 820. [CrossRef]
6. Yu, X.X.; Chi, S.X.; Wang, X.H.; Liu, B.H.; Wang, Y.; Zhang, Y.H. Preparation of fat substitute based on maize starch hydrolysates and application in reduced-fat acidified milk gel. *Int. J. Biol. Macromol.* **2023**, *244*, 125479. [CrossRef]
7. Emmambux, M.N.; Taylor, J.R.N. Morphology, physical, chemical, and functional properties of starches from cereals, legumes, and tubers cultivated in Africa: A review. *Starch—Stärke* **2013**, *65*, 715–729. [CrossRef]
8. Teklehaimanot, W.H.; Duodu, K.G.; Emmambux, M.N. Maize and teff starches modified with stearic acid as potential fat replacer in low calorie mayonnaise-type emulsions. *Starch—Stärke* **2013**, *65*, 773–781. [CrossRef]
9. Ren, S. Comparative analysis of some physicochemical properties of 19 kinds of native starches. *Starch—Stärke* **2017**, *69*, 1600367. [CrossRef]
10. Simanavičiūtė, D. Physical and chemical modification of the starch. A review. *Chem. Technol.* **2015**, *66*, 45–54. [CrossRef]
11. Mapengo, C.R.; Ray, S.; Emmambux, M.N. Pasting properties of hydrothermally treated maize starch with added stearic acid. *Food Chem.* **2019**, *289*, 396–403. [CrossRef] [PubMed]
12. D'Silva, T.V.; Taylor, J.R.; Emmambux, M.N. Enhancement of the pasting properties of teff and maize starches through wet–heat processing with added stearic acid. *J. Cereal Sci.* **2011**, *53*, 192–197. [CrossRef]
13. Martins, S.H.F.; Pontes, K.V.; Fialho, R.L.; Fakhouri, F.M. Extraction and characterization of the starch present in the avocado seed (*Persea americana* mill) for future applications. *J. Agric. Food Res* **2022**, *8*, 100303. [CrossRef]
14. Kaufman, R.; Wilson, J.; Bean, S.; Herald, T.; Shi, Y.C. Development of a 96-well plate iodine binding assay for amylose content determination. *Carbohydr. Polym.* **2015**, *115*, 444–447. [CrossRef]
15. Adewuni, A.O.; Robertson, M.D.; Grassby, T. The use of complexing index methods to examine the production of amylose-lipid complexes from potato starch. *Proc. Nutr. Soc.* **2023**, *82*, E4. [CrossRef]
16. González, L.C.; Loubes, M.A.; Bertotto, M.M.; Baeza, R.I.; Tolaba, M.P. Flow behavior and syneresis of ball milled rice starch and their correlations with starch structure. *Carbohydr. Polym. Technol. Appl.* **2021**, *2*, 100168. [CrossRef]
17. Wang, H.; Yang, Q.; Ferdinand, U.; Gong, X.; Qu, Y.; Gao, W.; Ivanistau, A.; Feng, B.; Liu, M. Isolation and characterization of starch from light yellow, orange, and purple sweet potatoes. *Int. J. Biol. Macromol.* **2020**, *160*, 660–668. [CrossRef] [PubMed]
18. Madeneni, M.N.; Faiza, S.; Ramaswamy, R.; Guha, M.; Pullabhatla, S. Physico-chemical and functional properties of starch isolated from ginger spent. *Starch—Stärke* **2011**, *63*, 570–578. [CrossRef]
19. Abegunde, O.K.; Mu, T.H.; Chen, J.W.; Deng, F.M. Physicochemical characterization of sweet potato starches popularly used in Chinese starch industry. *Food Hydrocoll.* **2013**, *33*, 169–177. [CrossRef]
20. Liu, Y.; Chen, X.; Xu, Y.; Xu, Z.; Li, H.; Sui, Z.; Corke, H. Gel texture and rheological properties of normal amylose and waxy potato starch blends with rice starches differing in amylose content. *Int. J. Food Sci.* **2020**, *56*, 1946–1958. [CrossRef]
21. Nhan, M.T.; Copeland, L. Effect of variety and growing environment on pasting and thermal properties of wheat starch. *Starch—Stärke* **2015**, *68*, 436–445. [CrossRef]
22. Bajaj, R.; Singh, N.; Kaur, A.; Inouchi, N. Structural, morphological, functional and digestibility properties of starches from cereals, tubers and legumes: A comparative study. *J. Food Sci. Technol.* **2018**, *55*, 3799–3808. [CrossRef] [PubMed]
23. Amani, N.G.; Tetchi, F.A.; Dufour, D.; Kamenan, A. A Comparative Study of the Syneresis of Yam Starches and other Modified Starches. *J. Food Technol. Afr.* **2002**, *7*, 4–8. [CrossRef]
24. Vamadevan, V.; Bertoft, E. Structure-function relationships of starch components. *Starch—Stärke* **2014**, *67*, 55–68. [CrossRef]

25. Panyoo, A.E.; Emmambux, M.N. Amylose–lipid complex production and potential health benefits: A mini-review. *Starch—Stärke* **2017**, *69*, 1600203. [CrossRef]
26. Chung, H.J.; Liu, Q. Impact of molecular structure of amylopectin and amylose on amylose chain association during cooling. *Carbohydr. Polym.* **2009**, *77*, 807–815. [CrossRef]

proceedings

MDPI

Proceeding Paper

Dynamics of Total Carotenoid Content of Yellow Root Cassava Varieties During *Gari* Processing †

Ibukunolu Oluwadamilola Udemba [1,2,*], Bunmi Olasanmi [1,*] and Peter Iluebbey [3]

1 Department of Crop and Horticultural Sciences, University of Ibadan, Ibadan 200005, Oyo State, Nigeria
2 Institute of Agricultural Research and Training, Obafemi Awolowo University, Moor Plantation, Ibadan 200273, Oyo State, Nigeria
3 International Institute of Tropical Agriculture, Ibadan 200001, Oyo State, Nigeria; p.iluebbey@cgiar.org
* Correspondence: idowuibukunolu2012@yahoo.com (I.O.U.); bunminadeco@yahoo.com (B.O.)
† Presented at the CORAF's 2023 Symposium on Processing and Transformation of Agricultural Products in West and Central Africa: Achievements and Opportunities for Private Sector Engagement, Lome, Togo, 21–23 November 2023.

Abstract: Changes in the total carotenoid content (TCC) of three yellow root cassava varieties, IBAI070593, IBAI011368, and IBAI070539, and a check white root variety (TMSI30572) as influenced by each *gari* processing stage were investigated in this study. In two cropping seasons, fresh storage roots were harvested 12 months after planting, analyzed for TCC (µg/g), and processed into *gari*. The TCCs in grated mash, fermented and dewatered mash, and *gari* were determined following the Bioanalyt procedure. Across processing stages and varieties, significant variations in TCC were found. The TCC levels followed the order of garification (10.46) > fermentation and dewatering (9.84) > peeling (8.78) > grating (7.62), with IBAI070593 exhibiting the highest TCC.

Keywords: *gari* processing; processing stages; total carotenoid content; yellow root cassava varieties

1. Introduction

Cassava (*Manihot esculenta* Crantz) is propagated principally for its storage roots, which serve as food, feed, and raw material for industries, though its leaves are also valued for vegetables. Globally, more than 800 million people depend on cassava for their staple calories [1]. In the developing world, cassava ranks fourth after wheat, maize, and rice as an energy source [2,3]. Specifically, about 700 million and 500 million of the tropics' populace derive about 500 and 100 cal/day, respectively, from cassava roots [4]. Notably, cassava is rated as the second most important crop after maize in Africa, serving as a major staple for more than 40% of its population [1]. The quantity of cassava produced in Nigeria ranks highest globally [5] with up to 80% of Nigerians consuming cassava on a daily basis [6]. An excerpt from the report of Okoye et al. [7] indicated that some Nigerians even eat cassava and its products more than once a day due to its comparative affordability in relation to other crops amidst the prevailing economic recession.

The frequent consumption of cassava and its products without complementing it with vitamin A-rich food is linked to vitamin A deficiency (VAD) in humans [8]. This explains the report of the prevalence of micronutrient deficiency (including vitamin A) among cassava producers and consumers who use the crop as a major staple [9]. The deficiency stems from no or negligible carotenoid and micronutrient contents of white root cassava varieties, which are the most prominent cultivars available to farmers for propagation, and consequently in the market for other stakeholders along the cassava value addition and distribution chain. In addition, animal products which contain pre-formed vitamin A are too expensive for lower socio-economic segments of the Nigerian population to buy. Meanwhile, the continuous intake of diets devoid of pro-vitamin A carotenoids (PVACs; precursor of vitamin A) or pre-formed vitamin A has been reported to lead to partial or total blindness [10]. Other consequences of VAD are the increased vulnerability to diseases, the slowed development and untimely demise of children, and the high mortality of pregnant women [11,12], hence the rising concern of possible sustainable and cost-effective measures of combating VAD in recent decades.

The use of intervention measures like immunization, food diversification, supplementation, and fortification to combat VAD in Nigeria had yielded limited results. This is due to the fact that the target population (usually rural dwellers with low income) can hardly access or afford these measures. This has driven the screening of germplasm by breeders and nutritionists in order to explore its potential for increasing micronutrient density in staples through bio-fortification [13]. Bio-fortification involves the addition of nutrients to staple crops of populations at risk of micronutrients (including vitamin A) deficiency [14]. The successful bio-fortification of cassava in Nigeria led to the development and release of some yellow root cassava varieties (YRCVs) with high total carotenoid content (TCC). This provides a means of making PVACs continuously accessible to malnourished people in rural areas (once the food is consumed) at no extra cost, considering the significant role cassava plays in their diet [6,15].

Unfortunately, cassava is rarely eaten raw by humans but processed into various forms prior to consumption. In Nigeria, about 90% of the total production is processed for consumption [16], with 70% of the processed cassava converted to *gari* [17,18]. Meanwhile, carotenoids are naturally intracellular and processing procedure involving multiple steps (like *gari*-processing techniques) can increase their availability and subsequent degradation [19]. The degradation is enhanced by the chemical instability of carotenoids due to the presence of dual bonds in their molecular structure, and the high sensitivity of β-carotene (the highest carotenoid constituent of YRCVs) to light, oxygen, and heat. These influence to a large extent the amount of TCC retained in bio-fortified cassava products for end users. Therefore, efforts harnessed in enhancing the TCC of YRCVs through bio-fortification will be jeopardized if a substantial quantity is not retained in final products like *gari*, which is the most relished and generally acceptable product of cassava after processing.

In Nigeria, six YRCVs were released in 2011 and 2014. However, there is a paucity of information on the effect of each step involved in processing the fresh storage roots of these novel cassava varieties into *gari* on their TCC. Understanding the trend of carotenoid changes during the processing of bio-fortified cassava varieties into *gari* can help set breeding targets appropriately, improve processing techniques to reduce TCC loss, and ultimately help in varietal selection for propagation and adoption. Hence, this study was conducted to quantify the TCC in fresh storage roots, intermediate products, and *gari* from some novel YRCVs during *gari* processing.

2. Materials and Methods

In the 2018 and 2019 cropping seasons, three YRCVs (IBAI011368, IBAI070539, and IBAI070593) and a white root variety (TMSI30572) as check were planted at the experimental site of the Department of Crop and Horticultural Sciences, University of Ibadan. The experiment was laid out on the field using a randomized complete block design and replicated thrice. Twelve (12) months after planting, the fresh storage roots were harvested and processed into *gari* at the processing unit of the International Institute of Tropical Agriculture (IITA) Ibadan, following the procedure described by James et al. [20]. About 20 kg of fresh storage roots of each variety (in their replicates) were peeled with stainless steel knives, rinsed with portable clean water, and grated mechanically with the aid of a petrol-powered rotating grating machine (Dandrea Agrimport brand). The grated mashes were packed into woven bags, placed inside plastic buckets, and left to ferment for 48 h. Subsequently, the bags containing the grated mashes were arranged in a metallic rack and pressed using a hydraulic jack to enable the simultaneous pressing and fermentation of the cassava mashes for 24 h. The pressed mashes were pulverized using the grater and garified in rectangular pans (heated by firewood) made from stainless steel with a chimney. Pictorial views of the fresh, peeled, grated storage roots and *gari* of the four cassava varieties are presented in Figure 1.

After each phase of *gari* processing, samples were analyzed for TCC at the i-check laboratory of IITA following the procedure described by Bioanalyt [21]. About 5 g of each sample was weighed into a mortar and macerated using a pestle. The macerated sample was made into slurry by mixing it with 20 mL of distilled water. The slurry was subsequently transferred into a well-labeled graduated falcon tube and shaken vigorously to create a uniform mixture. Using a syringe, 0.4 mL of each slurry sample was transferred into a reagent vial. The vials were shaken thoroughly for 10 s (to create a uniform mixture) and then allowed to stand for about 5 min. The absorbance of each sample was read on the i-check device and the TCC was estimated using the formula below:

$$\text{Total carotenoid content} = \text{Dilution factor} \times \text{Absorbance value} \quad (1)$$

The data collected were subjected to descriptive statistics, analysis of variance, and correlation analysis using SAS (9.0 version). Significant means were separated by the least significant difference of means at the 5% level of probability.

3. Results

The results of the combined analysis of variance for the effect of variety, processing stage, and season on the TCC of the cassava during *gari* processing are summarized in Table 1. Significant varietal, processing stage, seasonal, and factor interaction effects were observed for the TCC of the cassava varieties during processing into *gari* (Table 1). The variation in TCC of the cassava varieties after each stage of processing in the two seasons of evaluation is presented in Table 2. Notably, the white root variety (TMSI30572) had no TCC (Table 2). In the 2019 season, variety IBAI070593 had the highest TCC after each processing stage, which differed significantly from the values recorded for other varieties (Table 2). After peeling in the 2020 season, the TCC in the fresh storage roots of IBAI070539 (9.91 µg/g) was highest and statistically at par with IBAI070593 (9.61 µg/g) and IBAI011368 (8.78 µg/g) (Table 2). Conversely, the maximum TCC recorded after grating (9.20 µg/g) and dewatering (10.91 µg/g) in the 2020 season were obtained from variety IBAI070593, and varied significantly from the quantity recorded for other bio-fortified varieties after these processing stages (Table 2).

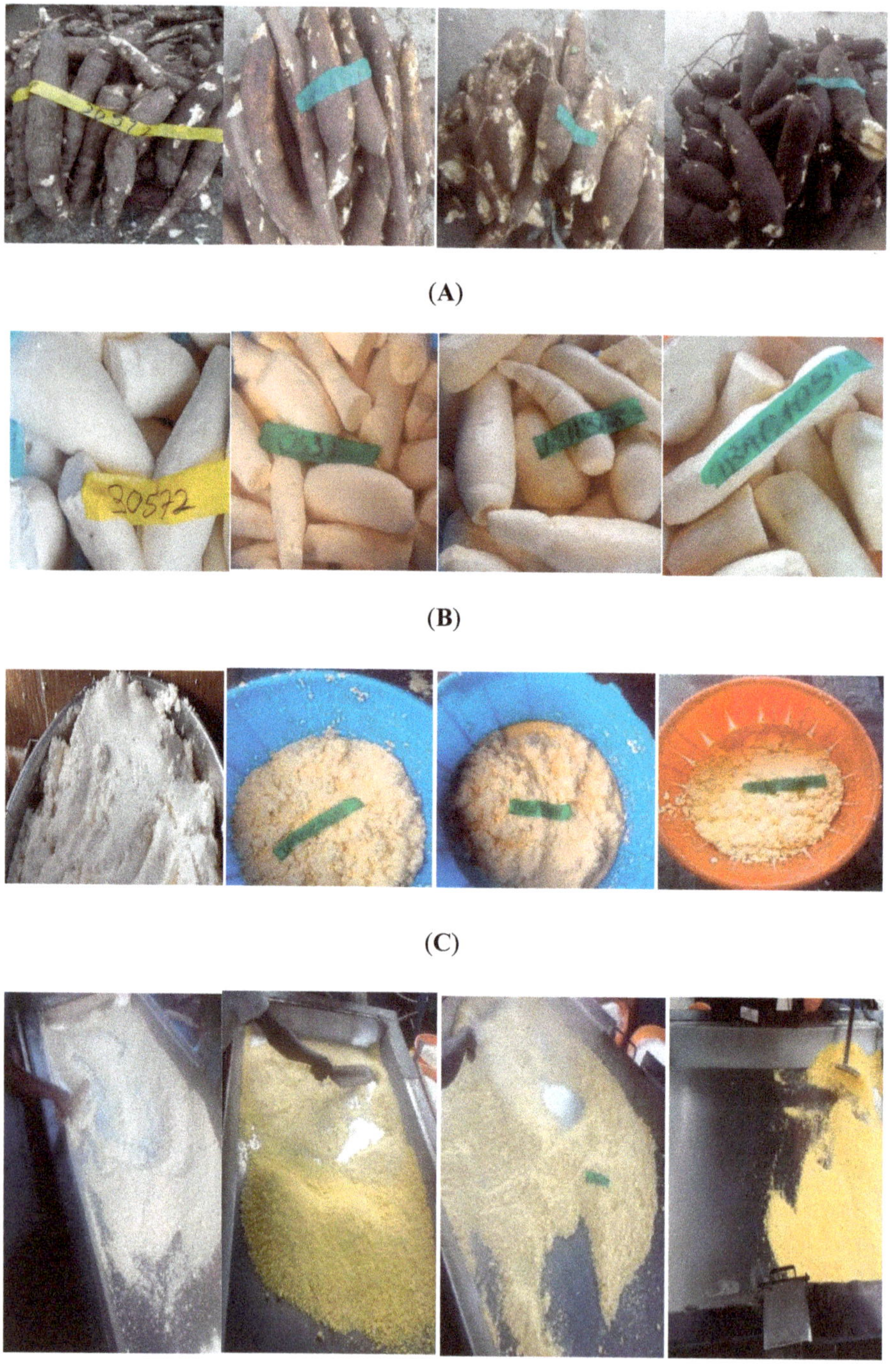

Figure 1. The fresh roots (**A**), peeled roots (**B**), grated mashes (**C**), and *gari* (**D**) of four cassava varieties: TMSI30572 (white root check) and yellow roots of IBAI070539, IBAI011368, and IBAI070593 (from the left).

Table 1. Combined analysis of variance for effect of variety, season, and processing stage on total carotenoid content of four cassava varieties during processing into *gari*.

Source of Variation	Degree of Freedom	Mean Square	Sum of Squares
Replication	2	0.17	0.35
Season	1	1.28 ***	1.28
Processing stage	3	18.68 ***	56.04
Variety	3	555.83 ***	1667.49
Processing stage × season	3	2.22 ***	6.67
Season × variety	3	4.11 ***	12.34
Processing stage × variety	9	14.64 ***	131.76
Processing stage × season× variety	9	2.37 ***	21.37

*** implies significance at 0.001 level of probability.

Table 2. Total carotenoid content (µg/g) of four varieties of cassava after different stages of *gari* processing in two seasons.

Variety	Fresh Storage Root	Grated Mash	Fermented and Dewatered Mash	*Gari*	Fresh Root	Grated Mash	Fermented and Dewatered Mash	*Gari*	Fresh Storage Root	Grated Mash	Fermented and Dewatered Mash	*Gari*
	2019				2020				Average			
IBAI010593	10.85	8.56	10.08	16.58	9.61	9.20	10.91	13.79	10.23	8.87	10.49	15.19
IBAI010539	6.69	6.37	8.07	8.52	9.91	8.03	10.26	8.72	8.30	7.20	9.17	8.62
TMSI30572	0.00	0.00	0.00	0.00	0.00	0.00	0.00	0.00	0.00	0.00	0.00	0.00
IBAI011368	6.85	6.16	9.93	6.86	8.78	5.39	9.80	8.13	7.81	5.78	9.86	7.50
*LSD ($p < 0.05$)	0.16	0.70	0.16	0.12	1.21	0.20	0.63	0.55	0.19	0.15	0.16	0.39

*LSD: least significant difference.

On average, variety IBAI070593 maintained the highest TCC across the processing steps and compared statistically with only IBAI011368 after dewatering. Meanwhile, mean TCC values of TMSI011368 were consistently and significantly lowest among the three YRCVs across the processing stages except after dewatering when it had a statistically higher TCC than IBAI070539 (Table 2). A critical appraisal of the trend of the change in TCC shows that the TCC in all the YRCVs became depleted after grating but increased after fermentation and dewatering (Table 2). However, garification resulted in an increase in TCC of IBAI070593 (in both seasons) and IBAI070539 (in 2019), but a depletion of TCC of IBAI011368 (in both seasons) and IBAI070539 (in 2020) (Table 2). The percentage changes in the mean TCC of the three YRCVs during processing into *gari* are presented in Table 3. After grating and dewatering, the rate of change in TCC ranged from −1.20% (IBAI070539) to −25.99% (IBAI011368) and 18.13% (IBAI070539) to 70.59% (IBAI011368), respectively (Table 3). About 47.93% increase in TCC of IBAI070593 was recorded after garification, while decline of about 6.00% and 24.04% were noted for IBAI070539 and IBAI011368, respectively (Table 3).

Table 3. Percentage changes in mean total carotenoid content of three yellow root cassava varieties during processing into *gari* in two seasons.

Variety	Peeling to Grating	Grating to Fermentation and Dewatering	Fermentation and Dewatering to Roasting
IBAI070593	−13.20	18.13	47.93
IBAI070539	−1.20	11.83	−6.00
TMSI011368	−25.99	70.59	−24.04

"−" indicates TCC loss/reduction.

On average, changes in the TCC of the YRCVs during processing into *gari* followed the order of garification (10.46 μg/g) > fermentation and dewatering (9.84 μg/g) > peeling (8.78 μg/g) > grating (7.62 μg/g) (Table 4). A significantly higher mean TCC was recorded in the second season (9.36 μg/g) compared to the first (Table 4). Meanwhile, TCC in fresh storage roots maintained a significantly strong positive correlation with the TCC of grated mash, dewatered mash, and *gari* (Table 5).

Table 4. Mean total carotenoid content (μg/g) of four cassava varieties in different *gari* processing stages in two seasons.

Processing Stage	Total Carotenoid Content (μg/g)
Peeling	8.78
Grating	7.62
Fermentation and dewatering	9.84
Garification	10.46
LSD ($p < 0.05$)	0.29
Season	
First	8.97
Second	9.36
LSD ($p < 0.05$)	0.20

Table 5. Pearson correlation coefficients between total carotenoid content of fresh storage root and products along *gari* processing chain.

Processing Stages	Fresh Storage Root TCC
Post-grating TCC	0.925 ***
Post-dewatering TCC	0.966 ***
Post-garification TCC	0.920 ***

*** implies significance at the 0.001 level of probability.

4. Discussion

The inedibility of cassava storage root in the raw state for humans and the chemical instability of carotenoids during processing make the benefits of bio-fortification of cassava to be dependent on the TCC status of products after typical processing. A record of 0 μg/g of TCC for the white root check variety affirms the submissions of [22,23] that most white varieties of cassava lack carotenoids. However, the presence of TCC in the fresh storage roots of the YRCVs suggests that YRCVs can proffer better health benefits than white root varieties by combating VAD when consumed, given the detrimental effect of VAD on humans. The TCC in the fresh storage roots of the three YRCVs, though lower than the target benchmark of 15 μg/g set for YRCVs [24], is within the range of 1.37 and 13 μg/g reported by some authors for the fresh storage root of some YRCVs from separate studies [25–29].

The observation of a significant varietal difference in TCC of the YRCVs during processing implies that the stability of carotenoids in the three YRCVs during processing into *gari* depends on their genetic potential for this trait. The observed reduction in TCC of all the YRCVs after grating agrees with the earlier observation of [29–31] who reported a decline in TCC of some YRCVs after grating. Carotenoids are naturally intracellular. Rupturing the YRCV tissues during grating exposed their inherent carotenoids, thereby enhancing their availability. Subsequently, the carotenoids were oxidized in the grater and environment, thus increasing their vulnerability to being lost through leaching from the grated mash. The increase in TCC of all dewatered bio-fortified cassava mashes after fermentation may not be unconnected to the production of additional carotenoids by the lactic bacteria and yeast in the cassava mash during fermentation and the enhanced availability of carotenoids locked up in cassava through softening [32–34]. Dewatering

Draft Preparation, I.O.U. with input from B.O. and P.I., Writing—Review and Editing, I.O.U., B.O. and P.I.; Visualization, I.O.U.; Supervision, B.O. and P.I.; Project Administration, I.O.U., B.O. and P.I. All authors have read and agreed to the published version of the manuscript.

Funding: This research received no external funding.

Data Availability Statement: The original contributions presented in this study are included in the article.

Acknowledgments: The authors wish to appreciate the HarvestPlus Project and the Cassava Breeding Unit staff of IITA, Ibadan, for providing the support for the total carotenoid analysis.

Conflicts of Interest: The authors declare no conflicts of interest.

Abbreviations

The following abbreviations are used in this manuscript:

TCC	Total carotenoid content
VAD	Vitamin A deficiency
YRCVs	Yellow root cassava varieties
PVACs	Pro-vitamin A carotenoids

References

1. FAO. *Protecting Cassava, a Neglected Crop, from Pests and Diseases*; FAO: Rome, Italy, 2019. Available online: https://www.fao.org/publications/card/en/c/CA7117EN (accessed on 13 February 2024).
2. McCallum, E.J.; Anjanappa, R.B.; Gruissem, W. Tackling agriculturally relevant diseases in the staple crop cassava (*Manihot esculenta*). *Curr. Opin. Plant Biol.* **2017**, *38*, 50–58. [CrossRef] [PubMed]
3. Amelework, A.B.; Bairu, M.W.; Maema, O.; Venter, S.L.; Laing, M. Adoption and promotion of resilient crops for climate risk mitigation and import substitution: A case analysis of cassava for South African Agriculture. *Front. Sustain. Food Syst.* **2021**, *5*, 617783. [CrossRef]
4. Byju, G.; Suja, G. Mineral nutrition of cassava. *Adv. Agron.* **2020**, *159*, 169–235.
5. FAOSTAT. Data on Yield of Crops and Area Harvested. 2022. Available online: http://www.fao.org/faostat/en/#data/QC (accessed on 23 December 2022).
6. Salau, S.A.; Nofiu, N.B.; Jimoh, A.T. Effect of cassava products consumption on food security of farming households in Kwara state, Nigeria. *J. Agric. Environ. Sci.* **2019**, *4*, 36–45.
7. Okoye, F.U.; Okoye, A.C.; Umeh, S.I. Consumption behaviour analyses of cassava products among rural household in Ebonyi state, Nigeria. *Agro-Sci.* **2021**, *20*, 14–19. [CrossRef]
8. Udemba, I.O.; Olasanmi, B.; Iluebbey, P. Tandem effect of bio-fortification on dry matter content and *gari* yield of some yellow root cassava (*Manihot esculenta* Crantz) varieties. *Asian J. Adv. Agric. Res.* **2023**, *23*, 11–19.
9. Okwuonu, I.C.; Narayanan, N.N.; Egesi, C.N.; Taylor, N.J. Opportunities and challenges for bio-fortification of cassava to address iron and zinc deficiency in Nigeria. *Glob. Food Secur.* **2021**, *21*, 100478. [CrossRef]
10. Martini, S.; Rizzello, A.; Corsini, I.; Romanin, B.; Fiorentino, M.; Grandi, S.; Bergamaschi, R. Vitamin A difficiency due to selective eating as a cause of blindness in a high income setting. *Pediatrics* **2018**, *141*, S439–S444. [CrossRef]
11. Maziya-Dixon, B.; Akinyele, B.I.; Sanusi, R.A.; Oguntona, T.E.; Nokoe, S.K.; Harris, E.W. Vitamin A deficiency is revalent in children less than 5 years of Age. *Niger. J. Nutr. Sci.* **2006**, *136*, 2255–2261.
12. UNICEF DATA. Vitamin A Deficiency in Children-UNICEF DATA. 2023. Available online: https://data.unicef.org/topic/nutrition/vitamin-a-deficiency/ (accessed on 9 February 2024).
13. Njoku, D.N.; Vernon, G.; Egesi, C.N.; Asante, I.; Offei, S.K.; Okogbenin, E.; Kulakow, P.; Eke-Okoro, O.N.; Ceballos, H. Breeding for enhanced ß-Carotene content in cassava: Constraints and accomplishments. *J. Crop Sci.* **2011**, *25*, 560–571.
14. Mandial, A.; Sood, S.; Chaudhary, D.R.; Jhorar, P.; Rana, R.S. Bio-fortification: An approach to eradicate micronutrient deficiency. *Front. Nutr.* **2023**, *10*, 1233070. [CrossRef]
15. Duah, E.A. Consumption Patterns, Perceptions and Total Carotenoids, Iron and Zinc Contents of Yellow Flesh Cassava. Master's Thesis, University of Ghana, Accra, Ghana, 2016.
16. Githunguri, C.; Gatheru, M.; Ragwa, S. Cassava production and utilization in coastal, eastern and western regions of Kenya. In *Handbook on Cassava Production, Potential Uses and Recent Advances*; Nova Science Publishers, Inc.: New York, NY, USA, 2017; pp. 41–54.

17. Onyemauwa, C.S. Analysis of household consumption of cassava products in Ohaozara, Ebonyi State, Southeast Nigeria. *Researcher* **2010**, *2*, 1–6.
18. Otekunrin, O.A.; Sawicka, B. Cassava a 21 century staple crop. How can Nigeria harness its enormous trade potentials. *Acta Sci. Agric.* **2019**, *3*, 194–202. [CrossRef]
19. Rodriguez-Amaya, D.B. Effects of processing and storage on food carotenoids. *Sight Life Newsl. (Spec. Issue)* **2002**, *3*, 25–35.
20. James, B.; Okechukwu, R.; Abass, A.; Fannah, S.; Maziya-Dixon, B.; Sanni, L.; Osei-Sarfoh, A.; Fomba, S.; Lukombo, S. *Producing Gari from Cassava: An Illustrated Guide for Small Holder Cassava Processors*; International Institute of Tropical Agriculture (IITA): Ibadan, Nigeria, 2012; p. 32.
21. Bioanalyt. *User Manual iCheck Carotene*; Bioanalyt: Teltow, Germany, 2014; 24p.
22. Gegios, A.; Amthor, R.; Maziya-Dixon, B.; Egesi, C.; Mallowa, S.; Nungo, R.; Gichuki, S.; Mbanaso, A.; Manary, M. Children consuming cassava as a staple food are at risk for inadequate zinc, iron, and vitamin A intake. *Plant Foods Hum. Nutr.* **2010**, *65*, 64–70. [CrossRef]
23. Gomes, R.S.; De Almedia, C.F.; Da Silva Costa, J.R.; Machado, J.R.; Delazari, F.T.; Da Silva Santos, F.C.; Da Silva Santos, D.J. Genetic diversity in sweet cassava from the Brazilian middle north region and selection of genotypes based on morpho-agronomical descriptors. *Afr. J. Agric. Res.* **2016**, *11*, 3710–3719.
24. CGIAR. Roots, Tubers and Banana: Planting for Greater Impact. 2013. Available online: https://www.rtb.cgiar.org (accessed on 15 March 2020).
25. Mulokozi, G.I. Content and In Vitro Accessibility of Provitamin A Carotenoids in Some Tanzania Vegetables and Fruits. Effect of Processing and Preparation and Influence on Vitamin A Status. Ph.D. Thesis, University of Technology, Goteborg, Sweden, 2003.
26. Kapinga, R.; Agili, S.; Tumegamires, A. *Development of High Beta Carotene and Other Agronomic Traits Through Conventional Breeding with Farmer's Participation: Annual Report*; International Potato Centre: Kampala, Uganda, 2004.
27. Omodamiro, R.M.; Oti, A.; Etudaiye, H.A.; Egesi, C.; Olasanmi, B.; Ukpabi, U.J. Production of fufu from yellow cassava roots using the odorless flour technique and the traditional method: Evaluation of carotenoids retention in the fufu. *Adv. Appl. Sci. Res.* **2012**, *3*, 2566–2572.
28. Eyinla, T.; Maziya-Dixon, B.; Alamu, E.O.; Sanusi, R.A. Retention of pro-vitamin a content in products from new biofortified cassava varieties. *Foods* **2019**, *8*, 177. [CrossRef]
29. Oliveira, R.G.; Lucia de Carvalho, M.J.; Nutti, R.M.; José de Carvalho, L.V.; Fukuda, W. Assessment and degradation study of total carotenoid and ß-carotene in bitter yellow cassava (*Manihot esculenta* Crantz) varieties. *Afr. J. Food Sci.* **2010**, *4*, 148–155.
30. Maziya-Dixon, B.; Awoyale, W.; Dixon, A. Effect of processing on the retention of total carotenoid, iron and zinc contents of yellow-fleshed cassava roots. *J. Food Nutr. Res.* **2015**, *3*, 483–488.
31. Bechoff, A.; Tomlins, K.I.; Chijioke, U.; Ilona, P.; Westby, A.; Boy, E. Physical losses could partially explain modest carotenoid retention in dried food products from biofortified cassava. *PLoS ONE* **2018**, *13*, e0194402. [CrossRef] [PubMed]
32. Bechoff, A.; Ugo, C.; Keith, I.; Pesila, G.; Paul, I.; Andrew, W.; Erick, B. Carotenoid stability during storage of yellow *gari* made from biofortified cassava or with palm oil. *J. Food Compos. Anal.* **2015**, *44*, 36–44. [CrossRef]
33. Ligia, A.C.; Karp, G.S.; Francielo, V.F.; Karen, Y.F.; Zoz, L.I.; Carvalho, J.C. *Carotenoids: Biotechnological Production of Carotenoids and Their Applications in Food and Pharmaceutical Products*; Intech Open Publisher: London, UK, 2017; pp. 125–143. Available online: http://dx.doi.org/10.5772/67725 (accessed on 15 March 2020).
34. Osagie, V.E.; Onimawo, I.A.; Alamu, O.E. Residual β-carotene and cyanide levels in *gari* produced from unfermented yellow cassava (*Manihot esculenta* Crantz) using local processing method. *JSRR* **2017**, *16*, 1–5. [CrossRef]
35. Dahdouh, L.; Escobar, A.; Rondet, E.; Ricci, J.; Fliedel, G.; Adinsi, L.; Dufour, D.; Cuq, B.; Delalonde, M. Role of dewatering and roasting parameters in the quality of handmade *gari*. *Int. J. Food Sci. Technol.* **2021**, *56*, 1298–1310. [CrossRef]
36. Aregbesola, M.O.; Olaosebikan, Y.O.; Sanni, L.A. Moisture sorption characteristics of garri produced using a mechanical dryer. *Eur. J. Eng. Technol.* **2016**, *4*, 59–67.
37. Gomes, S.; Torres, G.A.; Godoy, R.; Pacheco, S.; Carvalho, J.; Nutti, M. Effects of boiling and frying on the bioaccessibility of β-carotene in yellow fleshed cassava roots (*Manihot esculenta* Crantz cv BRSJari). *Food Nutr. Bull.* **2013**, *34*, 65–74. [CrossRef]
38. Lee, H.S.; Coates, G.A. Effect of thermal pasteurization on Valencia orange juice color and pigments. *LWT* **2003**, *36*, 153–156. [CrossRef]
39. La Frano, M.; Zhu, C.; Burri, B. Effects of processing, cooking, and storage on β-carotene retention and bioaccessability in biofortified cassava *(Manihot esculenta)*. *FASEB J.* **2014**, *28*, 646–654. [CrossRef]

proceedings

MDPI

Proceeding Paper

The Peanut Sector and Artisanal Wafer (Kwlikwli) and Oil (Agonlinmin) Production Units: Lessons for Policies to Support Small-Scale Agrifood Production in West Africa? †

Alida Adjilé * and Roch L. Mongbo

Laboratory for Analysis of Social Dynamics and Development Studies (LADYD), Faculty of Agronomic Sciences, University of Abomey-Calavi, Cotonou 01 BP 526, Benin; rochl_mongbo@yahoo.fr
* Correspondence: adjilealida01@gmail.com
† Presented at the CORAF's 2023 Symposium on Processing and Transformation of Agricultural Products in West and Central Africa: Achievements and Opportunities for Private Sector Engagement, Lome, Togo, 21–23 November 2023.

Abstract: Peanut production in Benin has experienced impressive growth, from 69,222 tons in 1989 to 156,901 tons in 2018, despite the liberalization of the sector, which has deprived it of significant public support. This ethnographic study conducted in Ouessè aims to understand this evolution using the path dependency theory. Semi-structured interviews were conducted with several actors, including producers, collectors, traders and processors. The results show that these actors and private microfinance have reappropriated the historical past of peanuts in the commune of Ouessè, as well as the previous orientations of agricultural policy, to reorganize the financing, collection and marketing of peanuts in order to meet the growing demand from processing units.

Keywords: peanuts; path dependency; policy

1. Introduction

From the colonial period until the early 1980s, agricultural policies implemented in Benin, as in most countries in the sub-region, favored cash crops such as oil palm, peanuts, cotton and, more recently, cashew nuts [1]. Regional specialization was thus established, with a predominance of oil palm in the south, peanuts and cotton in the center, and cotton in the north [2]. The structural reforms of the 1980s led to significant fluctuations in peanut production. However, since 2011, there has been a recovery in this production, particularly in the commune of Ouessè, in the center of the country, with an extension to the north. Paradoxically, this recovery in production has developed in the absence of specific public policies in favor of peanuts. This article examines the historical factors and current dynamics that influenced the evolution of peanut production between 1969 and 2017-2018. It also examines the strategies implemented by stakeholders in the sector to adapt to market constraints. The aim is to better understand current practices, identify ways to sustainably support peanut production in the main production areas and identify lessons for policies to support agrifood production.

2. Materials and Methods

This study is based on the theory of path dependence. The notion of economics of path dependence was developed in political science during the 1990s, to emphasize the weight of choices made in the past and that of political institutions on present options and decisions [3]. In general, the concept of path dependence corresponds to a situation where past progress in a given direction induces subsequent movements in the same direction [3]. Path dependence is defined as a sequential series of steps or events that results from antecedent historical conditions, which define a set of options available at a given moment, called a critical moment [3]. In the context of path dependence, two types of dominant sequences are often studied [4]. The first type involves self-reinforcing sequences, where an institutional pattern, once adopted, provides increasing benefits as it continues to be used. This leads to the pattern becoming more reinforced over time, making it difficult to change or adopt alternatives, even if these might be more effective. Path-dependent analysis may also involve the study of reactive sequences, which are chains of successive events linked by causal relationships, where each event reacts to the previous ones. The final outcome of the sequence is usually the object of study, and the chain of events is seen as a path leading to that outcome. For a sequence to be considered path-dependent, the initial event must be contingent, and the chain must exhibit "inherent sequentiality". However, not all sequences possess these characteristics, and those that do not should therefore not be considered path-dependent.

In the case of peanut production in Benin, the operationalization of the path dependence theory could mean the following:

- Past agricultural policies have created structures and institutions that still guide producers' practices and market dynamics today.
- Initial investments in certain regions, including the commune of Ouessè (Benin), encouraged specialization in peanut cultivation, which has been difficult to change since then.
- Social networks and cultural norms related to peanut production can be barriers to change and, conversely, forces that drive progress in the culture.

Based on this assumption, data were collected using qualitative and quantitative surveys. A diachronic approach made it possible to trace the evolution of agricultural policies from 1960 to the present day. The effects of these policies on production and areas cultivated with peanuts from 1969 to 2017–2018 were assessed using agricultural statistics obtained via FAOSTAT, MAEP compendiums and the yearbooks of the Directorate of Agricultural Statistics of Benin.

Information on national rural development policies and their local implementation, particularly in Ouessè, was collected through a documentary review and interviews with key informants. Fifteen (15) semi-structured interviews, including ten (10) individual and five (5) group interviews, were conducted with producers, collectors, traders, processors, microfinance managers and stakeholders from the technical services of the Ministry in charge of agriculture, livestock and fisheries (MAEP). In addition, a synchronic approach made it possible to document local initiatives developed by stakeholders after the State withdrew from the sector, as well as private financial support mechanisms. In-depth interviews were conducted with officials from the Village Savings Bank and Agricultural Credit (CAVECA) and the Ouessè Food Bank Group (GBVO). Structured surveys were conducted among 30 collectors, 60 producers and 60 traders (wholesalers) randomly selected and interviewed during market days. The data were processed and analyzed using Microsoft Excel 2016 and NVivo 10 software.

3. Results

3.1. *The Main Phases of the Development of Peanut Cultivation in Benin*

The evolution of peanut production in Benin since independence, presented in Figure 1, reveals a trajectory marked by significant fluctuations, closely linked to the agricultural policies implemented. Four major phases can be distinguished: (1) the period from 1969 to

1975; (2) the period from 1975 to 1990; (3) the period from 1990 to 2000; and (4) the period from 2000 to the present day (2017–2018).

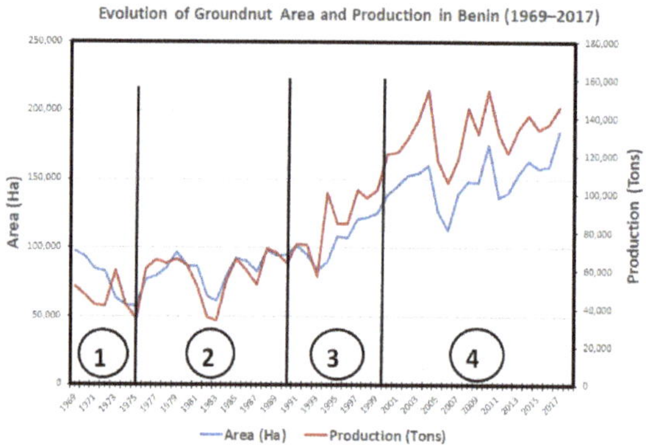

Figure 1. Evolution of national peanut areas and production from 1969 to 2017–2018.

- Period 1: from 1969 to 1975
  Following independence, Benin implemented policies to promote cash crops, including peanuts. A commodity-specific extension approach was adopted, focusing on marketing and supplying local industries [5]. Peanut cultivation areas and production were stable until 1970, before declining until 1975. This collapse was partly attributable to the end of the guaranteed price by France in 1965, which reduced the profitability of the crop for producers [6]. The drought that hit the region in the mid-1970s also contributed to the decline in peanut production [7,8]. Development projects, such as the Integrated Development Operation of Zou (ODIZ), were implemented to promote cash crops, including peanuts. Although the area cultivated with the improved peanut variety increased thanks to these projects, political priority was given to cotton [2]. The fall in cotton prices from 1975 onwards encouraged a slight recovery in peanut production, as producers turned to this crop to compensate for their losses with the cotton crop.
- Period 2: from 1975 to 1990
  Between 1975 and 1990, peanut production in Benin experienced significant fluctuations. The Regional Action Centers for Rural Development (CARDER), created in 1975, coordinated efforts in rural areas, contributing to an increase in peanut production from 34,655 tons over 57,945 hectares in 1975, to 63,931 tons over 94,892 hectares in 1990. However, a drastic decline in peanut production was observed in 1982 and 1983, reaching only 33,595 tons over 61,997 hectares. A recovery began in 1984 with 54,377 tons, peaking again at 63,931 tons in 1990. In parallel, in 1989, Benin suffered a financial crisis, marked by a decline in Gross Domestic Product (GDP) [9], a budget deficit and the collapse of the banking system, affecting the National Society of Fatty Body Industries (SONICOG), which was no longer able to finance the purchase of peanuts. In response to the economic crisis, the Benin government adopted structural adjustment programs in 1989 and 1991, reducing public spending and opening agricultural development enterprises to farmers' organizations and the private sector.
- Period 3: from 1990 to 2000
  Between 1990 and 2000, Benin reorganized its rural sector as part of the post-crisis reforms of 1989, in line with the 1991 rural development policy paper. This period saw a restructuring of the Regional Action Centers for Rural Development (CARDER) and agricultural sectors, with a gradual disengagement of the State and greater in-

volvement of the private sector. The area cultivated with peanuts increased by 46% between 1990 and 2000, with an average annual growth rate of 5.7%. Production increased by 89.5% over the same period, with an annual average of 9.9%. In 2000, 138,596 hectares were sown and production reached 121,166 tons, although the peanut sector did not benefit from the same support as cotton. In 2000, however, the Ministry of Rural Development developed a "Peanut Industry Revival Program" proposing solutions for its relaunch.

- Period 4: from 2000 to the present (2017–2018)
  From 2000 to 2018, peanut production and areas cultivated in Benin stagnated, despite a slight increase over the last five years, with an area of 199,527 hectares cultivated for a production of 156,901 tons. Significant declines were recorded in 2005 (−24% compared to 2004) and in 2011 (−14% compared to 2010). Some efforts to introduce new varieties and support artisanal units have been made, particularly in the Zou and Collines departments, without much success. The sector has suffered from a lack of supervision and a lack of suitable seeds and specific fertilizers, as well as climate degradation and conflicts between farmers and herders. Despite these challenges, recent progress in peanut processing offers potential for future improvement in the peanut sector.

3.2. *The Peanut Marketing System in Ouessè: From the Eve of Independence to Liberalization*

Between 1955 and 1958, the colonial administration set up peasant agricultural mutual societies in each village to supply European industries with peanuts. These mutual societies were federated into regional unions in 1958, notably in the communes of Dassa and Savè (Benin). The present-day commune of Ouessè, then a subdivision of Savè, was an important production area. The Mutual Union of Savè (UMAS) was created to improve the supply of essential products, access to credit for producers, and the organization of peanut marketing. In 1968, the shelling plant, initially installed in Savè, was transferred to Ouessè under pressure from Ouessè producers. UMAS managed the plant with a board of directors made up of collectors, paid on commission and benefiting from loans. UMAS received funds from the Dahomey Agricultural Marketing Company (SOCAD), and then from the National Society for Agricultural Promotion (SONAPRA) to prefinance the purchase of peanuts, which were shelled and sold at regulated prices. In 1975, accusations of mismanagement led to a change of office in 1976. The new office was said to have improved management, increasing the factory's profits. However, market liberalization in 1989 reduced the capacity of UMAS, as it could no longer purchase all of the production. Competition intensified with the arrival of private individual traders, offering more attractive prices to producers. UMAS's attempts to redress the situation by shelling the traders' goods failed. The factory eventually went bankrupt and ceased operations.

3.3. *The Rehabilitation of Old Support Instruments for Peanut Production in the Commune of Ouessè*

With the end of direct government intervention in agricultural production, actors in the Ouessè peanut sector have reactivated the old collection and financing mechanisms to reorganize the financing, collection and marketing of peanuts.

3.3.1. Marketing Organization

Since the liberalization of the peanut sector and the withdrawal of the State, Ouessè producers have had to adapt their marketing strategy. They now sell their peanuts directly or via collectors, mainly to traders in the Zou (Bohicon, Abomey) and Mono (Klouékamè, Azovè) departments of Benin. However, the lack of infrastructure and outlets has led to abusive practices and crises of trust between traders and producers. To address these problems, former UMAS collectors created a new association, the Ouessè Food Bank Group (GBVO). This association elects a president for a renewable three-year term, and organizes peanut collection in subsections at village, district and commune level. It has an average of 80 collectors per village. The GBVO has introduced reforms such as the setting of new measurement units (basin and 100 kg bag) to replace the jute bag. It sets a floor price (10,000 FCFA for a 100 kg bag) that is adjusted according to the regional market. Direct sales between producers and traders were banned to allow for better control. Traders often seek large volumes and establish business relationships with several collectors, directly or through their representatives. Collectors, when they receive orders at the beginning of the campaign, contact producers to reserve the peanuts before harvest. After the harvest, the producer is paid according to the current market price, as soon as the collector receives the goods. If the producer lacks the resources to finance production, the collector can, based on the floor price and the expected number of peanut bags, grant him a loan. Collectors receive 500 FCFA per bag sold, including 400 FCFA for their service and 100 FCFA for transport. They finance their activities mainly through prefinancing from traders, with secondary contributions from their own savings or microfinance loans.

3.3.2. Production and Marketing Support Systems

Financing agricultural production, particularly for food crops such as peanuts, has been a persistent challenge since the end of direct State support. Previously supported by the government with seeds and fertilizers, producers must now finance their production themselves, often with the help of traders, collectors or moneylenders. With the resumption of peanut production in recent years, several microfinance institutions have set up in Ouessè, though it is a landlocked place.

According to the results of our surveys, 87% of traders, 75% of collectors and 60% of producers use the services of these institutions. Producers mobilize, on average, 76.55% of their savings to finance production, mainly for seeds and labor. The presence of microfinance institutions, such as the Village Savings Bank and Agricultural Credit (CAVECA), the Local Agricultural Credit Mutual Fund Gross Domestic Product (CLCAM) and the Benin Agricultural Mutual Insurance (AMAB), has reduced the use of prefinancing by traders or usurers. CAVECA offers loans for agricultural production, covering a period of 12 to 18 months. Repayment begins after 6 months of activity and is spread over the following two quarters, with a decreasing interest rate of 2% per month. Regardless of the type of credit requested, the value of the financial guarantee, initially set at 25%, was reduced to 16% in 2011, and then to 10% in 2013, to better respond to customers' concerns and deal with competition between institutions.

In view of the data presented in Table 1, facilitating access conditions, in particular by reducing the financial guarantee and interest rate, has significantly increased the number of beneficiaries and encouraged producers and traders to get involved.

Table 1. Types and values (in FCFA) of loans granted by CAVECA GBEME KPASSA to finance agricultural activities in 2005, 2013 and 2014.

Types of Credit Granted	Years	Number of People Receiving Credit	Total Value of Credit Granted (in FCFA)
Loans granted for production	2005	5	1,000,000
	2013	688	148,570,000
	2014 (January to September)	198	90,910,000
Credits granted for the agricultural food trade	2003	19	2,621,600
	2013	1273	389,900,000
	2014 (January to September)	1307	367,390,000

In 2013, Village Savings Bank and Agricultural Credit of Gbeme Kpassa (Ouessè, Benin) granted 688 loans, with a total value of 148,570,000 FCFA, for agricultural production, and 1273 loans, with a value of 389,900,000 FCFA, for food trade.

> "Before the arrival of microfinance institutions, 8 out of every 10 producers took out loans with traders and loan sharks. But today there are barely 4 who still take out loans with these two actors. I think that this business will flourish in the years to come. Many farmers don't grow cotton, they grow peanuts. Some give up, but at the same time others continue to produce them. So, there's no shortage of goods", says a peanut grower.

However, loans granted for the marketing of agricultural products dominate those granted for agricultural production.

4. Discussion

The analysis of the evolution of peanut production in Benin, in light of the path dependency theory, shows a strong influence of previous agricultural policies on the current dynamics of this crop in the commune of Ouessè. The trajectory of peanut production has been shaped by several critical moments, such as the influence of colonial policies [10], structural adjustment programs, price incentives, technical support with the establishment of CARDER, the organization of marketing and climatic disturbances. After independence, the continuation of colonial policies focused on cash crops consolidated the place of the peanut in the Beninese economy. Many African countries continued with colonial policies of promoting export crops until 1970 [11]. However, the Benin era of revolutionary rule introduced new paradigms favoring food self-sufficiency, which initially boosted peanut production, before challenges such as the reorganization of the cotton sector and drought reversed this trend. The marketing system set up by the colonial administration, particularly UMAS, supported the sector until the end of the 1980s. However, with the disengagement of the State, producers were faced with new challenges linked to market liberalization, requiring reorganization to maintain the value chain [11–13]. In this new environment, the former leaders of UMAS collectors were able to adapt by creating a new organization (GBVO). This structure continues to play a key role in facilitating transactions between producers and traders and ensuring a secure marketing network. By establishing a minimum price for peanuts, these collectors not only strengthened their influence, but also helped to stabilize production within family farms. The specialization of the commune of Ouessè in peanut cultivation, despite the liberalization of the sector, is therefore the result of a series of historical choices and favorable initial conditions. The organization of the collection system, the social networks established between producers, collectors and traders, the facilitation of access to microcredits for the financing of production and collection activities, and then the local standards that have developed around this crop, constitute the essential elements of the continuity of this production system. Although local stakeholders have been able to adapt to certain conditions in their commercial environment to maintain the peanut sector, significant challenges persist, particularly in terms of resilience to climatic hazards and market fluctuations. It is essential to consider these historical dynamics to better understand the reluctance to certain changes proposed today, and to envisage development strategies that take into account not only structural constraints, but also the assets inherited from the past.

5. Conclusions

This study of the peanut sector in Ouessè demonstrates a remarkable capacity for adaptation of local actors in the face of political and economic developments. Despite historical fluctuations in production, linked to agricultural policies and different critical moments, peanut production remains dynamic thanks to the revival of local stakeholders, who have managed to adapt former practices to the current liberal and non-State context. They set informal governance rules and internal economic dynamics that shape the marketing system. The creation of the Ouessè Food Bank Group (GBVO) after the liberalization of the market is a significant example of the reappropriation of institutional instruments inherited from the colonial era. This organization played an important role in structuring the sector, guaranteeing a minimum income for producers and promoting market stability. However, several challenges remain. Producers are still vulnerable to price fluctuations and climatic hazards. In addition, reliance on microfinance institutions for production financing poses a risk in the event of economic shocks [14]. In conclusion, this study highlights the importance of taking into account local dynamics and historical trajectories in the design of agricultural development policies. An approach adapted to specific realities and past experiences is essential to strengthen the resilience of actors in agricultural sectors and ensure the long-term sustainability of these sectors in places that are major producers, such as the case of the Ouessè commune for peanut production.

Author Contributions: Conceptualization, A.A. and R.L.M.; methodology, A.A. and R.L.M.; software, A.A.; validation, A.A. and R.L.M.; formal analysis, R.L.M.; investigation, A.A.; resources, A.A.; data curation, A.A.; writing—original draft preparation, A.A.; writing—review and editing, A.A. and R.L.M.; visualization, A.A.; supervision, R.L.M.; project administration, A.A.; funding acquisition, R.L.M. All authors have read and agreed to the published version of the manuscript.

Funding: This research was funded by the project "Governance of Local Identity Product Hubs for the Prosperity of Agricultural and Agro-food Craft Enterprises in Southern and Central Benin (GOFIT)", financed under the Interdisciplinary and Interfaculty Competitive Research Fund Projects 2012–2014 of the University of Abomey-Calavi (UAC).

Institutional Review Board Statement: Not applicable (No ethical approval required).

Informed Consent Statement: Informed consent was obtained from all subjects involved in the study.

Data Availability Statement: The raw data supporting the conclusions of this article will be made available by the authors on request.

Conflicts of Interest: The authors declare no conflict of interest.

Abbreviations

AMAB	Assurance Mutuelle Agricole du Benin/Benin Agricultural Mutual Insurance
CARDER	Centres d'action régionaux pour le développement rural/Regional Action Centers for Rural Development
CAVECA	Caisse Villageoise d'Epargne et Crédit Agricole/Village Savings Bank and Agricultural Credit
CLCAM	Caisse Locale de Crédit Agricole Mutuelle/Local Agricultural Credit Mutual Fund
GPD	Gross Domestic Product
GBVO	Groupement de Banques de Vivres de Ouessè/Ouessè Food Bank Group
MAEP	Ministère de l'Agriculture de l'Elavage et de la Pêche/Ministry of Agriculture, Livestock and Fisheries
ODIZ	Opération de Développement Intégré du Zou/Integrated Development Operation of Zou
PIB	Produit Intérieur Brut/Gross Domestic Product
UMAS	Union de Mutuelle de Savè/Mutual Union of Savè
SOCAD	Société de commercialisation agricole du Dahomey/Dahomey Agricultural Marketing Company

SONAPRA	Société Nationale pour la Promotion Agricole/National Society for Agricultural Promotion
SONICOG	Société Nationale des Industries des Corps Gras/National Society of Fatty Body Industries

References

1. Nubukpo, K. The Political Economy of Reform in the Cotton Sector in Western and Central Africa: Toward the Convergence of Modes of Organization. *Mondes En Développement* **2011**, *3*, 93–109. [CrossRef]
2. Mongbo, R.L.; Dossou-Houessou, A. The rural sector in Benin: The colonial legacy and its management until 1990. In *Economy and Society in Benin from Yesterday to Tomorrow*; Hountondji, P., Ed.; L'Harmattan: Paris, France, 2000; Chapter 2; pp. 75–99.
3. Palier, B. Path dependence. In *Dictionary of Public Policies*, 4th ed.; Boussaguet, L., Jacquot, S., Ravinet, P., Eds.; Presses de Sciences Po: Paris, France, 2014; pp. 411–419. [CrossRef]
4. Mahoney, J. Path Dependence in Historical Sociology. *Theory Soc.* **2000**, *29*, 507–548. [CrossRef]
5. Otchoun, B. Case study on Benin. In *Study of the Situation and Evolution of Forest Extension and Animation Systems in Sahelian Africa*; FAO, Ed.; FAO: Rome, Italy, 2004.
6. Breusers, M.; Gibbon, D. Adja Agriculture. In *Peasant Dynamics on the Adja Plateau in Benin*; Daane, J., Breusers, M., Frederiks, E., Eds.; Karthala: Paris, France, 1997; pp. 189–240.
7. Koudamiloro, O.; Vissin, E.W.; Sintondji, L.O.; Housou, C.S. Socio-economic and environmental effects of hydroclimatic risks in the Ouémé River watershed at the Bétérou outlet in Benin (West Africa). In Proceedings of the XXVIII Colloquium of the International Climatological Association, Liège, Belgium, 1–4 July 2015.
8. Sircoulon, J. Drought in West Africa: Comparison of the years 1982–1984 with the years 1972–1973. *Cah. ORSTOM Ser. Hydrol.* **1985**, *21*, 75–86.
9. Soglohoun, G.; Satoguina, H. Domestic Savings and Economic Growth in Benin. Master's Thesis, Faculty of Economics and Management, University of Abomey-Calavi, Abomey Calavi, Benin, 2009.
10. Daviron, B. The Historical Integration of Africa in the International Food Trade: A Food Regime Perspective. In *Globalization and Restructuring of African Commodity Flows*; Fold, N., Nylandsted Larsen, M., Eds.; Nordiska Afrika Institutet: Uppsala, Sweden, 2008; pp. 44–78.
11. Daane, J.; Mongbo, R.L. Peasants' influence on development projects in Benin: A critical analysis. *Geneva-Africa* **1991**, *29*, 49–76.
12. Achancho, V. The role of peasant organizations in the professionalization of agriculture in sub-Saharan Africa: The case of Cameroon. Ph.D. Thesis, AgroParisTech, Paris, France, 17 December 2012.
13. Spinat, J.-B.; Traoré, B.; Saywell, A.-S. *Support and Advice to Farmer Organizations in the Office du Niger Area: From the Service Provision Center Project to the "Faranfasi So"*; Traverse No. 26; Les Éditions du Groupe Initiatives, Institute for Research and Application of Development Methods: Paris, France, 2006; p. 10.
14. Kpadé, P.C.; Boinon, J.-P. Dynamics of cotton policies in Benin: A reading through path dependence. *Rural. Econ.* **2011**, *321*, 58–72. Available online: http://journals.openedition.org/economierurale/2949 (accessed on 2 May 2019). [CrossRef]

proceedings

MDPI

Proceeding Paper

Endogenous Adaptation Strategies of Rice Millers to Water and Weather-Related Pressures in Benue State, Nigeria †

Fanen Terdoo [1,2]

[1] Department of Geography, Federal University Dutsin-Ma, Dutsin-Ma PMB 5001, Katsina State, Nigeria; fanenterdoo@gmail.com or fterdoo@bsum.edu.ng; Tel.: +234-9018138072

[2] Department of Geography, Benue State University, Makurdi PMB 102119, Benue State, Nigeria

† Presented at the CORAF's 2023 Symposium on Processing and Transformation of Agricultural Products in West and Central Africa: Achievements and Opportunities for Private Sector Engagement, Lome, Togo, 21–23 November 2023.

Abstract: Rice milling in Sub-Saharan Africa (SSA) remains highly dependent on water availability and favorable weather conditions, making local millers vulnerable to water- and weather-related pressures (WWrP). This study examines how rice millers in Makurdi and Adikpo, Benue State, Nigeria, adapt to these pressures. Data were collected through snowball sampling involving 21 rice millers and two workshops attended by 14 millers. Content and thematic analyses of data were carried out using NVivo 11. Findings indicate that rice milling operations are highly exposed and sensitive to flooding, excessive wetness, dryness, high humidity, and water scarcity. These pressures are likely to intensify due to climate change and pronounced climate variability. To adapt, millers employ endogenous strategies, including temporarily relocating or evacuating mills, rotating drying schedules, modifying drying methods, reducing or suspending parboiling during floods, and digging wells. In spite of these adaptation measures, locally milled rice continues to face issues of poor quality and diminished competitiveness. Consequently, this study highlights upgrading milling technologies and infrastructure, promoting proactive and long-term adaptation measures, and supporting collective adaptation strategies among millers to enhance resilience in the rice value chain.

Keywords: agri-food systems; rice value chains; rice processing; post-harvest management; climate change and variability; vulnerability; exposure; sensitivity; adaptation; resilience

1. Introduction

Nigeria is a prominent rice producer in Africa with a substantial installed capacity of over 3 million metric tons for rice milling each year [1,2]. This capacity encompasses both large-scale integrated rice mills and smaller mills spread throughout the country [2]. The smaller mills, often operated by local entrepreneurs or cooperatives, are dispersed across different regions such as the North-West, North-Central, and South-East and serve the local markets and communities [3]. Consequently, these mills play a significant role in ensuring food security and boosting the economy. They contribute substantially to rice consumption in Nigeria and generate employment opportunities along the value chains. By reducing reliance on rice imports, which exposes the country to external factors such as market fluctuations, currency exchange rates, and trade disruptions, domestic rice milling enhances food security and price stability [4,5]. Furthermore, local rice milling helps to stabilize the rice supply within the country by reducing transportation costs and logistical challenges associated with importing rice from distant locations [6]. This not only saves foreign reserves but also ensures a consistent supply of rice, particularly during global food crises or trade disruptions. Moreover, rice milling operations not only support local rice farmers by providing a reliable market for their produce but also create employment opportunities across different stages of the supply chain, thereby alleviating poverty, stimulating economic growth, and fostering rural development in agricultural regions [7].

Despite the crucial role of domestic rice milling in ensuring food security and economic stability, rice milling in Sub-Saharan Africa (SSA) heavily relies on water availability and favorable weather conditions [8,9]. Local millers often lack access to modern rice milling technologies, leading to a dependence on water for various stages of the milling process, including cleaning, soaking, and parboiling [10,11]. In regions with irregular or unreliable water availability due to seasonal variations in rainfall or inadequate irrigation infrastructure, small millers may face challenges in maintaining consistent milling operations [5]. Water scarcity can disrupt production schedules, cause delays in processing activities, and reduce milling capacity, resulting in decreased productivity and income for mill owners. In low-lying and flood-prone areas, excessive rainfall can lead to flooding, which can damage rice milling factories, causing waterlogging, soil erosion, and harm to milling facilities, machinery, and storage structures [12,13]. This disrupts milling operations, necessitates costly repairs, and leads to downtime for small-scale millers, impacting their profitability and competitiveness in the market [14].

Furthermore, high air humidity levels and moisture content during rainy seasons can worsen the growth of mold and spoilage of rice grains during storage and processing [15]. Small-scale rice millers may encounter difficulties in mitigating the risks of grain deterioration and quality degradation caused by fungal contamination, resulting in product losses, rejection by buyers, and damage to reputation [16]. The challenges of managing moisture-related issues in rice milling are further complicated by inadequate drying facilities and poor ventilation in storage areas. Moreover, the reliability of electricity supply for small-scale rice mills can be affected by extreme weather conditions, such as power outages during storms or damage to electrical infrastructure. This can lead small millers to rely on grid electricity, diesel generators, or other energy sources for their milling operations as any interruptions or fluctuations in energy supply can result in increased production costs, decreased efficiency, and hinder the overall profitability of rice milling enterprises [17].

In addition to the above, difficulties faced by small-size millers in addressing the overreliance of domestic milling on WWrP is due to a lack of adequate infrastructure for rice milling, including milling facilities, storage facilities, transportation networks, and power supply [17,18]. The absence of modern milling equipment and technology also hampers the efficiency and competitiveness of domestic rice milling operations. Additionally, small-scale rice millers often have limited access to credit and financial services, which constrains their ability to invest in modern milling equipment, infrastructure upgrades, and working capital [19]. This limits their capacity to expand their operations, improve productivity, and compete with larger-scale producers. Furthermore, the lack of skilled personnel and technical expertise in rice milling techniques, equipment maintenance, and quality control poses a significant challenge to domestic rice millers [18]. This affects their ability to optimize their milling processes, adhere to quality standards, and meet consumer preferences [20]. As a result, inadequate market access and competition arise, with domestic rice being outcompeted by imported rice that may be subsidized and offered at lower prices [1]. Inadequate market access, trade barriers, and lack of government support for domestic rice promotion further exacerbate the challenges faced by local millers in competing with imported rice.

In spite of the significant adverse impacts of WWrPs on domestic rice milling, there is limited documentation on the efforts made by indigenous millers in the region to adapt and build the resilience of their rice milling processes and operations to these pressures and continue to supply the majority of rice consumed across Nigeria. These efforts are referred to here as endogenous adaptation strategies. These are locally developed strategies based on indigenous knowledge, social structures, and cultural institutions [13,21,22]. These practices enable local actors and communities to manage or mitigate the impacts of climate change. Key characteristics of endogenous adaptation are that they are deeply rooted in local culture and shaped not by external influences but by the socioeconomic, political, and environmental conditions unique to each community [13,22]. Thus, given the importance of such practices to managing and moderating the impacts of climate change, this study aims to document the exposure and sensitivity of rice millers to WWrPs in Benue State, Nigeria. It also seeks to examine the current endogenous adaptation strategies used by rice millers in response to WWrPs and identify potential entry points for policy interventions to enhance the resilience of rice milling in Benue State.

2. Methodology

This section provides an overview of the study area, including the criteria for selecting data collection sites. It also describes the data collection process, highlighting the sampling techniques and procedures used for conducting interviews and workshops. Finally, it outlines the methods used for data analysis and the presentation of results.

2.1. Study Area

The study was carried out in Benue State, commonly known as the "Food Basket of the Nation" [23]. This state is situated in the North-Central geopolitical zone of Nigeria, between latitude 7°47′ N and 10°0′ N, as well as longitude 6°25′ E and 9°30′ E (Figure 1). It shares its boundaries with Nasarawa State to the north, Taraba State to the east, Cross River State to the southeast, Enugu State to the south, Ebonyi State to the southwest, and Kogi State to the west. Makurdi is the capital city of Benue State, where agriculture forms the backbone of the state's economy, encompassing the production of a diverse range of fruits, grains, and tuber crops. The state is known for its significant rice cultivation originating from the six primary rice production and processing belts or clusters, namely Lower Benue River Basin (Makurdi), Gboko, Katsina-Ala, Utukpo, Ukum, and Adikpo [24]. These collectively contribute to the state's substantial rice production output, playing a pivotal role in fulfilling local demand and making a notable contribution to the broader agricultural sector in Nigeria.

2.2. Data Collection

To account for the current exposures and adaptation responses of the miller to WWrP, a qualitative strategy that draws from a combination of participatory tools, including participants' observations, semi-structured interviews, and workshops, were used in two rice processing and milling clusters: Makurdi in the West and Adikpo in the North. These locations were purposefully chosen due to their contrasting geographical and political characteristics, as well as the presence of different types of rice mills, ranging from large-scale integrated mills to smaller ones.

A total of 21 face-to-face semi-structured interviews (10 in Adikpo and 11 in Makurdi) to obtain a comprehensive insight into the issue being studied. Although the sample size appears small, it is considered representative, as it reflects the characteristics of rice millers in the two study locations. In Adikpo, Kwande, 10 participants (22%) were selected from 45 active rice milling houses, while in Makurdi, 11 participants (64.7%) were chosen from a total of 17 milling houses. The interviews, each lasting 1 hour and 30 minutes or longer, were initially conducted between February and July 2017, with additional interviews added in August 2019. In addition, 2 workshops involving 14 rice millers (7 in Makurdi and 7 in Adikpo) were conducted to complement and validate the interviews in the respective locations. This approach engages participants in practical exercises and activities, enhancing the co-creation process while generating qualitative insights into endogenous adaptive practices and identifying entry points to reduce millers' exposure to water-related pressures [25].

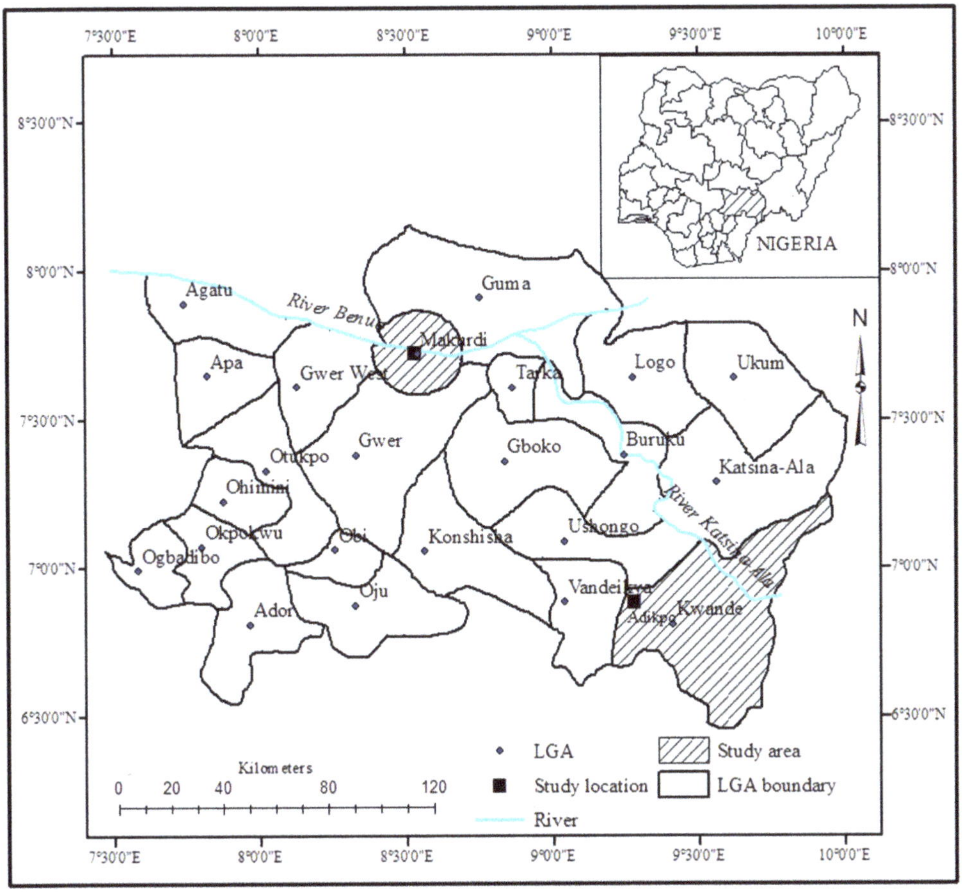

Figure 1. Map of study area, Benue State, Nigeria.

Given the complexity of exploring endogenous adaptation practices, the small sample size allows for a more in-depth examination and meaningful engagement with each participant. This enables researchers to uncover rich qualitative data that larger samples might overlook (see also, [26]). Thus, with participants' consent, both the interviews and workshops were recorded using an Olympus Digital Voice Recorder VN-741PC. Additionally, two research assistants took notes, which provided a foundation for analyzing the interviews and workshops in the later stages of the research.

Moreover, due to the millers' skepticism and reluctance to grant interviews to individuals they perceive as outsiders, such as government officials or representatives of organizations whom they feel have been unsupportive or have made unfulfilled promises of loans and safety nets to improve their milling quality but failed to deliver, a snowball sampling was implemented to overcome this challenge and address the issue of distrust. This involved having a leader introduce the interviewer to the next miller, and this process continued until all interviews were completed, reaching the saturation point where responses became consistently similar, and repetition was no longer necessary. Although rice processing activities typically fluctuate seasonally (within a year), the exposure, sensitivity, and adaptive practices are influenced by barriers and enabling factors that develop over a more extended period.

Consequently, data and information about rice millers' vulnerability (i.e., exposure, sensitivity, and adaptive strategies) to WWrPs, such as the types of pressure and impacts currently being experienced, as well as the strategies employed to manage or mitigate these impacts on rice milling processes and activities, were collected. Accordingly, during the interviews, rice millers were invited to share their experiences with exposure, sensitivity, and adaptation to WWrPs, particularly extreme events, and to describe how they had responded to these challenges over the past 5–10 years (refer to Figure 2 and Table 1 for the summary of data generated from the interviews).

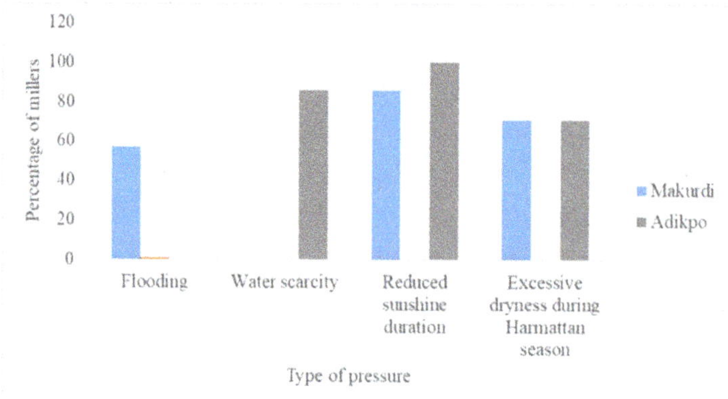

Figure 2. Types of pressures identified by rice millers at the study sites.

Table 1. Percentage of millers implementing various adaptation practices.

Type of Adaptation	Source of Pressure	Examples of Adaptation Strategies	Makurdi (n = 11)	Adikpo (n = 10)
Tactical reactive	Flooding	Temporary closure and relocation	7 (63.6%)	
	Limited sunshine in the rainy season	Changing rice drying technique (e.g., from thickly to thinly)	7 (63.6%)	6 (60%)
		Reducing the usual quantity processed	6 (54.5%)	7 (70%)
		Taking turns to dry	4 (36.4.6%)	5 (50)
	Excessive dryness in the harmattan	Changing rice drying technique (e.g., from thinly to thickly)	7 (63.6%)	4 (40%)
		Reducing the usual quantity processed	6 (54.5%)	7 (70%)
	Water scarcity	Going to nearby streams and rivers		4 (40%)
Strategic anticipatory		Digging a hand well		7 (70%)

n represents the total number of millers interviewed at each site, with the group percentage provided in parentheses.

2.3. Data Analysis

With the help of two research assistants, both native speakers of Tiv, one holding an MSc in English Language and the other an MSc in Geography, the researcher, who also has a good command of Tiv, translated the audio recordings of the interviews and workshop sessions from Tiv to English. The translations were then stored as Microsoft Word documents. To minimize bias during the transcription of interview data, verbatim translation was used to avoid "cleaning up" grammar, paraphrasing, or adding interpretation. Additionally, collaboration with two research assistants who are native speakers helped ensure accurate transcription and translation. In addition, peer debriefing sessions were conducted after each transcription to check for potential bias or misrepresentation. During this process, the field and workshop notes were utilized to supplement the audio recordings.

A computer-assisted NVivo 11 (QSR International, 17; [27]) was used during the analysis. The transcripts of the interviews were imported into the software and files were created to organize the data. Codes were generated progressively and refined or merged by selecting text and assigning it to nodes or themes. An inductive thematic and content analysis was conducted to identify code frequencies [28,29]. This involved counting how often specific codes appeared, particularly by noting how many participants in a set of interviews mentioned or alluded to a given issue or concept. Percentages were then calculated based on these frequencies. The utilization of content and inductive thematic analyses allowed for the identification of emerging themes, patterns, and the hiding and underlying insights within the qualitative data [30].

To ensure validity, the findings from the analysis of interview data were cross-checked against the workshop analysis results. If both methods produced similar outcomes, it could be inferred that the findings are robust. However, if there were disparities between the two results, the reasons behind such discrepancies were provided. Finally, the results obtained from the analysis are presented through charts or graphics and tables.

3. Results

The section presents the results on the types of WWrPs that rice millers are exposed and sensitive to and the endogenous strategies employed in response to the pressures.

3.1. Exposure and Sensitivity of Rice Millers to WWrPs in Benue State

The results of the analysis of data on millers' exposure and sensitivity to WWrPs are presented in Figure 2. Results showed that rice millers encountered frequent flooding of rice mills, water shortages, limited sunshine hours during peak rainy seasons, and extreme dryness caused by harmattan winds in the dry season (Figure 2).

A number of millers shared their experiences regarding these pressures, which are highlighted in Figure 2. According to the millers, flooding has posed a significant challenge in the region of Makurdi. In this particular area, both the Wurukum and Wadata rice mills are located on the banks of River Benue, allowing them to fulfill the water requirements during the rice parboiling and milling. Nevertheless, this location of the two mills makes them prone to frequent flooding, often disrupting key milling activities such as soaking, parboiling, and drying. This location also makes the two mills vulnerable to frequent flooding, which often disrupts milling activities, such as soaking, parboiling, and drying. These disruptions can go on for days, weeks, or even months, especially during the heavy rainfall period in August and September.

In contrast to Makurdi, scarcity of water was identified by millers in Adikpo as one of the challenges frequently encountered in the mills. The rice mills in these locations are strategically situated near water sources to ensure convenient access during the milling processes. Nevertheless, during the peak of the dry season, typically in March and April, these water sources tend to dry up, leading to greater challenges, including raising the costs of rice milling during this time. Since millers have no other options, they are forced to buy water from vendors, often at a high cost.

Another ongoing challenge for rice millers across the two study sites is the inconsistent duration of sunshine, caused by seasonal fluctuations and irregularities in weather and climate patterns. Firewood is the primary energy source for milling, but during the rainy season's peak (August and September), frequent cloud cover and reduced sunlight make drying difficult. This lack of adequate sunshine complicates the drying of parboiled rice to the desired moisture content, leading to a higher likelihood of grain shattering or preventing the complete removal of the chaff during the dehusking process (i.e., a process where the milling machines remove the chaff from the rice grains).

In contrast to the challenges posed by limited sunshine hours due to wet conditions, millers in all the study locations highlight excessive dryness during the harmattan season (i.e., between December and January) as a significant issue for milling. According to the millers, this period is characterized by strong dry winds leading to quicker drying. Consequently, firewood tends to become excessively dry and burns at a faster rate than intended. Additionally, the wind during steaming and parboiling often blows out the fire, leading to improperly parboiled rice. In addition, rice dried during this time becomes overly dry, making it susceptible to splitting into smaller pieces during dehusking.

3.2. *Adaptation of Millers to WWrPs*

This section explores the endogenous adaptation strategies rice millers use to cope with WWrPs. Table 1 presents a summary of the current adaptive practices adopted by millers in both locations to address the challenges posed by WWrPs. These strategies are classified into various types: tactical (i.e., short-term), reactive (e.g., implemented after pressure arises), strategic (i.e., long-term), anticipatory, or preventive (e.g., applied proactively before pressures occur).

In Makurdi (Table 1), for instance, when faced with frequent flooding, rice millers commonly employ the strategy of temporarily closing their mills and relocating to safer areas. While this practice helps them avoid total damage from the floods, it also places significant pressure on their income throughout the flooding period when all operations are suspended. Furthermore, the millers noted other challenging situations, such as extended scarcity of unmilled rice or severe milling machine malfunction, which includes difficulties in accessing necessary spare parts, for which temporary closure of milling activities is often the best approach (Table 1).

Furthermore, when millers encounter changes in sunshine duration caused by overcast weather or extreme dryness resulting from the harmattan wind, they are often prompted to respond by altering their rice drying methods. During the peak of the rainy season, spanning August to September, most millers change their usual drying method to drying the rice in thin layers of about a 7–10 cm bed depth, allowing for easy circulation of air and efficient heat penetration resulting from limited sunlight. As reported by the millers (Table 1), this process ensures that the rice's moisture content is reduced to appropriate levels for milling, preventing spoilage, and discoloration. In contrast, during the harmattan season, typically between December and January, the dry and intense winds cause the rice's moisture content to fall below the recommended 12–14% required for high-quality milling [24]. To address this, millers dry the rice in thick layers or clusters above the usual 7–10 cm bed depth, limiting air circulation and sunlight exposure to minimize shattering during milling.

In addition to altering the drying method, millers also adjust the quantity of rice milled during times of intense rainfall when hours of sunlight are reduced. As highlighted in Table 1, by reducing the amount of rice milled to a manageable level, they can minimize losses from improper handling and maintain the quality of the milled rice. Although this reduction in supply during heavy rainfall causes a temporary shock in the market, leading to increased prices, it is short-lived.

Additionally, millers often adopt a rotational system for drying parboiled rice (Table 1). This approach is necessary because during wet weather, the time needed to adequately dry the rice, which is critical for achieving high-quality milling, significantly increases. As a result, the available drying slabs or pavements become insufficient for the heightened drying demands, forcing millers to resort to drying rice in unsuitable locations, such as roadsides, to avoid spoilage. Unfortunately, this practice compromises the quality of the rice, often introducing broken stone fragments into the milled product. To address these issues, a coordinated approach is implemented, allowing only a limited number of millers to dry parboiled rice at a time while others wait their turn, thereby reducing these challenges.

Furthermore, due to the lack of nearby major rivers in Adikpo, rice millers often depend on well water as an alternative source. According to the millers (Table 1), this practice is especially common between January and May, when water scarcity is more pronounced. However, during extended dry periods, particularly when rainfall is delayed, many hand wells dry up, worsening the problem. In these cases, some millers report having to travel long distances to collect water from larger streams and rivers.

4. Discussion

This study was carried out to increase understanding of how the rice millers in Benue State, Nigeria are adapting to WWrPs. Accordingly, the findings revealed that rice milling is currently exposed and sensitive to prolonged periods of wet conditions, high moisture content, excessive dryness, and water scarcity. These put pressure on the local rice milling sector in Benue State, resulting in poor milling quality and loss of competitiveness of local rice in the market and ultimately its profitability. To respond to these pressures, thereby building resilience to keep producing and to continue to supply the end-use market in the face of these pressures, rice millers have implemented various practices. These include temporarily relocating rice mills, rotating drying shifts, managing the quantity and frequency of rice parboiled during flooding, digging wells, and modifying drying methods. These results are in agreement with those of 5, 8, 9, 12, 13, and 24, which demonstrate that the entire rice value chain's activities and processes (e.g., rice production, processing, and milling and marketing and distribution) are vulnerable to multiple climatic and economic pressures. Thus, enhancing the capacity of rice millers to adapt to these pressures requires investments in modern technology, capacity building of millers to use modern technology, and access to soft loans [17,18] to invest in demand-lifting strategies, including advertisement and branding to reverse urban biases toward locally milled rice and its market outcompetition [14,20,31,32].

The results of this study also highlight that the differences in adaptation strategies between the two study sites (i.e., Makurdi and Adikpo) may be influenced by factors such as the severity and frequency of environmental pressures, availability of resources, access to infrastructure, and socioeconomic considerations. This demonstrates the ability of the rice millers to innovate and adjust their practices to mitigate the impacts of climate variability [9]. While the adoption of strategic anticipatory measures (such as digging hand wells to address water scarcity) indicates forward-looking planning and proactive management of risks, the domination of tactical reactive strategies, such as temporary closure, relocation during flooding, and changing drying techniques in response to limited sunshine or excessive dryness, used by rice millers to address immediate threats shows that more effort is needed to address long-term pressures.

These findings demonstrate that the widespread adoption of endogenous adaptation practices in the study sites is largely driven by the small scale of rice milling operations and the low-income status of most millers, which limits their ability to attract investment [8,21,22]. While such adaptation measures help manage exposure to water-related pressures in the short term, their long-term effectiveness is constrained by a lack of technological advancement [5,8,32].

The findings also underscore the importance of context-specific adaptation planning and the need for targeted interventions to strengthen the resilience of rice milling communities across different locations. Policymakers and stakeholders can leverage these insights to design tailored adaptation strategies that address maladaptive outcomes of endogenous adaptation measures and enhance the adaptive capacity of rice millers to cope with water-related pressures, ultimately promoting more sustainable livelihoods [33].

4.1. Potential Entry Points to Reduce WWrPs and Enhance Resilience

This study highlights three key opportunities for reducing WWrPs and enhancing the resilience of locally milled rice to maintain its supply to the end market. These include upgrading rice milling technologies and infrastructure, adopting more proactive and long-term adaptive approaches, and implementing collective strategies for adaptation [9,14].

4.1.1. Upgrading Rice Milling Technologies and Infrastructure

Investment in modern milling machinery such as advanced rice hullers, polishers, driers, graders, and sorters can significantly improve the efficiency and quality of rice processing. These machines enable milling activities and processes during unfavorable weather conditions (e.g., limited sunshine to dry parboiled rice) and help reduce processing time, minimize grain breakage, and ensure higher milling yields. In addition, upgrading rice milling technologies and infrastructure to mechanized rice milling processes can help to streamline operations and reduce dependence on manual labor. This involves providing access to affordable mechanization technologies such as rice threshers, milling machines, and drying equipment. This view is shared by a number of previous studies [9,11,14,34,35].

4.1.2. Adoption of More Proactive and Long-Term Adaptive Practices

Similar to the above point, proactive measures such as proper drying with the use of modern driers and storage techniques can help minimize post-milling losses associated with mold growth, insect infestation, and moisture-related damage. This ensures that a higher percentage of the milled rice reaches the end-user market in good condition, thereby reducing economic losses for millers, traders, and consumers [5,8]. Furthermore, long-term adaptive practices enable rice millers to anticipate and respond effectively to various challenges and risks, such as fluctuations in market prices, changes in consumer preferences, and environmental factors like climate change. By diversifying their operations, investing in innovation, and building resilient supply chains, millers can mitigate risks and maintain business continuity even in uncertain conditions [9].

4.1.3. Implementation of Collective Adaptive Strategies

The implementation of collective adaptive strategies empowers rice millers to collaboratively address WWrPs; by working together, sharing knowledge and resources, and embracing innovative solutions, millers can adapt to changing environmental conditions and contribute to the resilience and prosperity of their communities. In particular, collective action allows rice millers to pool resources and access climate finance, insurance schemes, and risk-sharing/reduction mechanisms or measures that provide financial support in the event of weather-related losses or damage to crops and infrastructure. This may involve the construction of communal flood barriers, the rehabilitation of drainage systems, and the establishment of early warning and evacuation systems to mitigate the impacts of floods and storms on rice production and milling operations [5,36]. By participating in collective risk-pooling arrangements and leveraging their collective bargaining power, millers can learn from the system, reduce their vulnerability to weather-related shocks, and build their resilience, as well as ensure more sustainable livelihoods for themselves and their communities [33].

5. Conclusions

This study demonstrates that adapting rice milling to weather and climate-related pressures is key to building the resilience of the rice value chains (RVCs) in Sub-Saharan Africa in the face of pressures to respond to growing demands for high-quality milled rice. The findings highlight the significance of local adaptation strategies used by rice millers in addressing WWrPs, showcasing their capacity to innovate and modify their methods to reduce the effects of climate variability. As effective adaptation requires a combination of tactical reactive responses to immediate threats and strategic anticipatory measures to prepare for future challenges, the use of both types of adaptation strategies indicates a proactive approach to managing risks and building resilience.

Policymakers and stakeholders can use these insights to design targeted interventions and support mechanisms to enhance the adaptive capacity of rice millers, such as improving access to water resources, providing training on adaptive practices, and investing in climate-resilient infrastructure. Consequently, further research could explore the effectiveness and efficiency of different adaptation strategies, assess their socioeconomic implications, and identify barriers to their adoption to inform more tailored and context-specific interventions.

Funding: This study benefited from the Tertiary Education Trust Fund (TETfund), Federal University Dutsin-Ma, Katsina State, Nigeria (FUDMA/REG/PS 059/1/91).

Institutional Review Board Statement: The study was conducted in accordance with the Declaration of Helsinki and approved by the University Research Ethics Committee (UREC) of the University of Reading, UK, on 12 May 2015.

Informed Consent Statement: Informed consent was obtained from all participants involved in the study.

Data Availability Statement: The original data presented in the study are openly available in the Central Archive at the University of Reading at https://centaur.reading.ac.uk/84913/ (DOI: 10.48683/1926.00084913/).

Acknowledgments: This study was conducted while the author was a doctoral student at the University of Reading, United Kingdom. He is grateful to Dr. Giuseppe Feola for his advice on the study, and he thanks CORAF for sponsoring his attendance at the conference where this paper was presented and for publishing it in the conference proceedings.

Conflicts of Interest: The author declares no conflicts of interest.

References

1. Durand-Morat, A.; Bairagi, S. International Rice Outlook: International Rice Baseline Projections 2020-2030. Available online: https://scholarworks.uark.edu/aaesrb/49 (accessed on 20 July 2023).
2. USDA. Grain and Feed Annual Report for Nigeria. In Global Agricultural Information Network &United States Department of Agriculture (Vols. NI2022-000, Issue Prepared By: Ebenezer Boluwade, Agricultural Specialist Approved By: Gerald Smith Report). 2022. Available online: https://fas.usda.gov/data/nigeria-grain-and-feed-annual-5 (accessed on 20 July 2023).
3. Johnson, M.E.; Masias, I. Assessing the State of the Rice Milling Sector in Nigeria: The Role of Policy for Growth and Modernization. Feed the Future Innovation Lab for Food Security Policy Research Papers 259580, Michigan State University, Department

of Agricultural, Food, and Resource Economics, Feed the Future Innovation Lab for Food Security (FSP). Available online: https://ideas.repec.org/p/ags/miffrp/259580.html (accessed on 20 June 2023).

4. Okpiaifo, G.; Durand-Morat, A.; West, G.H.; Nalley, L.L.; Nayga, R.M.; Wailes, E.J. Consumers' preferences for sustainable rice practices in Nigeria. *Glob. Food Secur.* **2020**, *24*, 100345. [CrossRef]
5. Terdoo, F. Vulnerability and adaptation of rice value chains in Benue State, Nigeria. *Afr. J. Econ. Sustain. Dev.* **2024**, *9*, 177–200. Available online: https://www.inderscienceonline.com/doi/pdf/10.1504/AJESD.2024.136055 (accessed on 1 February 2024). [CrossRef]
6. Abbas, A.M.; Agada, I.G.; Kolade, O. Impacts of Rice Importation on Nigeria's Economy. *J. Sci. Agric.* **2018**, *2*, 71. [CrossRef]
7. Okodua, H. Assessing the Impact of Rice Sector Policy Reforms on the Income Mobility of Rural Households in Nigeria. *Open Agric. J.* **2018**, *12*, 174–184. [CrossRef]
8. Onyeneke, R.U.; Amadi, M.U.; Njoku, C.L. Climate Change Adaptation Strategies by Rice Processors in Ebonyi State, Nigeria. *Ekológia Bratisl.-J. Inst. Landsc. Ecol. Slovak Acad. Sci.* **2022**, *41*, 283–290. [CrossRef]
9. Terdoo, F.; Feola, G. The Vulnerability of Rice Value Chains in Sub-Saharan Africa: A Review. *Climate* **2016**, *4*, 47. [CrossRef]
10. Olayanju, A.T.; Okonkwo, C.E.; Ojediran, J.O.; Hussain, S.Z.; Dottie, E.P.; Ayoola, A.S. Interactive effects and modeling of some processing parameters on milling, cooking, and sensory properties for Nigerian rice using a one-step rice milling machine. *Heliyon* **2021**, *7*, 1–17. [CrossRef] [PubMed]
11. Omotayo, O.; Akinwumi, P.; Clement, A.; Samuel, J.; Ozioma, A.; Ezekiel, O. Economic Efficiency of Rice Milling Industry among Value Chain Processors in North West of Nigeria. *Nepal. J. Agric. Sci.* **2024**, *26*, 170–188.
12. Michael, P.S.; Sanga, H.G.; Shitindi, M.J.; Kwaslema, D.R.; Herzog, M.; Massawe, B.H.J. Floods stress in lowland rice production: Experiences of rice farmers in Kilombero and Lower-Rufiji floodplains, Tanzania. *Front. Sustain. Food Syst.* **2023**, *7*, 1206754. [CrossRef]
13. Onyeneke, R.U. Does climate change adaptation lead to increased productivity of rice production ? Lessons from Ebonyi State, Nigeria. *Renew. Agric. Food Syst.* **2020**, *36*, 54–68. [CrossRef]
14. Demont, M.; Ndour, M. Upgrading rice value chains: Experimental evidence from 11 African markets. *Glob. Food Secur.* **2014**, *5*, 70–76. [CrossRef]
15. Muchlisyiyah, J.; Shamsudin, R.; Basha, R.K.; Shukri, R.; How, S.; Niranjan, K.; Onwude, D. Parboiled Rice Processing Method, Rice Quality, Health Benefits, Environment, and Future Perspectives: A Review. *Agriculture* **2023**, *13*, 1390. [CrossRef]
16. Garcia-llobodanin, L.; Billiris, A. Effect of the drying air conditions on the drying rate and milling quality of a long-grain rice variety. *Food Sci. Technol.* **2023**, *43*, e65722. [CrossRef]
17. Danbaba, N.; Idakwo, P.Y.; Kassum, A.L.; Bristone, C.; Bakare, S.O.; Aliyu, U.; Kolo, I.N.; Abo, M.E.; Mohammed, A.; Abdulkadir, A.N.; et al. Rice Postharvest Technology in Nigeria: An Overview of Current Status, Constraints and Potentials for Sustainable Development. *Open Access Libr. J.* **2019**, *6*, 1. [CrossRef]
18. Ango, A.K.; Mamman, M.; Muhammad, A.; Hamza, B.A. Constraints to Rice Milling in Sokoto State: Are the Socio-Economics, Milling Technologies, and Lack of Trainings Responsible Factors? *Int. J. Agric. Environ. Biores.* **2021**, *6*, 65–77. [CrossRef]
19. Okeke, C.G.; Oluka, S.I. A Survey of Rice Production and Processing in South East Nigeria. *Niger. J. Technol.* **2017**, *36*, 227–234. [CrossRef]
20. Demont, M. Reversing urban bias in African rice markets: A review of 19 National Rice Development Strategies. *Glob. Food Secur.* **2013**, *2*, 172–181. [CrossRef]
21. Ansah, G.O.; Siaw, L.P. Indigenous Knowledge: Sources, Potency and Practices to Climate Adaptation in the Small-Scale Farming Sector. *J. Earth Sci. Clim. Change* **2017**, *8*, 431. [CrossRef]
22. Kimengsi, J.N.; Silberberger, M. How Endogenous Cultural Institutions May (Not) Shape Farmers' Climate Adaptation Practices: Learning from Rural Cameroon. *Soc. Nat. Resour.* **2023**, *36*, 460–478. [CrossRef]
23. Olayiwola, O.A. Nomadic Terrorism, Displacement, and Food Insecurity Challenge in the Food Basket of the Nation. In *Global Food Politics and Approaches to Sustainable Consumption: Emerging Research and Opportunities*; Amadi, L., Allen, F., Eds.; IGI Global Scientific Publishing: Hershey, PA, USA, 2020; pp. 101–117. [CrossRef]
24. Terdoo, F. Assessing Vulnerability of Rice Value Chains to Multiple, Combined Pressures in Benue State, Nigeria. Ph.D. Thesis, University of Reading, Reading, UK, 2019. Available online: https://centaur.reading.ac.uk/84913/ (accessed on 11 December 2023).
25. Storvang, P.; Mortensen, B.; Clarke, A.H. Using Workshops in Business Research: A Framework to Diagnose, Plan, Facilitate and Analyze Workshops. In *Collaborative Research Design: Working with Business for Meaningful Findings*; Springer: Singapore, 2018; pp. 155–174. [CrossRef]
26. Crouch, M.; McKenzie, H. The logic of small samples in interview-based qualitative research. *Soc. Sci. Inf.* **2006**, *45*, 483–499. [CrossRef]
27. Bazerley, P.; Jackson, K. *Qualitative Data Analysis with Nvivo*; Routledge: Oxfordshire, UK, 2013.
28. Braun, V.; Clarke, V. Using thematic analysis in psychology. *Qual. Res. Psychol.* **2006**, *3*, 77–101. [CrossRef]

29. Fereday, J.; Muir-Cochrane, E. Demonstrating Rigor Using Thematic Analysis: A Hybrid Approach of Inductive and Deductive Coding and Theme Development. *Int. J. Qual. Methods* **2006**, *5*, 80–92. [CrossRef]
30. Denscombe, M. *The Good Research Guide for Small-Scale Social Research Projects*, 4th ed.; Open University Press: Buckingham, UK, 2010.
31. Kwofie, E.M.; Ngadi, M. A review of rice parboiling systems, energy supply, and consumption. *Renew. Sustain. Energy Rev.* **2017**, *72*, 465–472. [CrossRef]
32. Demont, M.; Rutsaert, P.; Ndour, M.; Verbeke, W. Reversing Urban Bias in African Rice Markets: Evidence from Senegal. *World Dev.* **2013**, *45*, 63–74. [CrossRef]
33. Terdoo, F.; Feola, G. Rapid participatory system mapping builds agri- food system resilience: Evidence from Nigeria. *Afr. Geogr. Rev.* **2021**, *40*, 63–75. [CrossRef]
34. Adjao, R.T.; Staatz, J.M. Asian rice economy changes and implications for sub-Saharan Africa. *Glob. Food Secur.* **2014**, *5*, 50–55. [CrossRef]
35. Fiamohe, R.; Nakelse, T.; Diagne, A.; Seck, P.A. Assessing the effect of consumer purchasing criteria for types of rice in Togo: A choice modeling approach. *Agribusiness* **2014**, *31*, 433–452. [CrossRef]
36. Becker, L.; Yoboue, N.G. Rice producer-processor networks in Cote d'lvoire. *Am. Geogr. Soc.* **2009**, *99*, 164–185. [CrossRef]

proceedings

MDPI

Proceeding Paper

Unlocking the Role of Food Processing in Nutrition-Smart and Nutrition-Sensitive Agriculture in West Africa: Challenges, Opportunities, and a Framework for Deployment †

G. Esaïe Kpadonou [1], Caroline Makamto Sobgui [1], Rebeca Edoh [2], Kyky Komla Ganyo [1], Sedo Eudes L. Anihouvi [2,3,4,*] and Niéyidouba Lamien [1]

1 West and Central African Council for Agricultural Research and Development (CORAF), Dakar BP 48, Senegal; e.kpadonou@coraf.org (G.E.K.); c.sobgui@coraf.org (C.M.S.); k.ganyo@coraf.org (K.K.G.); n.lamien@coraf.org (N.L.)

2 Research and Development Section, CRI-Bridge, Cotonou BP 409, Benin; edoh.nephtalie@gmail.com

3 Earth, Environmental and Life Sciences Section, Research Institute, University of Bucharest, 030018 Bucharest, Romania

4 Faculty of Agronomic Sciences, University of Abomey-Calavi, Abomey-Calavi BP 526, Benin

* Correspondence: aeudeslandry@gmail.com or aeudeslandry@icloud.com

† Presented at the CORAF's 2023 Symposium on Processing and Transformation of Agricultural Products in West and Central Africa: Achievements and Opportunities for Private Sector Engagement, Lome, Togo, 21–23 November 2023.

Abstract: West Africa's agri-food systems face a triple burden of malnutrition, climate vulnerability, and structural inefficiencies that compromise nutrition and public health. Despite increased attention to food security, agricultural strategies often prioritize yield over dietary quality. This paper explores the critical role of food processing in advancing Nutrition-Sensitive Agriculture (NSA) and Nutrition-Smart Agriculture (NSmartAg) across West Africa. Drawing on a systems lens, it positions food processing not as a peripheral activity, but as a catalytic mechanism that connects nutrient-dense production with improved consumption outcomes. Food processing can reduce post-harvest losses, preserve micronutrients, extend food availability, and foster inclusive value chains particularly for women and youth. Yet, persistent challenges remain, including institutional fragmentation, infrastructure gaps, and limited financial and technical capacity. This paper proposes a conceptual framework linking food processing to NSA and NSmartAg objectives and outlines operational entry points for implementation. By integrating processing into agricultural policies, investment, education, and monitoring systems, stakeholders and policymakers can reimagine agriculture as a platform for resilience and nutritional equity. Strategic recommendations emphasize multisectoral collaboration, localized solutions, and evidence-informed interventions to drive the transformation toward sustainable, nutrition-oriented food systems.

Keywords: food security; nutrition; nutrition-smart agriculture; food fortification; balanced diet; public health

1. Introduction

Rapid population growth, climate variability, and structural weaknesses in agri-food systems are fueling a nutritional crisis in West Africa, manifesting as undernutrition, micronutrient deficiencies, and rising obesity [1–3]. Despite the region's substantial agricultural potential, persistent food and nutrition insecurity continue to undermine health and development indicators. Stunting rates in children under five remain alarmingly high

https://doi.org/10.3390/proceedings2025118017

in several countries, and micronutrient deficiencies such as vitamin A, iron, and zinc are widespread [4–6]. Likewise, dietary transitions toward energy-dense, nutrient-poor foods, especially in urban areas, are accelerating the burden of diet-related non-communicable diseases. These nutritional challenges are further exacerbated by food insecurity, seasonal variations in food availability, and limited access to diverse, affordable diets. The regional population, currently exceeding 460 million, is expected to double by 2050, placing unprecedented pressure on food systems [7]. This demographic shift, combined with rising urbanization and an emerging middle class, is reshaping consumption habits and the food demand in both volume and quality [8]. Yet, agricultural systems remain largely rainfed, low-input, and dominated by subsistence farming, with low productivity and limited infrastructure for storage, distribution, and processing. As a result, food environments across the region fail to deliver consistent access to safe, nutritious, and culturally appropriate diets.

There is growing recognition that transforming agriculture alone will not resolve the region's nutrition crisis unless it is explicitly oriented toward dietary quality. This has led to increased interest in Nutrition-Sensitive Agriculture (NSA) and Nutrition-Smart Agriculture (NSmartAg) approaches that reframe agriculture as a vehicle for improved health outcomes. NSA emphasizes agricultural investments and policies that contribute to nutritional well-being, particularly for vulnerable populations, while NSmartAg focuses more specifically on technical innovations that enhance the nutrient content and bioavailability of food [9,10]. In this context, food processing emerges as a critical yet underexplored mechanism to strengthen the nutrition–agriculture nexus. By reducing post-harvest losses, enhancing food safety, preserving nutrient contents, and increasing shelf life, food processing can serve both technical and socio-economic functions in the food system [11,12]. Its integration into NSA and NSmartAg frameworks offers opportunities to address structural barriers, such as gender inequality, infrastructure deficits, and fragmented value chains, while additionally promoting inclusive economic growth and dietary diversity. This study investigates the role of food processing in accelerating the implementation of NSA and NSmartAg in West Africa. It explores the region's current agricultural and nutritional landscape, evaluates challenges and opportunities for deploying food processing innovations, and proposes a conceptual framework to guide future interventions. In doing so, it aims to contribute to ongoing efforts to build resilient, nutrition-sensitive food systems that are aligned with regional development goals and global commitments.

2. Conceptual Foundations: Integrating Nutrition into Agriculture Through NSA, NSmartAg, and Food Processing

The persistent burden of malnutrition across West Africa underscores the inadequacy of conventional agricultural development paradigms to deliver nutritional outcomes. NSA and NSmartAg have gained prominence as comprehensive responses to this challenge. Converging them with food processing systems offers a valuable alternative to enhance the nutritional efficacy of agricultural interventions.

2.1. Nutrition-Sensitive Agriculture (NSA)

NSA is defined as an approach that addresses the underlying determinants of malnutrition by making the entire agricultural value chain, from production and processing to marketing, responsive to nutrition objectives. It integrates nutritional goals into agricultural policies, research, and practices to ensure that food systems contribute not only to caloric sufficiency but also to dietary quality and public health [7,13]. NSA emphasizes the cultivation and availability of nutrient-dense crops, dietary diversity, and the inclusion of gender-responsive and pro-poor interventions that directly or indirectly influence household nutrition, particularly among women and children.

2.2. Nutrition-Smart Agriculture (NSmartAg)

NSmartAg is a subset of NSA, representing a more targeted and innovation-driven strategy aimed at explicitly enhancing the nutritional quality of agricultural outputs. It targets practices and encompasses the development and dissemination of biofortified crops, the application of agronomic practices that retain or improve the nutrient content, and the use of precision technologies in crop and livestock management to optimize both yields and nutrient profiles [10,14,15]. Unlike NSA's broader focus, NSmartAg prioritizes the generation of measurable nutritional outcomes through technical and scientific interventions.

2.3. Food Processing as a Nexus for Nutrition and Agriculture

Food processing constitutes a critical but often underleveraged component within both NSA and NSmartAg frameworks. Defined broadly, it includes all methods and technologies that convert raw agricultural commodities into consumable food products while preserving or enhancing their nutritional value, safety, and accessibility. This spans from traditional methods, such as drying and fermentation, to modern approaches, including biofortification, fortification, and sustainable packaging solutions [11,12]. Beyond its nutritional implications, food processing contributes to reduced post-harvest losses, improved food safety, value addition, and economic empowerment, especially for smallholder farmers, women, and youth who dominate the informal food economy in West Africa. Critically, food processing can adapt the food supply to the changing demands of urbanization, dietary transitions, and year-round availability challenges, while supporting more stable, inclusive agri-food value chains.

2.4. Toward a Unified Framework

Integrating NSA, NSmartAg, and food processing offers a comprehensive framework for transforming West African food systems into engines of nutritional resilience, equity, and sustainability. Individually, each of these approaches targets specific entry points. NSA seeks to align agricultural policies with nutrition objectives; NSmartAg emphasizes innovation and technical efficiency in nutrient delivery; and food processing bridges production with consumption by ensuring the safety, availability, and acceptability of nutritious foods. Together, they offer a synergistic strategy to address both immediate dietary deficiencies and long-term structural drivers of malnutrition. This unified framework enables a systemic shift from traditional, yield-centered production models to inclusive food systems that prioritize dietary diversity, nutrient density, and equitable access. It underscores the importance of moving beyond isolated interventions toward integrated, multi-level strategies that connect farmers, processors, markets, and consumers in ways that support public health outcomes. Importantly, it provides a foundation for designing policy, technological, and institutional pathways that link agriculture more explicitly with nutrition and sustainable development goals. Operationalizing this framework requires rethinking how agricultural value chains are structured from the input supply and on-farm practices to post-harvest handling, processing, and distribution. It also necessitates aligning national food and nutrition strategies, research agendas, and investment priorities. Food processing becomes the pivot around which NSA and NSmartAg coalesce, enabling the valorization of nutrient-rich crops, reducing seasonal gaps in supply, and creating market incentives for healthier diets. In this context, agriculture is reimagined not merely as a source of income or subsistence, but as a proactive platform for addressing public health challenges, fostering inclusive economic growth, and advancing regional food sovereignty. This unified approach is essential for building food systems that can meet the complex and evolving nutritional needs of West Africa's growing population.

3. Nutrition and Agriculture in West Africa: Status, Trends, and Structural Drivers

3.1. Nutrition Landscape: Persistent Deficits Amidst Dietary Transitions

West Africa continues to bear a disproportionate share of the global nutrition burden, characterized by the coexistence of undernutrition, micronutrient deficiencies, and emerging overweight and obesity trends, a phenomenon referred to as the "triple burden of malnutrition" [1,16]. Stunting affects over 30% of children under five in several countries, including Niger, Nigeria, and Burkina Faso, reflecting chronic undernutrition during early life [17–19]. Deficiencies in essential micronutrients, particularly vitamin A, iron, and zinc, remain widespread and underdiagnosed, contributing to increased morbidity, cognitive impairment, and maternal mortality [20]. Concurrently, urbanization and rising income levels have triggered significant shifts in dietary patterns. Traditional diets rich in vegetables, legumes, and cereals are increasingly being replaced by ultra-processed, energy-dense foods that are high in sugar, fat, and sodium but low in essential nutrients [3,21]. These dietary transitions are driving an increase in overweight, obesity, and diet-related non-communicable diseases, particularly in urban populations [22].

3.2. Agricultural Systems: High Potential, Low Nutritional Yield

The agricultural sector remains the backbone of most West African economies, contributing significantly to the GDP and employing most of the rural population [23]. Yet, the prevailing production systems are largely subsistence-based, low-input, and rainfed. Key staples, such as maize, millet, sorghum, yam, and cassava, dominate the landscape, providing bulk calories but limited micronutrient diversity [24,25]. The limited dietary diversification at the production level translates into poor dietary quality at the consumption level, especially among low-income households. The absence of systematic investment in nutrient-rich crops, such as legumes, fruits, and vegetables, constrain the potential of agriculture to meet nutritional needs [26,27]. In addition, the reliance on seasonal rainfed systems exacerbates food insecurity during lean periods.

3.3. Structural Drivers of Nutrition Insecurity

West Africa's persistent food and nutrition insecurity is not merely the result of inadequate agricultural production but is driven by a set of closely linked structural challenges. These barriers operate across environmental, social, and infrastructural dimensions, collectively undermining the region's ability to ensure equitable and consistent access to safe and nutritious food.

One of the most pressing threats is climate change, which has intensified the frequency and severity of droughts, floods, and erratic rainfall patterns. These climatic disruptions compromise planting and harvesting cycles, lower agricultural yields, and reduce the reliability of food supplies. Smallholder farmers, who dominate the region's agricultural landscape, are particularly vulnerable due to their limited access to adaptive technologies and risk mitigation means [28,29].

Another critical constraint is the high level of post-harvest losses, with estimates suggesting that up to 40% of perishable food is lost between harvest and consumption. These losses stem from inadequate storage infrastructure, poor transport systems, and the limited access to food preservation and processing technologies [30,31]. The result is not only economic inefficiency but also a significant erosion of the food quantity and nutritional quality.

Gender inequality further compounds these issues. Women play pivotal roles in food production, processing, and household nutrition yet face systemic barriers in accessing land, credit, education, and decision-making authority. Their underrepresentation in agricultural leadership and value chain governance limits both the productivity and nutritional impact of agricultural interventions [32,33].

Closely related are infrastructure deficits that restrict the flow and preservation of nutrient-rich foods. Limited road connectivity, an inconsistent electricity supply, and the near absence of cold chain systems hamper the distribution of perishable commodities, particularly in remote and underserved regions. These constraints increase food spoilage, elevate prices, and reduce dietary diversity [34].

Lastly, population growth and urbanization are reshaping food demand and consumption patterns. The region's population is projected to double by 2050, accelerating urban expansion and straining already fragile food systems. Urbanization is also driving dietary transitions toward ultra-processed and convenience foods, which are often nutrient-poor and energy-dense, thereby increasing the prevalence of non-communicable diseases [35,36].

These structural constraints do not operate in isolation. Rather, they converge to form a self-reinforcing system of vulnerability, where climate shocks exacerbate post-harvest losses, gender disparities limit the adaptive capacity, and infrastructural gaps obstruct distribution and market access. As illustrated in Figure 1, the complex interlinkages among these factors erode both food availability and nutritional adequacy. Table 1 further contextualizes these dynamics by presenting comparative indicators from selected West African countries, highlighting disparities in child stunting, micronutrient deficiencies, post-harvest loss rates, and overall food insecurity. Together, these conditions underscore the urgent need for integrated interventions that move beyond yield-centric approaches. Nutrition-sensitive and nutrition-smart strategies must be embedded within broader efforts to strengthen infrastructure, close gender gaps, enhance climate resilience, and improve post-harvest systems. Only then can West African food systems be transformed into inclusive, sustainable platforms for health and development.

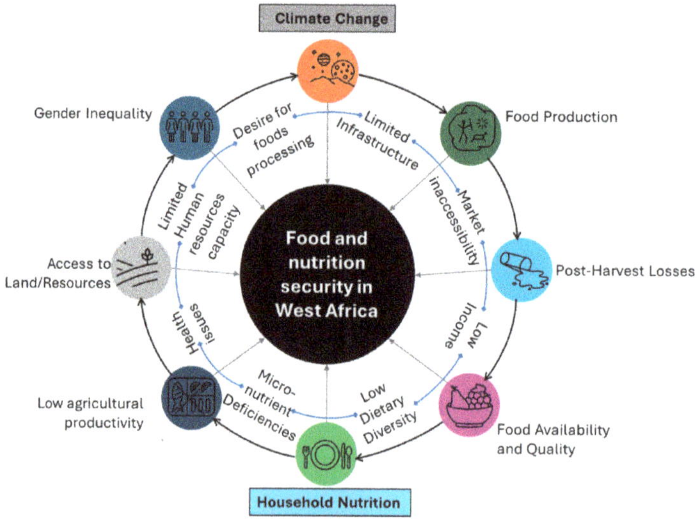

Figure 1. Interconnected drivers of agri-food system and nutrition insecurity in West Africa.

Table 1. Nutrition and agricultural indicators in some West African countries [ranked by stunting rate].

Country	Stunting Rate (%)	Vitamin A Deficiency (%)	Iron Deficiency (%)	Agriculture GDP Share	Major Crops	Urban Obesity	Food Insecurity Rate	Post-Harvest Losses	References
Niger	45.1	50	42	40	Millet, Sorghum, Cowpeas	Low but increasing	50	43	[6]
Nigeria	36.8	35	30	21	Maize, Cassava, Yam, Millet, Rice	Rising rapidly	42	Up to 45	[1,37,38]
Burkina Faso	33.1	42	36	31	Sorghum, Millet, Maize, Groundnut	Gradual rise	41	40	[29,39]
Mali	30.4	40	34	38	Sorghum, Millet, Maize, Rice	Moderate increase	36	35	[6]
Liberia	28.0	38	29	33	Rice, Cassava, Oil palm	Moderate rise	39	35	[1,37]
Sierra Leone	26.3	36	30	35	Cassava, Rice, Cocoa	Moderate increase	38	32	[6]
Guinea	25.0	34	31	27	Rice, Cassava, Groundnuts	Moderate rise	35	30	[9,29]
Gambia	23.4	31	28	22	Groundnuts, Rice, Millet	Slight increase	33	28	[37]
Côte d'Ivoire	21.6	30	27	24	Cocoa, Cassava, Yam, Rice	Urbanizing rapidly	30	28	[9,29]
Ghana	18.8	28	22	20	Cassava, Cocoa, Maize, Yam	Significant in cities	24	25	[37]
Togo	18.6	29	24	26	Maize, Cassava, Sorghum	Gradual urban rise	27	26	[1]
Senegal	17.3	25	21	15	Groundnut, Millet, Rice, Horticulture	Urban growth trend	28	30	[1]
Benin	16.2	26	23	20	Maize, Cassava, Rice, Yam	Moderate urban trend	29	27	[1,6]
Cabo Verde	8.5	14	15	10	Maize, Potatoes, Vegetables	High urbanization	10	15	[1,40]

4. Food Processing as a Link for Agricultural Production and Nutrition: Relevance, Challenges, and Opportunities

Food processing connects nutrient-rich agricultural outputs to end-users' nutritional needs by enhancing food safety, shelf life, palatability, and accessibility. In contexts marked by seasonal variability, a weak infrastructure, and dietary monotony, conditions typical in many parts of West Africa, food processing becomes indispensable for converting perishable crops into stable, nutritious products that can be consumed year-round. From a nutritional perspective, appropriate food processing techniques, such as fermentation, controlled drying, and fortification, can significantly reduce post-harvest nutrient losses and enhance bioavailability [41]. For instance, processing millet into fortified flours or transforming cassava through fermentation reduces anti-nutritional factors while improving digestibility. Additionally, processing enables the incorporation of underutilized and indigenous crops into mainstream diets, which is vital for dietary diversification and the preservation of agro-biodiversity. When embedded in inclusive business models, food processing also contributes to economic empowerment, especially for women and small-scale entrepreneurs, while fostering value additions across rural supply chains [11,12]. Despite its importance, the agri-food processing system in West Africa faces many challenges.

4.1. *Structural and Technological Constraints*

One of the foremost barriers to effective food processing in West Africa is the persistent lack of modern infrastructure. Many processing units, particularly those operated by small and medium enterprises, rely on outdated or artisanal equipment with limited capacity and efficiency [34]. Inadequate facilities lead to high post-harvest losses, reduced food safety, and compromised nutrient retention. Additionally, the scarcity of reliable electricity and cold storage further restricts the viability of processing perishable crops, especially in rural areas. Access to appropriate technology is also constrained by limited capital investment, prohibitive import costs, and weak maintenance ecosystems. Local manufacturers of food processing equipment are scarce with low mass reproductivity performance, and imported machines often face compatibility or repair issues due to the unavailability of spare parts. This technological stagnation impedes both innovation and scale, limiting the ability of processors to meet the rising urban demand for safe, shelf-stable, and nutritionally enhanced products [42].

4.2. *Financial and Market Barriers*

The development of the food processing sector is further constrained by financial exclusion. Smallholder farmers and agro-processors often lack access to affordable credit, as financial institutions perceive agricultural and food-related ventures as high-risk, particularly due to their dependence on seasonality and weather variability [10]. Even when financing is available, interest rates tend to be prohibitively high for small enterprises. In parallel, a weak market integration limits processors' ability to source quality raw materials or access consistent demand. Agricultural inputs are expensive or unavailable in many areas, while value chains remain fragmented and poorly coordinated. Limited transport infrastructure and high logistics costs erode competitiveness, often making imported processed foods cheaper and more accessible than local alternatives. These dynamics discourage private sector investment and widen the gap between rural producers and urban markets.

4.3. *Institutional and Policy Challenges*

Institutional fragmentation undermines the food processing sector's development potential. Ministries responsible for agriculture, health, trade, and industry frequently operate in silos, leading to disconnected policies and uncoordinated implementation [43]. Food processing is often overlooked in national agricultural or nutrition strategies, receiving little policy attention or budgetary allocations. This marginalization constrains opportunities to promote food fortification, enforce quality standards, or incentivize private-sector engagement. Regulatory environments also tend to be underdeveloped or inconsistently enforced. Food safety standards, labeling requirements, and certification processes are often unclear or costly to comply with, particularly for small processors. This regulatory vacuum not only poses health risks but also undermines consumer confidence and stifles regional trade in processed food products [44].

4.4. Socio-Economic Inequities and Gender Constraints

Women represent a significant share of the informal food processing workforce in West Africa, yet their potential remains underleveraged due to systemic gender-based constraints. Women typically face more limited access to land, capital, education, and productive resources than men. These inequalities restrict their ability to scale up operations or adopt improved processing technologies, despite their central role in household nutrition and informal food markets [32,33]. In addition, training programs, extension services, and financing mechanisms are often not tailored to the realities of women-led enterprises. Strengthening gender-responsive policies and capacity-building initiatives is therefore critical to expanding the reach and impact of food processing within NSA and NSmartAg frameworks.

4.5. Emerging Opportunities for Sectoral Transformation

Despite these challenges, West Africa's food processing sector presents considerable opportunities for driving inclusive growth and nutritional transformation. The region is rich in diverse raw materials, including cereals, legumes, roots, fruits, and vegetables, that can serve as inputs for nutrient-rich, culturally relevant food products [45]. Innovations in food preservation, such as solar drying, vacuum sealing, and fermentation, offer low-cost, scalable solutions for extending shelf life and improving food safety. The growing demand for fortified foods, driven by both public health priorities and consumer preferences, creates new market segments for processors. Regional initiatives, such as the African Continental Free Trade Area (AfCFTA), offer potential for expanding the intra-African trade in processed foods, provided harmonized standards and infrastructure investments accompany policy commitments [46].

Furthermore, advances in biofortification, sustainable packaging, and low-energy processing technologies open avenues for nutrition-enhancing innovations. Coupled with investments in youth entrepreneurship, cooperative models, and digital platforms for market linkage, these tools can unlock the full potential of food processing to support nutrition-sensitive and nutrition-smart agricultural systems.

5. Challenges to the Implementation of NSA and NSmartAg in West Africa

Despite the conceptual strengths and increasing policy recognition of NSA, NSmartAg, and food processing, the translation of these frameworks into actionable and sustainable programs in West Africa remains limited. Implementation efforts are frequently hindered by systemic, technical, financial, and sociocultural barriers that prevent these approaches from reaching scale or delivering measurable nutrition outcomes. Understanding these constraints is critical to designing interventions that are context-appropriate and scalable.

5.1. Institutional Fragmentation and Policy Incoherence

A major constraint is the fragmentation of institutional mandates across the agriculture, health, education, industry, and trade sectors. Ministries and agencies responsible for food systems often operate in silos, leading to disconnected policy frameworks and limited intersectoral coordination [43]. For instance, agricultural policies may prioritize staple crop productivity while public health strategies emphasize nutrition supplementation, with little overlap in strategy, resources, or objectives regarding agricultural production. Moreover, food processing is frequently marginalized in both nutrition and agricultural agendas, receiving insufficient investment or regulatory support. The absence of integrated food system governance inhibits synergies between sectors and reduces the effectiveness of multisectoral strategies intended to advance NSA or NSmartAg [47,48].

5.2. Underinvestment in Nutrition-Oriented Research and Infrastructure

NSA and NSmartAg remain chronically underfunded across most countries in the region. Agricultural investments continue to focus disproportionately on staple or export crops, such as maize, rice, cocoa, and cotton, with limited funding for nutrient-dense crops, underutilized species, or biofortified varieties. Similarly, research and extension services seldom prioritize nutrition outcomes, leading to the weak uptake of improved varieties and practices with demonstrable nutritional benefits [49]. Infrastructure deficits also constrain implementation. Inadequate rural road networks, unreliable electricity, and poor storage systems increase transaction costs and limit the distribution of perishable, nutrient-rich foods. The absence of cold chains, post-harvest technologies, and quality control mechanisms further undermines the reach and reliability of food processing enterprises [34].

5.3. Capacity Gaps and Limited Technical Expertise

At the grassroots level, smallholder farmers and local processors often lack the knowledge and training required to implement NSA and NSmartAg effectively. Agricultural extension services are under-resourced and seldom incorporate nutrition-sensitive content. They focus their interventions more on agricultural practices and innovations. Technical know-how in post-harvest handling, nutrient-preserving processing, and small-scale value addition remains limited, particularly in rural areas. This capacity gap is further compounded by the weak integration of nutrition curricula in agricultural education and by a general lack of training opportunities for women and youth who represent a significant portion of the informal food processing sector [39]. Without deliberate capacity strengthening, many nutrition-sensitive innovations will remain inaccessible or poorly adopted.

5.4. Financial Exclusion and Market Limitations

Access to finance remains a persistent barrier for small-scale farmers, processors, and entrepreneurs aiming to engage in nutrition-smart production or food processing. Commercial banks and microfinance institutions often perceive agriculture and food ventures as high-risk and rarely offer tailored financial products for nutrition-sensitive investments. Even when credit is available, high interest rates and collateral requirements deter uptake [50,51]. Market fragmentation and weak demand incentives also impede implementation. The lack of structured markets for biofortified or fortified foods disincentivizes producers and processors from investing in nutritional quality. In addition, rural consumers may face affordability and accessibility constraints, while urban markets may favor imported, ultra-processed foods over healthier, locally produced alternatives [3].

5.5. Sociocultural Barriers and Gender Inequality

Sociocultural dynamics, including food taboos, consumption norms, and perceptions around fortified or biofortified products, can hinder the adoption of nutrition-sensitive innovations. For example, in some communities, the consumption of orange-fleshed sweet potatoes or fortified flours may be resisted due to their unfamiliar appearance or taste, especially in the absence of adequate awareness campaigns [52,53]. Gender inequality further restricts the implementation capacity. Although women are key actors in production, processing, and nutrition management, they face limited access to land, extension services, agricultural inputs, and markets. Without gender-responsive policy frameworks and inclusive programming, NSA and NSmartAg interventions risk reinforcing existing inequities rather than addressing them [32,33].

5.6. Climate Vulnerability and Environmental Constraints

Climate variability poses an existential threat to the sustainability of Nutrition-Sensitive Agriculture in West Africa. Increasingly frequent and severe droughts, floods, and irregular rainfall patterns are destabilizing production systems, particularly those that rely on rainfed agriculture. These shocks not only affect yields but also the crop quality and nutrient composition [29]. Moreover, climate impacts often lead to reactive shifts in crop choice toward hardier but less nutritious staples, undermining dietary diversity. Without investments in climate-resilient infrastructure, early warning systems, and adaptive agronomic practices, the scalability of NSA and NSmartAg will remain highly constrained, and the timely availability of agricultural products used as inputs in processing systems will be compromised.

6. Leveraging Food Processing to Implement Nutrition-Sensitive and Nutrition-Smart Agriculture in West Africa

Food processing is not merely a downstream activity in the agri-food system. It is the central mechanism that helps to operationalize both NSA and NSmartAg in practical and scalable ways. Properly leveraged food processing can catalyze the transformation of nutrient-rich agricultural production into safe, affordable, and culturally appropriate diets, while reducing post-harvest losses and improving livelihoods. This section presents a conceptual framework for operationalizing NSA and NSmartAg through food processing principles. To guide the integration of food processing within NSA and NSmartAg strategies, Figure 2 presents a deployment framework that maps the essential components and linkages needed to operationalize nutrition-oriented food systems through processing interventions.

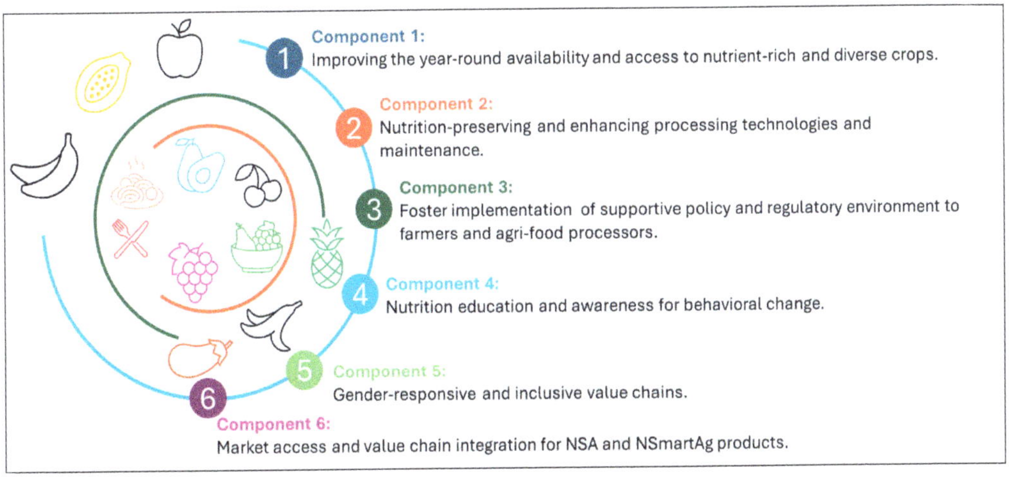

Figure 2. The deployment framework of food processing in the implementation of Nutrition-Sensitive Agriculture and Nutrition-Smart Agriculture.

6.1. Conceptual Framework for Nutrition-Oriented Food Processing Actions

Building on the agriculture-to-nutrition pathways articulated by [54], we propose an adapted framework that positions food processing as a central mechanism linking production, value addition, and consumption. The framework identifies six interdependent components:

6.1.1. Year-Round Availability of Nutrient-Dense Raw Materials

The foundation of nutrition-sensitive food systems is the consistent access to diverse, nutrient-rich crops [50]. This requires transitioning from seasonal, rainfed production to year-round systems supported by irrigation, climate-smart agriculture, crop diversification, and adequate processing systems. Scaling the adoption of biofortified and underutilized crops can expand the nutrient base of processed foods.

6.1.2. Nutrition-Preserving and -Enhancing Processing Technologies

Food processing must retain or enhance the nutritional value of raw materials. This can be achieved through methods such as solar drying, fermentation, extrusion, and vacuum packaging [50]. Fortification techniques, whether mandatory or voluntary, can address specific micronutrient deficiencies (e.g., vitamin A and iron). Importantly, technologies should be accessible to small-scale processors and tailored to local dietary preferences [50]. Systems maintenance as well as replacement parts of the processing equipment should be made available and affordable.

6.1.3. Supportive Policy and Regulatory Environment

Effective implementation demands enabling policies that align agricultural, health, and trade sectors. These include investment incentives for processors producing nutritious foods, quality assurance systems, streamlined food safety regulations, and subsidies for equipment or inputs. National food and nutrition strategies should formally integrate food processing within NSA and NSmartAg programming.

6.1.4. Nutrition Education and Behavioral Change

Supply-side innovations are insufficient without corresponding shifts in demand. Strategic communication campaigns, nutrition labeling, and school-based interventions can increase consumer awareness and acceptance of fortified and biofortified foods. Training for extension workers and food processors is also essential to embed nutritional goals into processing decisions.

6.1.5. Gender-Responsive and Inclusive Value Chains

Women and youth dominate informal processing sectors but face systemic barriers to land, finance, and training. Strengthening their participation in food value chains via cooperatives, technical support, or access to finance can enhance food security and equity while unlocking economic multipliers.

6.1.6. Market Access and Value Chain Integration

Robust distribution systems are critical for ensuring that processed nutritious foods reach diverse consumers. This includes developing local agro-processing hubs, linking rural producers to urban markets, and promoting regional trade under frameworks like the African Continental Free Trade Area (AfCFTA). Digital platforms can also support logistics, traceability, and price transparency.

Together, these components create a virtuous cycle (Figure 2) in which nutritious foods are consistently produced, preserved, and delivered across geographies and seasons, improving diet quality and reducing food waste. Table 2 provides some operational pathways for leveraging food processing to advance NSA and NSmartAg in West Africa.

Table 2. Operational pathways for leveraging food processing to advance NSA and NSmartAg in West Africa *.

Strategic Entry Point	Recommended Actions	Expected Outcomes
Availability of nutrient-dense raw materials	Promote climate-resilient and biofortified crop varieties Support crop diversification programs	Year-round availability of nutritious crops Expanded supply base for processors
Nutrition-preserving processing technologies	Introduce low-cost, nutrient-retaining equipment [e.g., solar dryers, extruders] Train processors in safe and hygienic techniques	Improved nutrient quality and food safety Reduced post-harvest nutrient loss
Policy and regulatory support	Integrate food processing in national food and nutrition strategies Simplify food safety compliance for SMES	Institutional alignment Increased formalization and product quality assurance
Nutrition education and behavior change	Implement consumer awareness campaigns Introduce school-based nutrition interventions	Increased demand for fortified and diverse foods Greater dietary diversity
Gender-responsive and inclusive value chains	Provide finance, training, and cooperative support to women/youth-led processors Reduce barriers to land and credit access	Increased inclusion and empowerment Enhanced productivity and enterprise growth
Market access and value chain integration	Develop rural–urban supply linkages Promote regional trade under afcfta Leverage digital platforms for logistics and sales	Expanded access to affordable nutritious foods Strengthened regional food systems
Monitoring and adaptive learning	Establish metrics for nutrition and equity impact Use monitoring and evaluation data to refine programs	Evidence-based policy adjustments Scalable, replicable interventions

* Adapted from [10,11,54].

6.2. Operationalizing the Framework: Pathways and Entry Points

Translating this framework into practice requires coordinated, multisectoral actions, as depicted in Table 2. Below are priority entry points for implementation, each corresponding to components of the framework:

- **Crop-Processing Alignment**: Link biofortified crop value chains to local processors through input subsidies and aggregation models. For example, supporting farmer co-operatives to grow iron-rich pearl millet that is processed into fortified porridge mixes.
- **Investment in Appropriate Technology**: Promote affordable, energy-efficient processing tools suitable for small and medium enterprises [e.g., solar dryers and low-cost extruders]. These technologies reduce nutrient loss while enhancing food safety.
- **Nutrition-Sensitive Procurement Policies**: Integrate locally processed, fortified foods into institutional feeding programs, such as school meals or maternal health initiatives, to stabilize demand and improve nutritional outcomes.
- **Capacity Building for Processors**: Provide training in nutrient-preserving processing, regulatory compliance, and business development. This includes incubator programs for youth- and women-led food ventures.

- **Public–Private Partnerships [PPPs]**: Incentivize partnerships between governments, NGOs, and private sector actors to co-develop value chains for nutrition-oriented products, especially in underserved areas.
- **Monitoring and Evaluation Systems**: Embed nutrition and gender metrics into food system assessments to guide policy and investment. Indicators should track not just production or income, but access to the market, actual dietary outcomes, and the affordability of nutritious processed foods.

7. Conclusions

West Africa faces an urgent need to reorient its food systems toward nutrition outcomes. The region continues to struggle with a triple burden of malnutrition, climate-induced agricultural stress, demographic pressure, and systemic infrastructure gaps highlighting the limitations of yield- and export-focused agricultural models. This article argues that food processing, strategically embedded within Nutrition-Sensitive Agriculture (NSA) and Nutrition-Smart Agriculture (NSmartAg), offers a transformative pathway to align food systems with nutritional, equity, and sustainability goals. When guided by nutritional objectives, food processing can significantly enhance the food system performance. It reduces post-harvest losses, preserves micronutrients, improves food safety, and increases the seasonal availability of diverse, culturally acceptable foods. Additionally, it supports value addition, creates markets for biofortified and underutilized crops, and offers inclusive economic opportunities especially for women and youth active in informal food economies. Yet, these benefits remain underrealized due to persistent policy, infrastructure, financing, and knowledge barriers. Unlocking this potential demands a shift in how food processing is valued in policy and practice. Governments should institutionalize food processing within agriculture and nutrition strategies, promote cross-sectoral alignment, and invest in nutrient-dense value chains and decentralized infrastructure. Access to affordable, efficient technologies and rural transport systems must also be expanded. Human capital development is equally critical. Training for small-scale processors, extension agents, and nutrition educators should integrate technical, business, and nutrition content, with an emphasis on gender equity. Generating consumer demand through behavior change, communication and clear food labeling can reinforce supply-side progress. Robust monitoring systems must track both nutritional and socio-economic outcomes. These insights should inform adaptive policy and investment strategies. Ultimately, embedding food processing within NSA and NSmartAg represents a paradigm shift that reimagines food systems as platforms for health, dignity, and resilience.

Author Contributions: Conceptualization, G.E.K.; methodology, G.E.K., S.E.L.A., K.K.G.; investigation, R.E., G.E.K., S.E.L.A., K.K.G.; writing—original draft preparation, R.E., G.E.K.; writing—review and editing, S.E.L.A., C.M.S., K.K.G.; supervision, N.L. All authors have read and agreed to the published version of the manuscript.

Funding: This research received no external funding.

Institutional Review Board Statement: Not applicable.

Informed Consent Statement: Not applicable.

Data Availability Statement: Data are contained within the article.

Conflicts of Interest: The authors declare no conflict of interest.

References

1. FAO; IFAD; UNICEF; WFP; WHO. *In Brief to The State of Food Security and Nutrition in the World 2023. Urbanization, Agrifood Systems Transformation and Healthy Diets Across the Rural–Urban Continuum*; FAO: Rome, Italy, 2023. [CrossRef]
2. Ahinkorah, B.O.; Amadu, I.; Seidu, A.A.; Okyere, J.; Duku, E.; Hagan, J.E., Jr.; Budu, E.; Archer, A.G.; Yaya, S. Prevalence and factors associated with the triple burden of malnutrition among mother-child pairs in sub-Saharan Africa. *Nutrients* **2021**, *13*, 2050. [CrossRef]
3. Popkin, B.M.; Ng, S.W. The nutrition transition to a stage of high obesity and noncommunicable disease prevalence dominated by ultra-processed foods is not inevitable. *Obes. Rev.* **2022**, *23*, e13366. [CrossRef] [PubMed]
4. John, C.; Poh, B.K.; Jalaludin, M.Y.; Michael, G.; Adedeji, I.; Oyenusi, E.E.; Akor, B.; Charles, N.C.; Buthmanaban, V.; Muhardi, L. Exploring disparities in malnutrition among under-five children in Nigeria and potential solutions: A scoping review. *Front. Nutr.* **2024**, *10*, 1279130. [CrossRef]
5. WHO. *Global Nutrition Report 2023: Addressing Malnutrition in Fragile Contexts*; World Health Organization: Geneva, Switzerland, 2023.
6. UNICEF. *Child Nutrition Situation in West and Central Africa: Regional Overview and Country Profiles*; United Nations Children's Fund: New York, NY, USA, 2022.
7. Simane, B.; Kapwata, T.; Naidoo, N.; Cissé, G.; Wright, C.Y.; Berhane, K. Ensuring Africa's Food Security by 2050: The Role of Population Growth, Climate-Resilient Strategies, and Putative Pathways to Resilience. *Foods* **2025**, *14*, 262. [CrossRef] [PubMed]
8. Hussain, M.A.; Li, L.; Kalu, A.; Wu, X.; Naumovski, N. Sustainable Food Security and Nutritional Challenges. *Sustainability* **2025**, *17*, 874. [CrossRef]
9. FAO. *Nutrition-Sensitive Agriculture and Food Systems in Practice: Options for Intervention*; Food and Agriculture Organization of the United Nations: Rome, Italy, 2017.
10. World Bank. Nutrition-Smart Agriculture in Haiti; 2020; Volume 1. Available online: http://documents.worldbank.org/curated/en/222031597126417223 (accessed on 14 May 2025).
11. Bilali, H.E.; Rokka, S.; Calabrese, G.; Borelli, T.; Grazioli, F.; Tietiambou, S.R.F.; Nanema, J.; Guimbo, I.D.; Dambo, L.; Nouhou, B.; et al. Conservation and promotion of neglected and underutilized crop species in West Africa: Policy and governance. *Sustainability* **2024**, *16*, 6194. [CrossRef]
12. Adeyeye, S.A.O. The role of food processing and appropriate storage technologies in ensuring food security and food availability in Africa. *Nutr. Food Sci.* **2017**, *47*, 122–139. [CrossRef]
13. Ruel, M.T.; Quisumbing, A.R.; Balagamwala, M. Nutrition-sensitive agriculture: What have we learned so far? Glob. *Food Secur.* **2018**, *17*, 128–153. [CrossRef]
14. Avnee; Sood, S.; Chaudhary, D.R.; Jhorar, P.; Rana, R.S. Biofortification: An approach to eradicate micronutrient deficiency. *Front. Nutr.* **2023**, *10*, 1233070. [CrossRef]
15. Sheoran, S.; Kumar, S.; Ramtekey, V.; Kar, P.; Meena, R.S.; Jangir, C.K. Current status and potential of biofortification to enhance crop nutritional quality: An overview. *Sustainability* **2022**, *14*, 3301. [CrossRef]
16. Prentice, A.M. The triple burden of malnutrition in the era of globalization. In *Nestlé Nutrition Institute Workshop Series*; Karger: Basel, Switzerland, 2023; Volume 97, pp. 51–61. [CrossRef]
17. Compaore, E.W.R.; Ouedraogo, O.; Bationo, P.E.; Zerbo, P. Determinants of stunting in children aged 0–59 months in three regions of Burkina Faso. *BMC Public Health* **2024**, *24*, 2392. [CrossRef]
18. Tamir, T.T.; Gezhegn, S.A.; Dagnew, D.T.; Mekonenne, A.T.; Aweke, G.T.; Lakew, A.M. Prevalence of childhood stunting and determinants in low and lower-middle income African countries: Evidence from standard demographic and health survey. *PLoS ONE* **2024**, *19*, e0302212. [CrossRef] [PubMed]
19. UNICEF. *Child Nutrition Situation in West and Central Africa: Regional Overview and Country Profiles*; United Nations Children's Fund: New York, NY, USA, 2023.
20. Bailey, R.L.; West, K.P.; Black, R.E. The epidemiology of global micronutrient deficiencies. *Ann. Nutr. Metab.* **2015**, *66* (Suppl. S2), 22–33. [CrossRef]
21. Casari, S.; Di Paola, M.; Banci, E.; Diallo, S.; Scarallo, L.; Renzo, S.; Gori, A.; Renzi, S.; Paci, M.; de Mast, Q.; et al. Changing dietary habits: The impact of urbanization and rising socio-economic status in families from Burkina Faso in sub-Saharan Africa. *Nutrients* **2022**, *14*, 1782. [CrossRef] [PubMed]
22. Muonde, M.; Olorunsogo, T.O.; Ogugua, J.O.; Maduka, C.P.; Omotayo, O. Global nutrition challenges: A public health review of dietary risks and interventions. *World J. Adv. Res. Rev.* **2024**, *21*, 1467–1478. [CrossRef]
23. FAO; IFAD; UNICEF; WFP; WHO. *The State of Food Security and Nutrition in the World 2022. Repurposing Food and Agricultural Policies to Make Healthy Diets More Affordable*; FAO: Rome, Italy, 2022. [CrossRef]
24. Tura, M.; Tappi, S.; Valli, E.; Salas, P.G.; Suleiman, R.; Thabet, C.; Essaidi, I.; Muyonga, J.H.; Akumu, G.; Bahri, W.R.; et al. A selection of African traditional and emerging foods: Processing, composition, nutritional, sensory and nutraceutical aspects. *Curr. Res. Food Sci.* **2025**, 101112, *in press*. [CrossRef]

25. Mnisi, C.M.; Oyeagu, C.E.; Akuru, E.A.; Ruzvidzo, O.; Lewu, F.B. Sorghum, millet and cassava as alternative dietary energy sources for sustainable quail production—A review. *Front. Anim. Sci.* **2023**, *4*. [CrossRef]
26. Vilakazi, B.; Mafongoya, P.L.; Odindo, A.O.; Phophi, M.M. The Role of Neglected Grain Legumes in Food and Nutrition Security and Human Health. *Sustainability* **2025**, *17*, 350. [CrossRef]
27. Bhardwaj, R.L.; Parashar, A.; Parewa, H.P.; Vyas, L. An alarming decline in the nutritional quality of foods: The biggest challenge for future generations' health. *Foods* **2024**, *13*, 877. [CrossRef]
28. Furtak, K.; Wolińska, A. The impact of extreme weather events as a consequence of climate change on the soil moisture and on the quality of the soil environment and agriculture—A review. *Catena* **2023**, *231*, 107378. [CrossRef]
29. FAO. *The State of Food and Agriculture 2019: Moving Forward on Food Loss and Waste Reduction*; Food and Agriculture Organization of the United Nations: Rome, Italy, 2019.
30. Kumar, D.; Kalita, P. Reducing postharvest losses during storage of grain crops to strengthen food security in developing countries. *Foods* **2017**, *6*, 8. [CrossRef] [PubMed]
31. Awulachew, M.T. Evaluation of pot technology preservation techniques for tomato fruit in East Arsi, Ethiopia. *Int. J. Res. Agric. Food Sci.* **2019**, *5*, 24–31.
32. Bryan, E.; Alvi, M.; Huyer, S.; Ringler, C. Addressing gender inequalities and strengthening women's agency to create more climate-resilient and sustainable food systems. *Glob. Food Secur.* **2024**, *40*, 100731. [CrossRef]
33. Gnisci, D. *Women's Roles in the West African Food System: Implications and Prospects for Food Security and Resilience*; OECD Publishing: Paris, France, 2016. [CrossRef]
34. Lawrence, M. Fundamentals of a healthy and sustainable diet. *Nutr. J.* **2024**, *23*, 150. [CrossRef]
35. de Bruin, S.; Dengerink, J.; van Vliet, J. Urbanisation as driver of food system transformation and opportunities for rural livelihoods. *Food Sec.* **2021**, *13*, 781–798. [CrossRef]
36. Hawkes, C.; Harris, J.; Gillespie, S. Changing diets: Urbanization and the nutrition transition. In *2017 Global Food Policy Report*; International Food Policy Research Institute (IFPRI): Washington, DC, USA, 2017; Chapter 4; pp. 34–41. [CrossRef]
37. Baby-friendly Hospital Initiative (BFHI) tools. Nutrition and Food Safety. Available online: www.who.int/teams/nutrition-and-food-safety/food-and-nutrition-actions-in-health-systems/ten-steps-to-successful-breastfeeding (accessed on 18 May 2023).
38. WHO. *Report of the Commission on Ending Childhood Obesity*; World Health Organization: Geneva, Switzerland, 2016.
39. IFPRI. *2021 Global Food Policy Report: Transforming Food Systems After COVID-19*; International Food Policy Research Institute: Washington, DC, USA, 2021. [CrossRef]
40. World Bank. *Macro Poverty Outlook. Country-by-Country Analysis and Projections for the Developing World*; Word Bank: Washington, DC, USA, 2022.
41. Knez, E.; Kadac-Czapska, K.; Grembecka, M. Effect of fermentation on the nutritional quality of the selected vegetables and legumes and their health effects. *Life* **2023**, *13*, 655. [CrossRef]
42. Wudil, A.H.; Usman, M.; Rosak-Szyrocka, J.; Pilař, L.; Boye, M. Reversing Years for Global Food Security: A Review of the FoodSecurity Situation of theSub-Saharan Africa (SSA). *Int. J. Environ. Res. Public Health* **2022**, *19*, 14836. [CrossRef]
43. OECD. *Making Better Policies for Food Systems*; OECD Publishing: Paris, France, 2021. [CrossRef]
44. Corigliano, O.; Algieri, A. A comprehensive investigation on energy consumptions, impacts, and challenges of the food industry. *Energy Convers. Manag.* **2024**, *23*, 100661. [CrossRef]
45. Zhou, Y.; Staatz, J. Projected demand and supply for various foods in West Africa: Implications for investments and food policy. *Food Policy* **2016**, *61*, 198–212. [CrossRef]
46. Kariuki, A.N. Upgrading strategies and food security implications on smallholder farmers in Sub-Saharan Africa: A value chain review. *J. Food Secur.* **2018**, *6*, 141–154.
47. Pingali, P. Agricultural policy and nutrition outcomes—Getting beyond the preoccupation with staple grains. *Food Secur.* **2015**, *7*, 583–591. [CrossRef]
48. Balz, A.G.; Heil, E.A.; Jordan, I. Nutrition-sensitive agriculture: New term or new concept? *Agric. Food Secur.* **2015**, *4*, 6. [CrossRef]
49. Global Nutrition Report. Action on Equity to End Malnutrition; Development Initiatives. 2020. Available online: https://globalnutritionreport.org/reports/2020-global-nutrition-report/ (accessed on 19 May 2025).
50. World Bank. Financing Agribusiness in Sub-Saharan Africa: Opportunities, Challenges, and Investment Models. 2016. Available online: https://www.findevgateway.org/sites/default/files/publications/files/africa_agrifinance_2016.pdf (accessed on 14 May 2025).
51. Langyintuo, A. Smallholder farmers' access to inputs and finance in Africa. In *The Role of Smallholder Farms in Food and Nutrition Security*; Gomez y Paloma, S., Riesgo, L., Louhichi, K., Eds.; Springer: Berlin/Heidelberg, Germany, 2020; pp. 133–152. [CrossRef]
52. Thompson, B.; Amoroso, L. *Combating Micronutrient Deficiencies: Food-Based Approaches*; Food and Agriculture Organization of the United Nations and CAB International: Rome, Italy, 2011.

53. Masset, E.; Haddad, L.; Cornelius, A.; Isaza-Castro, J. *A Systematic Review of Agricultural Interventions That Aim to Improve Nutritional Status of Children*; EPPI-Centre, Social Science Research Unit, Institute of Education, University of London: London, UK, 2011.
54. Herforth, A.; Ballard, T.J. Nutrition indicators in agriculture projects: Current measurement, priorities, and gaps. *Glob. Food Secur.* **2016**, *10*, 1–10. [CrossRef]

proceedings

Proceeding Paper

Strategic Pathways for Applying Food Processing Principles in the Implementation of Nutrition-Smart and Nutrition-Sensitive Agriculture in West Africa †

Sedo Eudes L. Anihouvi [1,2,3], Kyky Komla Ganyo [4], G. Esaïe Kpadonou [4,*], Rebeca Edoh [2], Caroline Makamto Sobgui [4] and Niéyidouba Lamien [4]

1 Research and Development Section, CRI-Bridge, Cotonou 03 BP 409, Benin; aeudeslandry@gmail.com or aeudeslandry@icloud.com
2 Earth, Environmental and Life Sciences Section, Research Institute, University of Bucharest, 030018 Bucharest, Romania; edoh.nephtalie@gmail.com
3 Faculty of Agronomic Sciences, University of Abomey-Calavi, Abomey-Calavi BP 526, Benin
4 West and Central African Council for Agricultural Research and Development (CORAF), Dakar BP 48, Senegal; k.ganyo@coraf.org (K.K.G.); c.sobgui@coraf.org (C.M.S.); n.lamien@coraf.org (N.L.)
* Correspondence: e.kpadonou@yahoo.com or e.kpadonou@coraf.org
† Presented at the CORAF's 2023 Symposium on Processing and Transformation of Agricultural Products in West and Central Africa: Achievements and Opportunities for Private Sector Engagement, Lome, Togo, 21–23 November 2023.

Abstract: West Africa faces persistent food and nutrition insecurity despite agricultural efforts, exacerbated by population growth, climate change, and socio-economic vulnerabilities. This study argues that integrating food processing principles with nutrition-sensitive agriculture (NSA) and nutrition-smart agriculture (NSmartAg) offers a transformative solution for human health. Therefore, we delineate these interconnected concepts and highlight their synergistic potential for a nutrition-focused food system. Likewise, critical analysis of key regional challenges, including infrastructural weaknesses, policy gaps, and gender inequities, was made prior to identifying significant opportunities for leveraging food processing as a strategic entry point to accelerate the implementation of NSA and NSmartAg. Based on these insights, six strategic pathways are proposed to achieve this objective: (i) integrating food processing into policies; (ii) investing in interdisciplinary R&D that puts nutrition and health benefits at the forefront of desired outcomes along with others; (iii) strengthening farmer and food processor capacities; (iv) improving agri-food infrastructure; (v) fostering multi-sectoral collaboration; and (vi) prioritizing youth engagement and market development. By adopting these integrated strategies, West African countries can build more resilient, equitable, and nutrition-centered food systems, ultimately improving public health outcomes and fostering sustainable regional development.

Keywords: nutrition and food security; agri-food systems; nutrition-smart agriculture; agricultural transformation; smart food

1. Introduction

Many West African countries have made food security a policy priority. However, the region continues to face high rates of malnutrition, foodborne illnesses, and hunger [1]. A study by [2], using six nutrition indicators including stunting, wasting, anemia, and overweight, classified West African countries into four trend groups and projected minimal progress toward the 2030 WHO targets. Similarly, Ref. [3] estimates that 20% of the population in the region remains undernourished, with limited

 https://doi.org/10.3390/proceedings2025118018

likelihood of achieving the goal of zero hunger by 2030. This situation underscores the need to move beyond increasing food quantity to enhancing food quality, safety, and nutritional value. Nutrition is not just about calorie intake but also involves the adequate consumption of proteins, fats, fiber, vitamins, and minerals, all essential to human health and immune function [4]. Food processing can support achieving this goal by preserving nutrient quality, extending shelf life, and reducing losses along the food chain. Yet, West Africa's food processing systems remain largely traditional and underdeveloped, leading to significant inefficiencies. Moreover, food loss and waste (FLW) continue to increase in the region. For instance, Ref. [5] estimated food losses of over 2000 kcal per person per day in countries like Benin, Mali, and Niger, with more than 50% attributed to post-harvest handling and processing. Therefore, improving food processing practices offers multiple benefits, including reducing waste, enhancing food safety, and supporting rural livelihoods, particularly for women and youth. While modernization is needed, such improvements should build on the advantages of traditional systems to ensure sustainability and local relevance.

In this context, two complementary frameworks have gained prominence: nutrition-smart agriculture (NSmartAg) and nutrition-sensitive agriculture (NSA). NSmartAg emphasizes nutrition-focused farming practices, including biofortification and climate-smart agriculture, while NSA takes a broader approach that connects agriculture to nutrition, health, and gender empowerment. Despite their potential, implementing NSmartAg and NSA in West Africa is constrained by limited investment, weak infrastructure, and inadequate institutional support [6]. Meanwhile, food processing offers an actionable entry point for operationalizing these frameworks by linking agriculture to nutrition and health outcomes. Sustainable food processing plays a crucial role in accelerating the implementation of NSA and NSmartAg in West Africa [7]. This study explores and identifies actionable entry points as strategic pathways to strengthen the integration of food, agriculture, nutrition, and health to accelerate the implementation of NSmartAg and NSA in West Africa. These strategies are intended to inform policies and interventions that address the region's pressing food and nutrition security challenges while supporting sustainable development.

2. Food Processing Principles, Nutrition-Sensitive Agriculture, and Nutrition-Smart Agriculture Linkage

Addressing food and nutrition insecurity in West Africa requires an integrated understanding of how agriculture, food processing, and health systems intersect. Three concepts are central to this integration: food processing, nutrition-sensitive agriculture (NSA), and nutrition-smart agriculture (NSmartAg). While distinct, these approaches are deeply interconnected and together form the foundation of a nutrition-focused food system. NSA provides a broader framework that links agriculture to nutrition and human health through policies and cross-sector collaboration. NSmartAg contributes by improving the nutrient quality of agricultural outputs, while food processing ensures those nutrients are preserved, enhanced, and delivered effectively to consumers. Together, they offer a comprehensive pathway to be explored for strengthening food systems, improving diets, and reducing malnutrition in West Africa. Table 1 summarizes the core principles of each approach.

Table 1. Key principles/descriptions of food processing and preservation, NSmartAg, and NSA.

	Food Processing and Preservation	Nutrition Smart Agriculture (NSmartAg)	Nutrition Sensitive Agriculture (NSA)
Focus and principles	Applying processes for transforming agricultural products into food products for: - Value addition - Nutrient retention - Shelf-life extension	Innovation-driven strategy for enhancing the nutritional quality of agricultural products. Focus on practices and integration: - Biofortification - ICT tools - Climate resilience	System-driven approach that influences policies and strategies for: - Food diversification - Gender empowerment - Cross-sectoral linkages
Description	Minimize nutrient loss through drying, fermentation, and low-heat cooking; extend shelf-life with hygienic packaging and storage; improve palatability and marketability of foods	Integrates nutrition goals with productivity through technology; promotes climate-resilient, nutrient-rich crops; utilizes ICT and precision farming tools to enhance outcomes	Focuses on increasing availability and affordability of diverse diets; promotes women's empowerment and household nutrition decision-making; links agriculture to health and education
References	[8,9]	[10,11]	[12,13]

3. Agri-Food Systems in West Africa: Status and Constraints

Despite various regional and national initiatives aimed at improving agriculture and nutrition, West Africa continues to grapple with severe systemic challenges that impede the effective implementation of NSA and NSmartAg. These multifaceted challenges are deeply rooted in structural, institutional, and socio-economic constraints that affect the entire system from primary production through processing to final consumption. A nuanced understanding of these complex barriers is essential for designing impactful and sustainable interventions. This section elaborates on these critical challenges, drawing upon the identified need for a more robust discussion of implementation barriers.

3.1. Structural Constraints in Agricultural Systems

Agricultural production in West Africa is severely constrained by environmental and socio-economic factors. Soil fertility is in continuous decline, a problem exacerbated by unsustainable farming practices, such as monocropping and inadequate fallow periods, and the escalating impacts of climate change, which contribute to widespread land degradation [14]. This situation is further compounded by increasing water scarcity, particularly in the arid and semi-arid Sahelian regions, leaving agriculture heavily reliant on erratic and unpredictable rainfall patterns [15]. The high cost of agricultural inputs, including improved seeds, fertilizers, and pesticides, represents a significant financial burden for the region's predominantly smallholder farmers. These economic and environmental pressures directly contribute to reduced overall yields and actively discourage the cultivation of nutrient-rich crops, which are frequently more resource-intensive and expensive to produce than traditional staple crops [16]. Consequently, dietary diversity is compromised right from the production stage, impacting the nutritional quality of available food. Furthermore, inadequate access to essential agricultural extension services poses a persistent problem. Farmers often lack the necessary knowledge and skills regarding modern farming techniques, sustainable soil management practices, or improved post-harvest handling methods. The dissemination of agricultural innovations is notably slow, and available technical advice is not always adapted to specific local contexts or the unique needs of diverse farming communities [17]. Access to finance also represents a barrier; limited availability of credit and formal financial services discourages farmers from making crucial investments required to enhance productivity or adopt new and more efficient technologies [18]. These structural constraints disproportionately affect smallholder farmers, especially women and youth, who typically face even greater restrictions in accessing resources and information, thereby perpetuating a cycle of low productivity and food insecurity across the region.

3.2. Post-Harvest Losses and Weak Infrastructure

Post-harvest losses (PHL) constitute a pervasive and substantial threat to food and nutrition security in West Africa, leading to significant quantitative and qualitative losses of foodstuffs and income for farmers. Estimates, such as those highlighted by [5], indicate that in countries like Benin, Mali, and Niger, over 50% of food losses occur during critical stages of post-harvest handling, processing, and storage. These losses are primarily attributable to underdeveloped or non-existent infrastructure. The road networks, often in poor condition and sparse in rural areas, render the transportation of agricultural products to markets and processing centers difficult, costly, and prone to delays that lead to the spoilage of perishable goods. The near-total absence of cold chains and modern storage facilities, such as refrigerated warehouses and silos, means that vast quantities of fruits, vegetables, dairy products, and grains are lost due to heat, humidity, pests, and mold [19]. In parallel, traditional food processing methods, while culturally embedded, are often inefficient in preserving nutrients or ensuring food safety at a larger scale [20]. These methods can result in significant nutritional losses and frequently fall short of the hygiene and quality standards required for commercialization or export. The pervasive lack of access to reliable and affordable energy, particularly electricity, in many rural areas further limits the modernization of processing and preservation techniques. This critical infrastructural deficit, combined with sub-optimal traditional processing methods, not only reduces the quantity of food available but also critically compromises its nutritional quality and safety. This exacerbates malnutrition problems and diminishes the income of actors throughout the value chain, creating a significant impediment to sustainable food systems.

3.3. Institutional and Policy Gaps

While many West African countries have commendably developed national food security strategies or agricultural development plans, the effective integration of explicit nutrition goals into these policies remains a challenge. Frequently, agricultural policies disproportionately prioritize staple crop yields (such as maize, rice, or cassava) and export potential over the crucial objectives of dietary diversity, the nutritional quality of food products, and the promotion of underutilized but nutrient-rich crops [21,22]. This narrow focus can exacerbate reliance on a limited number of staple crops, thereby reducing the resilience of food systems to shocks and constraining the availability of essential micronutrients vital for public health. Furthermore, the critical role of food processing is often overlooked or severely underestimated within national strategies. It is frequently perceived as a purely industrial concern, disconnected from primary agricultural production and broader nutritional outcomes, rather than being recognized as an essential link in the agri-food value chain that can add significant value, reduce losses, and enhance the year-round availability of nutritious foods [23,24]. This fragmented perception hinders the development of holistic policies that would support the entire food value chain, from farm to fork. A foundational issue also lies in inter-sectoral coordination, which remains persistently weak among ministries and agencies responsible for agriculture, health, education, trade, and national planning. This fragmentation of responsibilities leads to institutional silos, duplicated efforts, gaps in policy implementation, and a lack of synergy among various initiatives. A truly multi-sectoral approach is indispensable for the effective implementation of NSA and NSmartAg, as these concepts inherently demand coordinated interventions across multiple levels of the food system to achieve meaningful impacts on nutrition and food security.

3.4. Gender and Social Inequities

Women occupy a central and multifaceted role within West African food systems. They are pivotal agricultural producers, food processors, traders, and key decision-makers regarding household nutrition. Despite these indispensable contributions, they consistently confront persistent and deeply entrenched barriers that limit their full potential and, consequently, the effectiveness of nutrition-sensitive interventions [25]. Women's access to land ownership is frequently restricted by customary laws or traditional practices, thereby depriving them of essential collateral needed to access credit, invest in their farms, or fully benefit from agricultural development programs [26]. Their access to formal financial services, such as bank loans or insurance, is also significantly limited, which curtails their capacity to purchase inputs, equipment, or expand their economic activities. Similarly, opportunities for training and skill development are often less accessible to women due to time constraints imposed by heavy domestic and agricultural workloads, prevailing social norms, or geographical distance from training centers [27]. These gender disparities directly restrict the reach and efficacy of nutrition-sensitive interventions, as women are less able to adopt new farming or processing practices that could enhance the nutritional quality of foods. Moreover, entrenched cultural norms and low levels of nutrition literacy, particularly prevalent in many rural areas, significantly influence both production decisions (favoring cash crops or less nutritious staple grains) and household food consumption practices, perpetuating sub-optimal diets. Understanding and actively working to dismantle these systemic inequalities is therefore fundamental for any strategy aimed at sustainably improving nutrition across the region.

3.5. Limited Investment in Nutrition-Oriented Innovation

Agricultural research and development (R&D) in West Africa suffers from underfunding, which significantly hampers the emergence and widespread dissemination of innovations crucial for enhancing food and nutrition security [28]. This underinvestment is particularly acute in specific, yet vital, areas such as the biofortification of staple crops (the enhancement of micronutrient content through conventional breeding), the development of sustainable and nutrient-preserving food processing techniques, and the establishment of agricultural extension services specifically tailored to address nutritional outcomes. Consequently, promising innovations developed by national research institutes or international centers often fail to achieve widespread adoption among farmers. Several compounding factors explain this low adoption rate. The dissemination of research findings is frequently inefficient, with limited communication channels that struggle to effectively reach smallholder farmers and other stakeholders. A lack of awareness regarding the potential benefits of these innovations also constitutes a major obstacle. Farmers are often uninformed about the existence of new crop varieties or improved techniques, or they may not immediately perceive their direct relevance or economic viability for their own farming operations [29,30]. Furthermore, innovations might not be sufficiently adapted to specific local contexts, traditional farming practices, or the resource availability of diverse farming communities, rendering their implementation difficult or economically unfeasible. This critical gap in investment and the subsequent ineffective diffusion of innovations severely limit the region's capacity to transform its food systems and sustainably address the persistent challenges of malnutrition and food insecurity.

4. Leveraging Food Processing Principles and Past Initiatives to Overcome Constraints and Improve the Agri-Food System in West Africa

Leveraging food processing offers a multifaceted approach to address food and nutrition insecurity in West Africa, providing significant opportunities to advance NSA and NSmartAg agendas. This involves innovative practices, enhanced regional collaboration, and a strategic re-evaluation of local resources. Food processing appears as a particularly promising entry point, capable of significantly improving nutrition outcomes, drastically reducing post-harvest losses, and concurrently enhancing livelihoods across the diverse sub-region. Various NSA and NSmartAg initiatives already underway in several West African countries, exemplified in Table 2, demonstrate the tangible potential and serve as valuable case studies for broader implementation. Furthermore, a study by [31] identified a range of agricultural technologies and innovations that are inherently gender- and nutrition-sensitive, underscoring their potential for further improvement and integration into NSA and NSmartAg frameworks.

Table 2. Successful NSA and NSmartAg initiatives in West Africa.

Country	Innovation Type	Description	Outcomes	References
Nigeria	Iron-rich millet fortification	Biofortified millet processed into flours, retaining nutritional content	Reduced anemia among women and children	[32,33]
Ghana	Orange-fleshed sweet potato processing	Community processors convert sweet potatoes into flour and bakery items	Improved vitamin A intake, local economic empowerment	[32,34]
Senegal	Women's Agro-processing cooperatives	Female-led groups engaged in milling, drying, and packaging local crops	Increased income, improved diet diversity, community food resilience	[35]
Burkina Faso and Benin	Dried leafy vegetable packaging	Solar drying and hygienic packaging of moringa and other vegetables	Extended shelf life and reduced nutrient degradation	[36,37]
Mali	Rice fortification pilot projects	Fortified rice distributed through public school feeding programs	Enhanced nutrition outcomes for children	[38]
Benin	Cassava chips with fortified powder	Women's groups produce cassava chips enriched with protein powder	Boosted protein intake and rural entrepreneurship	[39]
Togo	Small-scale solar dehydrators	Use of solar dryers for fruits and vegetables	Reduced spoilage and extended access to seasonal produce	[40]

4.1. Building on Previous Initiatives for Scaling NSA and NSmartAg in West Africa

Food processing technologies present substantial potential to fundamentally improve the nutritional quality and extend the shelf life of staple foods, thereby increasing their availability and accessibility. Indigenous and traditional crops, such as millet and sorghum, which are inherently rich in essential micronutrients, can be valorized and transformed through advanced processing techniques like fortification, roasting, or extrusion [41,42]. These methods are crucial for enhancing nutrient bioavailability and improving the palatability and marketability of these vital grains. For instance, in Nigeria, significant strides have been made in fortifying millet with iron and processing it into stable, shelf-ready food products specifically designed to combat anemia [43,44]. Similarly, in Ghana, orange-fleshed sweet potatoes, a rich source of Vitamin A, are being successfully utilized in community-based projects to produce nutritious, Vitamin A-rich flour and various snack items [45]. These targeted efforts not only directly address critical micronutrient deficiencies prevalent in the population but also provide support for local food enterprises and empower women-led cooperatives, fostering inclusive economic growth.

Moreover, regional institutions and food innovation centers play an increasingly active and pivotal role in championing NSA and NSmartAg initiatives. Organizations such as the Economic Community of West African States (ECOWAS), the West and Central African Council for Agricultural Research and Development (CORAF), and the Permanent Interstate Committee for Drought Control in the Sahel (CILSS) are instrumental in promoting these agendas [46]. These crucial platforms facilitate vital knowledge exchange, foster collaborative research efforts, and ensure policy alignment across member states, which is essential for effectively scaling successful interventions regionally. Digital tools and mobile platforms are also being strategically leveraged to disseminate critical information to farmers on nutrition-sensitive practices, fluctuating market prices, and climate-smart agricultural techniques [47,48]. While these existing actions and partnerships are commendable, their impact can be significantly amplified through strengthened efforts and expanded support to accelerate the comprehensive implementation of NSA and NSmartAg throughout West Africa.

Indigenous knowledge systems offer a unique and often underutilized opportunity to enhance agricultural biodiversity, resilience, and sustainability within the region. Traditional practices, including diverse intercropping systems, agroforestry, and the use of local seed varieties, not only bolster the resilience of agricultural systems to climate stressors but also inherently support more diversified and nutritious diets. By systematically integrating these time-tested practices into NSA frameworks, communities can effectively preserve their valuable cultural heritage while simultaneously and substantially improving their nutrition security [49,50]. Furthermore, strengthening food supply chains through local and regional market integration can serve as a powerful catalyst for advancing NSA and NSmartAg. Smallholder farmers significantly benefit from structured market linkages, which not only enhance their income generation but also crucially reduce post-harvest losses and ensure that nutritious food products reach both urban and rural consumers more efficiently. Strategic investments in small-scale processing units and improved storage facilities at the community level empower farmers to engage in value addition activities, thereby significantly enhancing their economic standing while simultaneously boosting local food availability and food security.

4.2. Building on Food Processing Principles: NSmartAg and NSA for Enhancing Nutritious Food Availability and Shelf Life

Advanced food processing technologies are fundamental to addressing the complex food safety and security challenges confronting West Africa, especially in the context of rapidly growing populations [51]. These sophisticated technologies play a pivotal role in extending the shelf life of diverse food products, thereby mitigating food waste across the value chain and ensuring the consistent availability of nutritious foods throughout the year. Effective preservation techniques, encompassing novel approaches such as active and intelligent antimicrobial packaging [52] and the strategic incorporation of natural antioxidants derived from local sources [53], have emerged as critical strategies for maintaining optimal food quality and substantially prolonging usability. Further advancements in integrated food processing and packaging technologies, outlined by [54], clearly demonstrate a proactive commitment to addressing evolving consumer demands for safe, highly nutritious, and conveniently packaged food products. Additionally, the strategic integration of cutting-edge preservation methods, including non-thermal processing techniques like pulsed electric field (PEF) processing and the application of nanotechnology in food packaging [55], further enhances food shelf life while concurrently minimizing post-harvest losses that previously represented immense economic and nutritional drains. The widespread adoption of sustainable food preservation techniques, which are specifically designed to reduce environmental impact while maintaining superior food safety and nutritional integrity, is also gaining significant prominence and traction [56]. Ultimately, modern food processing serves as a foundational cornerstone of food security in West Africa. Its impact extends beyond merely extending shelf life; it fundamentally ensures the steady and reliable availability of diverse, safe, and nutritious food options for a growing population. The strategic integration of cutting-edge technologies and biotechnological advancements within the food processing sector strengthens the overall resilience of the regional food system, effectively addressing both immediate and long-term food security challenges across West Africa.

4.3. Building on Food Processing Principles: NSmartAg and NSA for Maintaining Food Safety and Increasing Economic Development in West Africa

As highlighted by [57], African policymakers are increasingly recognizing food safety not merely as a public health imperative but as a pivotal element in achieving comprehensive food security and, crucially, unlocking the immense economic potential of both intra- and extra-continental food trade. In this context, the role of modern food processing in systematically mitigating food safety risks is paramount, particularly in a region where traditional food handling practices can inherently contribute to significant safety concerns, as previously noted by [38]. By meticulously applying controlled processing techniques such as pasteurization, sterilization, or precise drying, harmful pathogens can be substantially reduced or eliminated, thereby significantly lowering the incidence of debilitating foodborne diseases that burden healthcare systems and reduce productivity. Therefore, sustainable and technologically advanced food processing stands as an indispensable cornerstone for concurrently enhancing public health safety and driving economic development across West Africa.

The economic implications of a well-developed and strategically implemented food processing sector are profound and far-reaching. The transformation of raw agricultural commodities into higher value-added products generates substantial employment opportunities across the value chain, from direct labor in processing plants to indirect jobs in logistics, marketing, and support services. This value addition vigorously stimulates local economies and fosters the organic growth of ancillary industries, creating a ripple effect of economic prosperity. As underscored by [58], increasing agricultural productivity in conjunction with significantly improving food processing and distribution systems is a fundamental prerequisite for supporting the region's rapidly expanding food economy. Targeted investments in critical infrastructure (e.g., cold chains, modern processing facilities), advanced technology adoption, and comprehensive training programs for the workforce can further enhance economic diversification and accelerate overall growth.

A particularly crucial aspect of this transformative process is the indispensable role of women within the West African food system. Women play an overwhelmingly significant part in both informal and formal food processing and trade activities, yet they frequently encounter systemic challenges that severely limit their productivity and economic empowerment. These barriers include restricted access to essential credit, appropriate technologies, relevant training, and meaningful decision-making opportunities. Ref. [39] highlighted the vital importance of enhancing women's roles across the food system, advocating for targeted policies that directly improve their access to critical resources and expand their economic opportunities. Similar insights were reported by [35], who identified highly profitable activities, such as the production of "atta" (a traditional cowpea fritter), predominantly carried out by women in Benin. However, their study also critically identified persistent barriers such as limited access to finance, improved processing technologies, and official government recognition. Addressing these challenges would not only empower countless women entrepreneurs but also contribute profoundly to poverty reduction and significantly enhance overall food security and resilience across the West African region. The strategic development of a well-structured and highly efficient food processing sector is therefore not merely beneficial but essential for West Africa. It not only safeguards public health by minimizing foodborne risks but also acts as a catalyst for sustainable economic growth, widespread job creation, and substantial poverty reduction. By investing strategically in food processing infrastructure and innovative technology and developing inclusive policies that support women's full participation, West African nations can collectively forge a more resilient, healthier, and prosperous food system for all.

5. Strategic Pathways for Advancing Implementation of NSmartAg and NSA in West Africa

To fully harness the transformative potential of NSA and NSmartAg, West Africa must adopt and vigorously pursue strategic, highly coordinated, and genuinely multi-sectoral development pathways. These integrated strategies are not merely critical for addressing persistent technical and infrastructural barriers but are equally vital for systematically embedding nutrition goals into the region's long-term agricultural and overarching economic planning. NSA and NSmartAg offer complementary and mutually reinforcing approaches essential for building food systems that are resilient, equitable, and inherently nutrition-focused. Based on a comprehensive synthesis of regional best practices and scientific insights, six strategic pillars emerge as foundational for achieving substantive progress (Figure 1). These pathways must be pursued simultaneously, consistently underpinned by unwavering political will, adequate resource mobilization, and inclusive governance mechanisms.

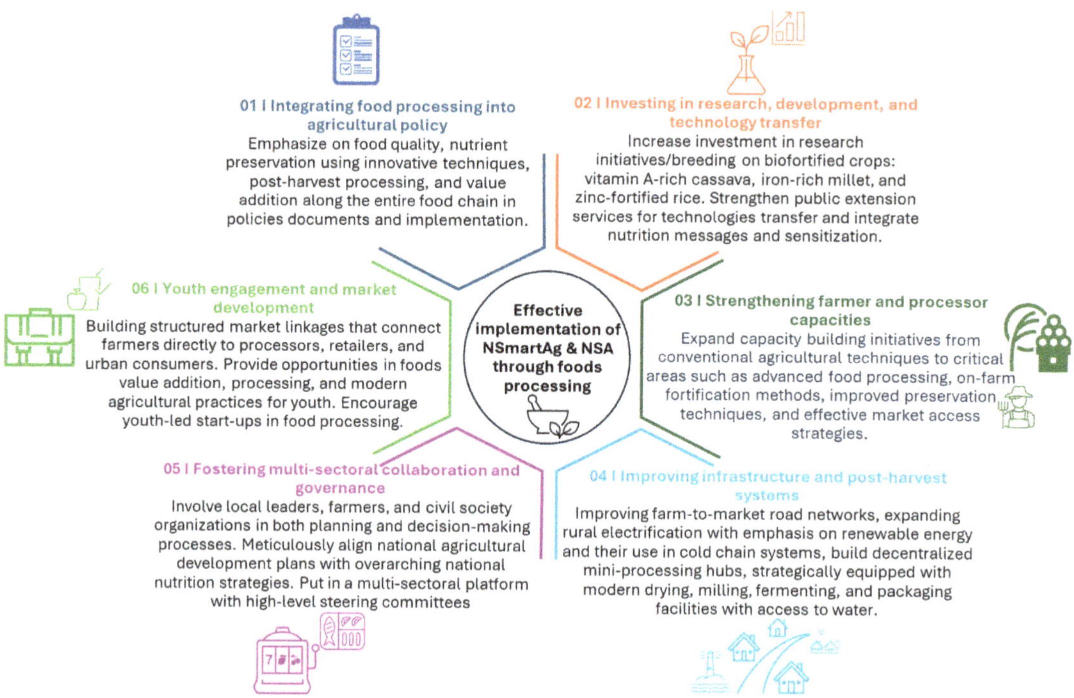

Figure 1. Strategic pathways for advancing implementation of NSmartAg and NSA.

5.1. Integrating Food Processing into Agricultural Policy

In many West African countries, existing agricultural policies predominantly emphasize narrow objectives such as increasing yield, boosting export volumes, or achieving food security primarily defined in terms of quantitative output. Consequently, crucial aspects like food quality, nutrient preservation, and value addition through processing often receive inadequate attention. The explicit integration of food processing into national agricultural strategies is, therefore, an indispensable step toward achieving comprehensive nutrition outcomes. Countries must update their agricultural and nutrition policies to incorporate clear and measurable objectives pertaining to post-harvest processing, innovative food preservation techniques, and value addition along the entire food chain. These policy updates should be meticulously informed by updated data on nutrient losses at various stages, evolving consumption trends, and pressing food safety issues specific to the region. Public investment should be strategically prioritized for the development of modern food processing infrastructure, with a particular focus on rural areas where perishable crops are abundant and where localized processing can significantly reduce spoilage and transportation costs. Investment models that cater specifically to small and medium-sized enterprises (SMEs) are critical for fostering inclusive growth. Concurrently, regulatory bodies must develop and rigorously enforce comprehensive food safety and nutritional labeling standards to ensure quality control across both local and regional value chains. Furthermore, targeted incentives, such as tax breaks or direct subsidies, can effectively encourage private sector investment in the processing of nutritious staples. Crucially, policies should also support decentralized processing models that empower community-level operations, thereby not only minimizing transport expenditures but also significantly enhancing local employment opportunities and fostering community resilience.

5.2. Investing in Research, Development, and Technology Transfer

A robust and dynamic research ecosystem is an indispensable prerequisite for effectively supporting and advancing NSA and NSmartAg initiatives. Current investment levels in agricultural research across the West African region remain unacceptably low, severely limiting the systematic development of innovative solutions specifically designed to address pervasive regional nutritional deficiencies. A substantial increase in funding should be strategically directed toward the accelerated breeding and widespread dissemination of biofortified crops, including but not limited to vitamin A-rich cassava, iron-rich millet, and zinc-fortified rice, which offer cost-effective pathways to combating micronutrient malnutrition. National agricultural research systems must actively foster deeper collaborations with leading universities and international research institutions. This collaborative approach is vital for designing context-specific technologies, conducting rigorous adaptive trials across diverse agro-ecological zones, and ensuring that innovations are locally relevant and scalable. Equally paramount is the efficient and widespread transfer of these technologies to end-users. Many promising biofortified seeds and affordable, nutrition-enhancing processing tools remain significantly underutilized due to a persistent lack of awareness or insufficient capacity at the farm level. Governments should thus establish and strengthen public extension services that seamlessly integrate nutritional messaging with evidence-based agricultural practices. Digital platforms can further enhance dissemination efforts, offering farmers mobile-based advisory services on optimal biofortified seed use, integrated pest management, and emerging value chain opportunities. Critically, gender-responsive technology dissemination strategies are essential: women, who are often marginalized by conventional extension services, must be explicitly included and prioritized through targeted outreach and tailored training initiatives to ensure equitable access and impactful adoption.

5.3. Strengthening Farmer and Processor Capacities

Smallholder farmers, particularly women, constitute the very heart of food production and household dietary decision-making processes across West Africa. Consequently, significantly enhancing their capacity to effectively engage in NSA and NSmartAg practices is pivotal for achieving sustainable success. Capacity building initiatives must transcend conventional training solely focused on planting and harvesting techniques; they must expand to comprehensively cover critical areas such as basic food processing, on-farm fortification methods, improved preservation techniques, and effective market access strategies. Programs must be designed to holistically address both knowledge gaps and practical constraints, including pervasive issues related to limited access to credit, appropriate inputs, and, critically, restricted decision-making power within households and broader community structures. Capacity development initiatives must also be meticulously tailored to specific local contexts and thoughtfully consider the intricate socio-cultural dynamics that influence adoption. Community-based training models utilizing peer-to-peer learning have demonstrated considerable promise in rural settings, especially when women are actively empowered to lead and facilitate these sessions, fostering greater trust and relevance. Furthermore, robust nutrition education should be seamlessly incorporated into all training curricula, helping farmers and processors understand not only the agricultural benefits but also the profound health advantages of producing and consuming diversified and fortified foods. Strategic partnerships with civil society organizations and local non-governmental organizations can significantly amplify outreach efforts, fostering sustained behavioral change and wider adoption. In the long term, establishing comprehensive vocational education programs focused on agri-food entrepreneurship for youth and women can unlock entirely new opportunities for employment, foster innovation, and cultivate future leadership within West Africa's evolving food systems.

5.4. Improving Infrastructure and Post-Harvest Systems

Persistent post-harvest losses continue to pose a formidable threat to food security and nutrition across West Africa. The pervasive inadequacy of infrastructure—ranging from poorly maintained rural roads to a critical lack of reliable energy for refrigeration—leads to massive spoilage of perishable goods and significant economic losses throughout the value chain. Therefore, strategic and targeted investments in rural infrastructure must constitute a paramount policy priority. Governments should commit substantial funding to crucial areas such as improving farm-to-market road networks, expanding rural electrification (with a strong emphasis on sustainable, renewable options like solar power), and enhancing access to clean water for both irrigation and food processing operations. Cold chain systems are of particular importance for preserving highly perishable, nutrient-rich foods, including fresh fruits, vegetables, dairy products, and animal proteins. Establishing affordable and energy-efficient cold storage units at strategic aggregation centers can dramatically reduce spoilage and extend the marketable shelf life of high-value crops, thereby stabilizing markets and increasing farmer incomes. Additionally, the development of decentralized mini-processing hubs, strategically equipped with modern drying, milling, fermenting, and packaging facilities, can effectively reduce post-harvest losses directly at the source and significantly improve the nutritional value and safety of food products. These community-level facilities should be readily accessible to farmer cooperatives, women's groups, and youth entrepreneurs, actively supported by tailored capacity-building programs and accessible microfinance schemes to ensure their sustainability and broad impact.

5.5. Fostering Multi-Sectoral Collaboration and Governance

The inherently complex and multifaceted nature of malnutrition, coupled with the ambitious scope of agricultural transformation, necessitates profound collaboration across diverse sectors and at multiple levels of governance. Agriculture, in isolation, cannot effectively resolve nutrition challenges without deeply coordinated action with the health, education, finance, trade, and environment sectors. Governments must establish and empower robust multi-sectoral platforms and high-level steering committees specifically tasked with coordinating the planning, funding allocation, implementation, and rigorous monitoring of NSA and NSmartAg initiatives. Joint policy frameworks are indispensable; they should meticulously align national agricultural development plans with overarching national nutrition strategies, relevant education curricula (e.g., through nutrition-sensitive school feeding programs), and comprehensive social protection systems. Collaborative monitoring and evaluation systems can significantly aid in tracking progress against shared, multi-dimensional indicators such as dietary diversity improvements, food safety adherence, and school nutrition outcomes. International donors, non-governmental organizations, and the private sector should be actively engaged early in the planning process to ensure coherence, optimize resource utilization, and enhance the long-term sustainability of interventions. Local government and community engagement are equally vital, as decentralized governance structures can effectively support context-specific interventions and build strong local accountability. By genuinely involving local leaders, farmers, and civil society organizations in both planning and decision-making processes, multi-sectoral collaboration can be rendered far more inclusive, responsive, and ultimately, more effective. Ultimately, sustained political commitment and visionary leadership are paramount to driving systemic change and ensuring that nutrition remains a central, non-negotiable tenet of West Africa's agricultural transformation agenda.

5.6. Youth Engagement and Market Development

Harnessing the demographic dividend of West Africa's burgeoning youth population and strategically developing robust market linkages are critical, yet often overlooked, pathways for advancing NSA and NSmartAg. The region's youth represent a dynamic force for innovation and entrepreneurship within the agri-food sector. Investing in their skills and providing opportunities in value addition, processing, and modern agricultural practices can transform them from being mere job seekers into catalysts for economic growth and nutritional improvement. This involves establishing vocational education programs focused on agri-food entrepreneurship that equip young people with technical skills in food science, business management, and marketing. Encouraging youth-led start-ups in food processing, particularly those utilizing local nutritious crops, can create new employment avenues and introduce innovative, healthy food products to the market.

Furthermore, strengthening market development is essential to ensure that nutritious foods produced through NSA and NSmartAg initiatives reach consumers efficiently and affordably. Many smallholder farmers struggle with limited access to reliable markets, leading to post-harvest losses and discouraging the production of diverse, nutrient-rich crops. Strategic interventions should focus on building structured market linkages that connect farmers directly to processors, retailers, and urban consumers. This includes developing digital market platforms that provide real-time price information, facilitating aggregation models for smallholders, and investing in market infrastructure such as collection centers and efficient transportation networks. Promoting local and regional trade in nutritious processed foods can further stimulate demand and provide stable outlets for agricultural produce. By empowering youth to drive innovation in these emerging markets and fostering more efficient and equitable market systems, West Africa can ensure that its agricultural productivity translates directly into improved nutrition and sustainable economic development for all. Table 3 summarizes insights into these strategic pathways and their expected outcomes.

Table 3. Strategic pathways and expected outcomes.

Strategic Pathway	Detailed Actions	Expected Outcomes
Integrate food processing into policy	Develop standards, invest in SME processors, mainstream nutrition targets.	Reduced losses, improved shelf life and nutrient retention
Invest in R&D and tech transfer	Support research institutes, fund fortification trials, foster public-private partnerships	Widespread access to enhanced crops and processing innovations
Strengthen farmer and processor capacity	Train on food safety, drying, fortification, and business models	Increased adoption of best practices and diversification
Improve infrastructure	Cold chains, rural roads, processing units powered by renewable energy	Minimized post-harvest losses, reduced rural poverty
Multi-sectoral collaboration	Create inter-ministerial working groups, nutrition task forces, joint monitoring systems	Coherent and sustainable nutrition–agriculture governance
Youth engagement	Support agri-preneurship and digital literacy programs	Boosted innovation and job creation in the agri-food sector
Market development	Facilitate access to urban markets for rural processed products	Greater value chain integration and smallholder revenue growth

6. Conclusions

West Africa faces a critical and complex food and nutrition security challenge, demanding a fundamental shift from merely increasing yields to prioritizing food quality, safety, and nutritional value. This review has established that food processing, NSA, and NSmartAg are intrinsically interconnected pillars, crucial for building resilient and nutrition-centered food systems. Food processing, notably, stands out as a pivotal and actionable entry point, offering diverse pathways to integrate agriculture, food systems, nutrition, and human health, thereby drastically accelerating the implementation of NSA and NSmartAg across the region. Our analysis not only highlights persistent challenges like infrastructural deficits and institutional gaps but also reveals significant, yet underexploited, opportunities rooted in local knowledge, technological advancements, and collaborative frameworks. The strategic imperative for a cohesive nutrition–agriculture alignment is undeniable, advocating for a deliberate and integrated approach. The proposed strategic pathways, ranging from policy integration and robust R&D investment to enhanced capacities, infrastructural improvements, multi-sectoral collaboration, and youth-driven market development, are not mere recommendations but interconnected components of a transformative agenda.

Achieving this vision is imperative for West Africa's future. It demands unwavering political commitment, innovative financing, and inclusive governance that empowers all stakeholders. By systematically embedding food processing as a central, catalytic element within the broader NSA and NSmartAg frameworks, West African nations can transcend current limitations, fostering more resilient, equitable, and nutrition-centered food systems that ensure healthier populations and sustainable regional development for generations to come.

Author Contributions: Conceptualization, G.E.K. and S.E.L.A.; methodology, G.E.K., S.E.L.A. and K.K.G.; investigation, R.E., G.E.K., S.E.L.A. and K.K.G.; writing—original draft preparation, S.E.L.A. and K.K.G.; investigation, R.E., G.E.K., S.E.L.A. and K.K.G.; writing—original draft preparation, R.E., G.E.K. and K.K.G.; writing—review and editing, S.E.L.A., C.M.S. and K.K.G.; supervision, N.L. All authors have read and agreed to the published version of the manuscript.

Funding: This research received no external funding.

Institutional Review Board Statement: Not applicable.

Informed Consent Statement: Not applicable.

Data Availability Statement: Data are contained within the article.

Conflicts of Interest: The authors declare no conflict of interest.

References

1. Hussaini, M.A.; Adegbite, O.; Yusuf, I. Strengthening food system resilience in Nigeria: Pathways for integrated approaches. *J. Food Secur. Resil.* **2024**, *8*, 210–225.
2. Chadare, F.J.; Idohou, R.; Agossou, D.J.; Azokpota, P.; Fanou-Fogny, N. Progress and prospects for nutrition improvement in West Africa: A classification based on historical trends and future projections. *Glob. Food Secur.* **2022**, *33*, 100618.
3. FAO; IFAD; UNICEF; WFP; WHO. *The State of Food Security and Nutrition in the World 2019: Safeguarding Against Economic Slowdowns and Downturns*; FAO: Rome, Italy, 2019.
4. Touch, V.; Tan, D.K.Y.; Cook, B.; Liu, D.L.; Cross, R.; Tran, T.A.; Utomo, A.; Yous, S.; Grunbuhel, C.; Cowie, A.L. Smallholder farmers' challenges and opportunities: Implications for agricultural production, environment and food security. *J. Environ. Manag.* **2024**, *370*, 122536. [CrossRef] [PubMed]
5. Aragie, E.A.; Balié, J.; Morales-Opazo, C. *Estimating Food Loss and Waste Along the Food Supply Chain: Evidence from Benin, Mali, and Niger*; FAO Agricultural Development Economics Working Paper 18-02; FAO: Rome, Italy, 2018. Available online: https://www.fao.org/3/I9526EN/i9526en.pdf (accessed on 2 June 2025).
6. Boadi, R.; Boakye, A.A.; Yaro, A.S. Implementation challenges of nutrition-sensitive agriculture in Ghana: Insights from agricultural officers. *J. Nutr. Agric.* **2024**, *10*, 45–56.
7. Kpadonou, G.E.; Sobgui, C.M.; Edoh, R.; Ganyo, K.K.; Anihouvi, S.E.L.; Lamien, N. Unlocking the Role of Food Processing in Nutrition-Smart and Nutrition-Sensitive Agriculture in West Africa: Challenges, Opportunities, and a Framework for Deployment. *Proceedings* **2025**, *118*, 17. [CrossRef]
8. Mafe, A.N.; Edo, G.I.; Makia, R.S.; Joshua, O.A.; Akpoghelie, P.O.; Gaaz, T.S.; Jikah, A.N.; Yousif, E.; Isoje, E.F.; Igbuku, U.A.; et al. A review on food spoilage mechanisms, foodborne diseases and commercial aspects of food preservation and processing. *Food Chem. Adv.* **2024**, *5*, 100852. [CrossRef]
9. Gómez, I.; Janardhanan, R.; Ibañez, F.C.; Beriain, M.J. The effects of processing and preservation technologies on meat quality: Sensory and nutritional aspects. *Foods* **2020**, *9*, 1416. [CrossRef] [PubMed]
10. Bhatnagar, S.; Chaudhary, R.; Sharma, S.; Janjhua, Y.; Thakur, P.; Sharma, P.; Keprate, A. Exploring the dynamics of climate-smart agricultural practices for sustainable resilience in a changing climate. *Environ. Sustain. Indic.* **2024**, *24*, 100535. [CrossRef]
11. Mešić, A.; Jurić, M.; Donsì, F.; Zavadlav, S.; Čolović, R.; Režek Jambrak, A. Advancing climate resilience: Technological innovations in plant-based, alternative and sustainable food production systems. *Discov. Sustain.* **2024**, *5*, 423. [CrossRef]
12. Kumari, K.; Mirzakhani Nafchi, A.; Mirzaee, S.; Abdalla, A. AI-driven future farming: Achieving climate-smart and sustainable agriculture. *AgriEngineering* **2025**, *7*, 89. [CrossRef]
13. Bryan, E.; Alvi, M.; Huyer, S.; Ringler, C. Addressing gender inequalities and strengthening women's agency to create more climate-resilient and sustainable food systems. *Glob. Food Secur.* **2024**, *40*, 100731. [CrossRef]
14. Dhakad, U.; Meena, S. Chapter-4 Soil Fertility Restoration: Challenges and Opportunities. *Adv. Agric. Sci.* **2024**, *47*, 51–72.
15. Adamou, O.I. Adapting to Climate Risks in the Sahel: Strategies for Smallholder Farmers in Niger. Ph.D. Thesis, Cornell University, Ithaca, NY, USA, 2024.
16. Papadopoulos, G.; Mavroeidis, A.; Stavropoulos, P.; Anastasopoulos, V.; Beslemes, D.; Tigka, E.; Kakabouki, I. Tritordeum: A versatile and resilient cereal for Mediterranean agriculture and sustainable food production. *Cereal Res. Commun.* **2024**, *52*, 323–331. [CrossRef]
17. Gamage, S.K.N.; Niranjala, S.A.U.; Upulwehera, J.M.H.M.; Piratdin, A.; Bandara, K.B.T.U.K.; Bandara, H.G.K.N.; Hettiarachchi, H.A.B.W.; Adikari, A.M.P.; Sumanapala, S.D.A.; Pabasara, M.K.N.; et al. Climate Change Impacts on Household Food Security in Sri Lanka's Dry Zones: A Qualitative Analysis. *Challenges* **2025**, *16*, 20. [CrossRef]
18. Khan, F.U.; Nouman, M.; Negrut, L.; Abban, J.; Cismas, L.M.; Siddiqi, M.F. Constraints to agricultural finance in underdeveloped and developing countries: A systematic literature review. *Int. J. Agric. Sustain.* **2024**, *22*, 2329388. [CrossRef]
19. Food and Agriculture Organization (FAO). *Nutrition-Sensitive Agriculture and Food Systems in Practice: Options for Intervention*; FAO: Rome, Italy, 2019. Available online: https://www.fao.org/3/ca5642en/ca5642en.pdf (accessed on 23 May 2025).
20. Ghosh, S.; Meyer-Rochow, V.B.; Jung, C. Embracing tradition: The vital role of traditional foods in achieving nutrition security. *Foods* **2023**, *12*, 4220. [CrossRef] [PubMed]
21. Kidane, B.; Urugo, M.M.; Hirpha, H.H.; Paulos, T.; Hundea, W.; Tessema, F. Nutritional challenges of staple crops due to increasing atmospheric carbon dioxide levels: Case of Sub-Saharan Africa. *J. Agric. Food Res.* **2025**, *19*, 101592. [CrossRef]

22. Noort, M.W.; Renzetti, S.; Linderhof, V.; du Rand, G.E.; Marx-Pienaar, N.J.; de Kock, H.L.; Magano, N.; Taylor, J.R. Towards sustainable shifts to healthy diets and food security in sub-Saharan Africa with climate-resilient crops in bread-type products: A food system analysis. *Foods* **2022**, *11*, 135. [CrossRef] [PubMed]
23. Maestre, M.; Poole, N.; Henson, S. Assessing food value chain pathways, linkages and impacts for better nutrition of vulnerable groups. *Food Policy* **2017**, *68*, 31–39. [CrossRef]
24. Apeh, O.O.; Nwulu, N.I. The water-energy-food-ecosystem nexus scenario in Africa: Perspective and policy implementations. *Energy Rep.* **2024**, *11*, 5947–5962. [CrossRef]
25. Perelli, G.; Cardarelli, M.; Bernini, R.; Lucarini, M.; Durazzo, A. Mediterranean plants-based dietary supplements: Focus on classification and coding. *Nat. Prod. Res.* **2025**, 1–8. [CrossRef]
26. Doss, C.; Meinzen-Dick, R.; Quisumbing, A.; Theis, S. Women in agriculture: Four myths. *Glob. Food Secur.* **2018**, *16*, 69–74. [CrossRef] [PubMed]
27. Fletschner, D.; Kenney, L. Rural women's access to financial services: Credit, savings, and insurance. In *Gender in Agriculture: Closing the Knowledge Gap*; Springer: Dordrecht, The Netherlands, 2014; pp. 187–208.
28. Wudil, A.H.; Usman, M.; Rosak-Szyrocka, J.; Pilař, L.; Boye, M. Reversing Years for Global Food Security: A Review of the FoodSecurity Situation of theSub-Saharan Africa (SSA). *Int. J. Environ. Res. Public Health* **2022**, *19*, 14836. [CrossRef] [PubMed]
29. Ndibalema, G. Effective Communication for Disseminating Soil and Water Management Practices to Bring About Sustainable Smallholder Agriculture in Tanzania: Case Study of Manyara and Dodoma Regions. Ph.D. Thesis, Wageningen University & Research, Wageningen, The Netherlands, 2019.
30. Funke, O.M.; Munyaradzi, C. The Adoption and Scaling of Climate-Smart Agriculture Innovation by Smallholder Farmers in South Africa: A Review of Institutional Mechanisms, Policy Frameworks and Market Dynamics. *World* **2025**, *6*, 51. [CrossRef]
31. Ganyo, K.K.; Kpadonou, G.E.; Lamien, N.; Maiga, M.; Ngaido, A.; Diallo, Y.; Sobgui, C.M.; Segnon, A.C.; Zougmoré, R.B. Promoting Gender-and Nutrition-Sensitive Agricultural Technologies and Innovations in West and Central Africa; 2023. Available online: https://cgspace.cgiar.org/bitstreams/10409e11-3d09-4a4f-a873-275378b58d3a/download (accessed on 24 May 2025).
32. FAO. *The State of Food and Agriculture 2019: Moving Forward on Food Loss and Waste Reduction*; Food and Agriculture Organization of the United Nations: Rome, Italy, 2019.
33. World Health Organization (WHO). *Double-Duty Actions for Nutrition: Policy Brief*; World Health Organization (WHO): Geneva, Switzerland, 2017.
34. United Nations Development Programme (UNDP). *Agribusiness for Africa's Prosperity*; UNDP Publications: Seoul, Republic of Korea, 2018.
35. Kpossilande, C.E.; Honfonga, B.G.; Ferre, T. Economic potentials of artisanal food processing microenterprises in West Africa: Case of "atta" production in Cotonou (Benin). *Agric. Food Econ.* **2020**, *8*, 24. [CrossRef]
36. Moussa, M. Innovative Millet Foods to Improve Nutrition and Expand Markets in West Africa. Ph.D. Thesis, Purdue University, West Lafayette, IN, USA, 2019.
37. Baco, M.N. Ethno-ecological variability in the consumption of leafy green plants in the republic of Benin. *Asian J. Agric. Ext. Econ. Sociol.* **2019**, *36*, 1–15. [CrossRef]
38. Adeyeye, S.A.O. Food safety in developing countries: An overview. *Nutr. Food Sci.* **2016**, *46*, 150–156. [CrossRef]
39. Gnisci, D. *Women and Food Security in Africa: Barriers and Opportunities*; European Centre for Development Policy Management (ECDPM): Maastricht, The Netherlands, 2016. Available online: https://ecdpm.org/publications/women-food-security-africa (accessed on 24 May 2025).
40. Kariuki, A.N. Upgrading strategies and food security implications on smallholder farmers in Sub-Saharan Africa: A value chain review. *J. Food Secur.* **2018**, *6*, 141–154.
41. Akinola, A.A.; George, O.A.; Ogbemhe, J.; Ipinnimo, O.; Oribayo, O. Quantitative review and machine learning application of refractance window drying of tuber slices. *Int. J. Food Eng.* **2024**, *20*, 125–140. [CrossRef]
42. Mgwenya, L.I.; Agholor, I.A.; Ludidi, N.; Morepje, M.T.; Sithole, M.Z.; Msweli, N.S.; Thabane, V.N. Unpacking the Multifaceted Benefits of Indigenous Crops for Food Security: A Review of Nutritional, Economic and Environmental Impacts in Southern Africa. *World* **2025**, *6*, 16. [CrossRef]
43. Nkhata, S.G.; Chilungo, S.; Memba, A.; Mponela, P. Biofortification of maize and sweet potatoes with provitamin A carotenoids and implication on eradicating vitamin A deficiency in developing countries. *J. Agric. Food Res.* **2020**, *2*, 100068.
44. Ravi, R.; Rana, M. Fortification and value addition of millet-based products: Innovations from Nigeria. *Int. J. Food Technol. Nutr.* **2024**, *19*, 112–124.
45. Low, J.W.; Mwanga, R.O.; Andrade, M.; Carey, E.; Ball, A.M. Tackling vitamin A deficiency with biofortified sweetpotato in sub-Saharan Africa. *Glob. Food Secur.* **2017**, *14*, 23–30. [CrossRef] [PubMed]
46. Dorji, T.; Rinchen, K.; Morrison-Saunders, A.; Blake, D.; Banham, V.; Pelden, S. Understanding how indigenous knowledge contributes to climate change adaptation and resilience: A systematic literature review. *Environ. Manag.* **2024**, *74*, 1101–1123. [CrossRef] [PubMed]

47. Abiri, R.; Rizan, N.; Balasundram, S.K.; Shahbazi, A.B.; Abdul-Hamid, H. Application of digital technologies for ensuring agricultural productivity. *Heliyon* **2023**, *9*, e22601. [CrossRef] [PubMed]
48. Choruma, D.J.; Dirwai, T.L.; Mutenje, M.J.; Mustafa, M.; Chimonyo, V.G.P.; Jacobs-Mata, I.; Mabhaudhi, T. Digitalisation in agriculture: A scoping review of technologies in practice, challenges, and opportunities for smallholder farmers in sub-saharan africa. *J. Agric. Food Res.* **2024**, *18*, 101286. [CrossRef]
49. Gonçalves, C.d.B.Q.; Schlindwein, M.M.; Martinelli, G.d.C. Agroforestry systems: A systematic review focusing on traditional indigenous practices, food and nutrition security, economic viability, and the role of women. *Sustainability* **2021**, *13*, 11397. [CrossRef]
50. Imoro, Z.A.; Zarouk, I.A.; Ballu, D.A.; Ammal, A. Harnessing indigenous technologies for sustainable management of land, water, and food resources amidst climate change. *Front. Sustain. Food Syst.* **2021**, *5*, 691603. [CrossRef]
51. Alamu, E.O.; Mooya, M.M. The role of food processing technologies in addressing food security and safety in West Africa. *J. Food Process. Technol.* **2017**, *8*, 1–6.
52. Fadiji, T.; Rashvand, M.; Daramola, M.O.; Iwarere, S.A. A review on antimicrobial packaging for extending the shelf life of food. *Processes* **2023**, *11*, 590. [CrossRef]
53. Petcu, C.; Borda, D.; Stănciuc, N. Natural antioxidants in food preservation: Trends and future perspectives. *Antioxidants* **2023**, *12*, 384.
54. Fadiji, T.; Pathare, P. Smart packaging innovations for enhancing food safety and quality in Sub-Saharan Africa. *Trends Food Sci. Technol.* **2023**, *134*, 75–85.
55. Sridhar, K.; Ali, A.; Pandiselvam, R. Emerging food processing technologies: A review of recent developments and applications. *J. Food Eng.* **2021**, *292*, 110368.
56. Lisboa, C.R.; da Silva, D.S.; de Almeida, D.T. Sustainable food preservation techniques: Implications for food safety and environmental protection. *Sustainability* **2024**, *16*, 158. [CrossRef]
57. Molnar, J.J.; Godefroy, S.B. Food safety as a driver for Africa's food trade and food security. *Food Control* **2020**, *118*, 107378.
58. Allen, T.; Heinrigs, P. *Emerging Opportunities in the West African Food Economy*; OECD Publishing: Paris, France, 2016. [CrossRef]

MDPI AG
Grosspeteranlage 5
4052 Basel
Switzerland
Tel.: +41 61 683 77 34

Proceedings Editorial Office
E-mail: proceedings@mdpi.com
www.mdpi.com/journal/proceedings

www.ingramcontent.com/pod-product-compliance
Lightning Source LLC
Chambersburg PA
CBHW041018270926
51850CB00024B/1467
* 9 7 8 3 7 2 5 8 5 0 4 1 9 *